CONTEMPORARY OPTICS

FOR

SCIENTISTS AND ENGINEERS

ALLEN NUSSBAUM and **RICHARD A. PHILLIPS**

Electrical Engineering Department
University of Minnesota

Research and Development
Foster Grant Co., Inc.

PRENTICE-HALL, INC., *Englewood Cliffs, New Jersey*

Library of Congress Cataloging in Publication Data

NUSSBAUM, ALLEN.
  Optics for scientists and engineers.

  Includes bibliographical references.
  1. Optics. I. Phillips, Richard A., joint author.
II. Title.
QC355.2.N87   535   74-30210
ISBN 0-13-170183-5

PRENTICE-HALL SOLID STATE PHYSICAL ELECTRONICS SERIES

Printed in the United States of America

PRENTICE-HALL INTERNATIONAL, INC., *London*
PRENTICE-HALL OF AUSTRALIA, PTY. LTD., *Sydney*
PRENTICE-HALL OF CANADA, LTD., *Toronto*
PRENTICE-HALL OF INDIA PRIVATE LIMITED, *New Delhi*
PRENTICE-HALL OF JAPAN, INC., *Tokyo*
PRENTICE-HALL OF SOUTHEAST ASIA (PTE.) LTD., *Singapore*

CONTENTS

PART III   THE FOURIER ANALYSIS APPROACH TO PHYSICAL OPTICS, *213*

PART IV THE INTERACTION OF LIGHT AND MATTER, *319*

14 NONLINEAR OPTICS, *370*

15 SOURCES, *417*

PREFACE

During the last decade, revolutionary changes have taken place in the field of optics. Foremost among these has been the development of the laser and the widespread application of lasers in many areas of science and engineering. Everyone who has made predictions of the future has concluded that the existing applications are only a beginning and that the greatest growth of this field lies ahead. The laser has made applications possible which had previously only been conjectured.

Two entirely new areas exploiting the unique properties of laser light have emerged: these are holography (three-dimensional photography) and nonlinear optics (frequency shifting and modulation of laser light). Other developments are the filtering of optical images to enhance selected features and character recognitition. Advances in detectors and detection techniques have made the infrared region more accessible.

Geometrical optics has undergone a similar advance during the decade. Matrix methods have greatly simplified the design of lens and optical systems. The use of computers in the interactive mode has made possible a great improvement in the design of optical systems and the tailoring of a system to a particular job. Further, it permits the elimination of a large amount of the tedious algebra associated with geometric optics.

All of these factors have indicated to us that a new and modern treatment of optics would serve a definite need. This book aims at a fairly broad coverage in both its subject matter and its audience, and it has been written using the modular approach. Each major topic is introduced separately and is self-contained. This permits different portions to be combined to serve different course needs. There is more material than can be incorporated in a one semester course, so that the instructor should choose the topics for his particular course. Traditionally, optics has been taught in physics departments. The pedagogical function of the optics course was to familiarize students with wave phenomena using light waves that they would see. This need is met in this book by the sections on interference, diffraction, and polarization. The modern aspects of these topics (coherence, spatial filtering, holography, and nonlinear optics) have been woven into the treatment. The initial development of each topic has been made as simple mathematically as possible. This provides a discussion of the fundamental aspects of the phenomena. The treatment of each topic is extended using more advanced mathematical techniques. Students preparing for graduate school can pursue the treatments in greater depth while others will still obtain a sound knowledge of phenomena. An example of this is the section on nonlinear optics. Each effect is first described simply, then a more advanced treatment using eigenvalues and eigenvectors is presented. Another example is our treatment of Fraunhofer diffraction, which is based on Huyghens' principle. Diffraction patterns for simple objects are calculated. Then a more advanced treatment leading to the optical transfer function is developed.

Recently, physics departments have renewed their interest in applied areas of physics, and the need to train physicists in applied areas is now widely recognized. At the same time, engineering departments are developing optics courses because of engineering applications of optics. The sections on holography, optical data processing, sources (including laser theory), detectors, crystal optics, and electro-optics meet this need.

Engineers are also particularly concerned with geometrical optics and systems design. The sections on matrix methods, the optical transfer function, sources, and detectors are well suited for this.

The following table contains suggestions for three different courses:

Geometrical Optics	Physical Optics	Applied or Engineering Optics
Matrix Methods	Introduction to Waves	Introduction to Waves
Introduction to Waves	Interference	Interference
Diffraction	Diffraction	Diffraction
Spatial Filtering	Holography	Holography
Optical Transfer Function	Absorption and Dispersion	Absorption and Dispersion
Sources	Crystal Optics	Crystal Optics
Detectors	Polarization	Polarization
	Nonlinear Optics (full treatment)	Nonlinear Optics (introductory parts)

The material in this book has been presented by the authors in regular courses, short intensive courses, and over television. We wish to express our gratitude for comments, from students and colleagues, that have improved our treatment. In addition, the material on the extension of matrix methods to non-paraxial optics (Chapters 2 and 3) was originally worked out under the auspices of an NSF Curriculum Development Grant, for which we are grateful.

ALLEN NUSSBAUM
RICHARD A. PHILLIPS

PART

I

ELEMENTARY AND ADVANCED
GEOMETRICAL OPTICS

The first four chapters of this book provide a fairly detailed introduction to geometrical optics. Chapter 1, which describes the matrix approach used by many teachers of optics, covers the material generally presented in an introductory physics course. This, plus some of the discussion of aberrations in the next two chapters, is all the background needed to understand the effect of lenses on light waves, as considered in Chapter 10. On the other hand, readers interested in the functioning or design of optical instruments such as cameras, telescopes, and spectroscopes should go through Chapters 1–3 thoroughly and select those parts of Chapter 4 which are appropriate. It will be noted that geometrical optics involves fairly extensive use of a computer or a scientific electronic calculator; it is the combination of matrix and numerical methods which takes what was once a very tedious and complicated discipline and makes it both simple and interesting.

1

PARAXIAL MATRIX OPTICS

1-1 The Newton and Gauss Lens Equations

Our ultimate purpose is to learn to calculate exactly how very complicated optical systems affect the light rays passing through them, but we shall start our approach to geometric optics in a very intuitive fashion. In Fig. 1-1, we show a simple magnifier—a double convex lens—used as a burning glass. The sun's rays, assumed to be traveling parallel to the axis of the lens, meet at the *focal point* or *focus F*, producing a small, intense spot of light capable of igniting paper. The lens is symmetric and if the sun's rays came from the opposite direction there would be another such point F' on the other side of the lens at an equal distance from the corresponding surface.

Next, let us use this lens to magnify. We find that when we view an object located a few centimeters away, it is enlarged significantly (Fig. 1-2(a)), but if the lens is brought closer to the object, the magnification is not very great (Fig. 1-2(b)). In fact, if the lens is placed very close to the print, the magnification is only perceptibly greater than unity. (The reader should try this with an ordinary magnifier.) This leads us to suspect that there is a position of the object, perhaps touching the lens, which would actually give an image of identical size and oriented in the same way. Denoting the object position for this situation by H and the image position by H', Fig. 1-3 shows how

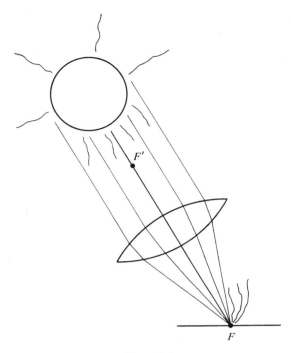

Figure 1-1

object and image might be related. (Actually, Fig. 1-3 is not quite a true representation of this one-to-one relation, for reasons we shall consider shortly.) The points H and H' are called the *unit* or *principal points* of the lens. We shall use the former term since it is self-explanatory. The four points F, F', H, and H' are collectively known as *cardinal points*, and the planes through these points normal to the axis are the *cardinal planes*.

Let us now place an object of height x at a distance s to the left of the symmetrical convex lens of Fig. 1-4. The axis of the lens is AA' and the cardinal points F, H, H', and F' lie on the axis in the order given, since the unit points are fairly close to the lens and the focal points are a little farther out. We consider a ray PQ which is parallel to AA'. Such a ray does not actually pass through the lens, so we cannot trace it from Q to R'. However, if the lens were larger or the object smaller, then this ray would emerge from the lens in such a way that it would cross the unit plane determined by H' at a distance $\overline{H'R'} = x$ from the axis. This follows from the fact that $\overline{HQ} = x$ and its image $\overline{H'R'}$ must be identical in size and orientation. Further, the ray which leaves P and passes through Q and R' must also pass through the focal point F'. This is a consequence of the focusing property discussed in connection with Fig. 1-1.

(a)

(b)

Figure 1-2

We shall assume for the moment that we know the location of the image point P'. We then take a ray $P'Q'$ parallel to the axis and, as before, trace it to the object space unit plane H, and then through the object space focal point F back to P. We may now reverse its direction, since the effect of the lens on a ray going from right to left is the same as on one going in the opposite sense. Thus, the ray PFR will emerge at Q' at a distance x' from the axis and will pass through the image point P'. The intersection of rays PQ and PR at P' in image space determines the height and position of the image.

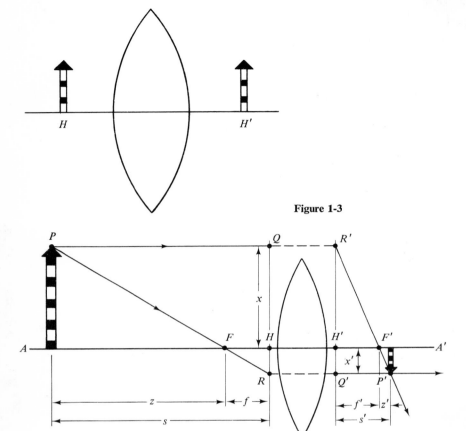

Figure 1-3

Figure 1-4

Denote the various distances in Fig. 1-4 by the symbols shown, where
$\overline{FH} = f$ is known as the *object space focal length* (with a similar definition
for $\overline{F'H'} = f'$). On the object side, we have three similar triangles, giving the
relations

$$\frac{x + x'}{s} = \frac{x}{z} = \frac{x'}{f} \tag{1-1a}$$

For the image side, we obtain in the same way

$$\frac{x + x'}{s'} = \frac{x'}{z'} = \frac{x}{f'} \tag{1-1b}$$

from which

$$x = \frac{zx'}{f} = \frac{x'f'}{z'}$$

or

$$zz' = ff' \tag{1-2}$$

which is *Newton's form* of the *lens equation*.

If we define the *magnification* β as

$$\beta = \frac{x'}{x} \tag{1-3}$$

then by Eqs. (1-1) and (1-2)

$$\frac{f}{z} = \frac{x'}{x} = \beta = \frac{z'}{f'} \tag{1-4}$$

and

$$x + x' = \frac{sx'}{f} = \frac{s'x}{f'}$$

Hence

$$\beta = \frac{x'}{x} = \frac{s'f}{sf'} \tag{1-5}$$

Further

$$\frac{x + x'}{x'} = \frac{s}{f} = 1 + \frac{1}{\beta} = 1 + \frac{sf'}{s'f}$$

or

$$\frac{f'}{s'} + \frac{f}{s} = 1 \tag{1-6}$$

This is the *Gauss lens equation*, which is equivalent to Eq. (1-2). Either of these relations implies that if an object point P produces an image point P' as shown in Fig. 1-4, then *all* rays from P which actually pass through the lens will intersect at P'. We shall examine the validity of this in the next section.

1-2 Snell's Law and the Lens Matrices

We have derived the Newton and Gauss relations on the basis of simple assumptions involving the unit and focal planes. We realize, then, that the explicit effect of the lens on a ray of light was not considered in obtaining Eqs. (1-2) or (1-6). Before we can proceed further, we must take note of the behavior of light at a glass-air interface, and this is specified by *Snell's law*. We define the *index of refraction n* of a material medium as the ratio of the velocity of light c in a vacuum to the velocity v in the medium, so that

$$n = \frac{c}{v} \tag{1-7}$$

(The velocity of light in air is approximately the same as in a vacuum.) Consider the ray of light in Fig. 1-5 passing from a medium of index n_1 to a medium of index n_2. In the early 1600s, Snell discovered that the angles θ_1

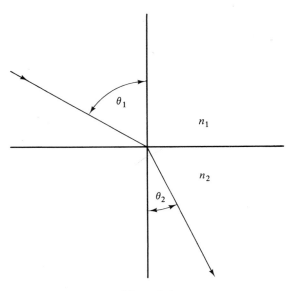

Figure 1-5

and θ_2 between the rays shown and the normal to the interface obey a law
of the form

$$n_1 \sin \theta_1 = n_2 \sin \theta_2 \qquad (1\text{-}8)$$

By Eq. (1-7), this may be written as

$$\frac{v_1}{v_2} = \frac{\sin \theta_1}{\sin \theta_2} \qquad (1\text{-}9)$$

Problem 1-1

Fermat's principle states that light originating at some arbitrary point in
a medium of index of refraction n_1 and terminating at another point in a
medium of the index n_2 will follow the path for which the travel time is a
minimum. Show from Fig. 1-5 that Snell's law corresponds to this condition.

         ▬

    Let us use Snell's law to treat the behavior of a ray of light passing
through the lens of Fig. 1-6. The axis of the lens coincides with the z axis of a
cartesian system, and the incident ray QP lies in the $x\text{-}z$ plane. The center of
curvature of the first surface is at C and its radius is r_1. If P lies a distance
x_1 from the axis (ignoring the second symbol x_1' for the moment), then the
angle PCO or φ is specified by

$$\sin \varphi = \frac{x_1}{r_1} \qquad (1\text{-}10)$$

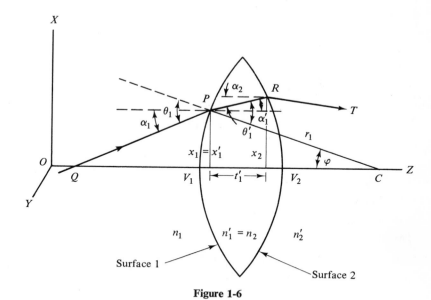

Figure 1-6

It is known that the sine, cosine, and tangent of a small angle x can be expressed by the first few terms of a Taylor series as

$$\left.\begin{array}{l} \sin x = x - \dfrac{x^3}{3!} + \dfrac{x^5}{5!} - \cdots \\[2mm] \cos x = 1 - \dfrac{x^2}{2!} + \dfrac{x^4}{4!} - \cdots \\[2mm] \tan x = x + \dfrac{x^3}{3} + \dfrac{2x^5}{15} + \cdots \end{array}\right\} \qquad (1\text{-}11)$$

For angles of about 5° or less, we may simplify these expressions to the approximate forms

$$\left.\begin{array}{l} \sin x = x \\ \cos x = 1 \\ \tan x = x \end{array}\right\} \qquad (1\text{-}12)$$

with an accuracy of about 1 %, and this restriction in optics is the *paraxial* (meaning "close to the axis") approximation.

For Eq. (1-10), we then have

$$\varphi = \frac{x_1}{r_1} \qquad (1\text{-}13)$$

and Snell's law Eq. (1-8) in paraxial form is

$$n_1\theta_1 = n_2\theta_2 \qquad (1\text{-}14)$$

From Fig. 1-6, we may express the inclination α_1 of the incident ray and α'_1

of the refracted ray as

$$\theta_1 = \alpha_1 + \varphi$$
$$\theta'_1 = \alpha'_1 + \varphi \tag{1-15}$$

Hence, Eq. (1-14) becomes

$$n_1(\alpha_1 + \varphi) = n'_1(\alpha'_1 + \varphi)$$

and by Eq. (1-13)

$$n_1\left(\alpha_1 + \frac{x_1}{r_1}\right) = n'_1\left(\alpha'_1 + \frac{x_1}{r_1}\right)$$

or

$$n'_1\alpha'_1 = \left(\frac{n_1 - n'_1}{r_1}\right)x_1 + n_1\alpha_1 \tag{1-16}$$

For a reason which will become apparent shortly, we shall designate the distance of P from the z axis as either x_1 or x'_1, leading to the relation

$$x'_1 = x_1 \tag{1-17}$$

Equations (1-16) and (1-17) may be combined into a single matrix equation

$$\begin{pmatrix} n'_1\alpha'_1 \\ x'_1 \end{pmatrix} = \begin{pmatrix} 1 & -k_1 \\ 0 & 1 \end{pmatrix}\begin{pmatrix} n_1\alpha_1 \\ x_1 \end{pmatrix} \tag{1-18}$$

where the constant k_1, called the *power* (or *refracting power*) of surface 1, is defined by

$$k_1 = \frac{n'_1 - n_1}{r_1} \tag{1-19a}$$

We may also write this as

$$k_1 = (n'_1 - n_1)c_1 \tag{1-19b}$$

where

$$c_1 = \frac{1}{r_1}$$

is the *curvature* of surface 1. Hence, k_1 is proportional to the curvature at the interface and the change in n as the light crosses it.

The 2×2 matrix on the right of Eq. (1-18), designated as R_1, is the *refraction matrix* for surface 1. Thus

$$R_1 = \begin{pmatrix} 1 & -k_1 \\ 0 & 1 \end{pmatrix} \tag{1-20}$$

Consider next the refracted ray, PR, which intersects surface 2 at a distance $x_2 = x'_2$ from the axis. Then

$$x_2 = x'_1 + t'_1 \tan \alpha'_1 \tag{1-21}$$

where t'_1 is the component of PR along the symmetry axis. In paraxial form, this becomes

$$x_2 = x'_1 + t'_1\alpha'_1 \tag{1-22}$$

Under the same small-angle restriction, t_1' is approximately equal to the lens thickness V_1V_2. We may combine this with the fact that the inclination α_1' of the ray PR is identical to α_2, or

$$\alpha_2 = \alpha_1' \tag{1-23}$$

to obtain a second matrix equation of the form

$$\begin{pmatrix} n_2\alpha_2 \\ x_2 \end{pmatrix} = \begin{pmatrix} 1 & 0 \\ \dfrac{t_1'}{n_1'} & 1 \end{pmatrix}\begin{pmatrix} n_1'\alpha_1' \\ x_1' \end{pmatrix} \tag{1-24}$$

where the *translation matrix* T_{21} specifying the propagation from surface 1 to surface 2 is

$$T_{21} = \begin{pmatrix} 1 & 0 \\ \dfrac{t_1'}{n_1'} & 1 \end{pmatrix} \tag{1-25}$$

It follows that we may trace the behavior of the ray from Q to P, where it is refracted, and then to R by substituting Eq. (1-18) into Eq. (1-24). We obtain

$$\begin{pmatrix} n_2\alpha_2 \\ x_2 \end{pmatrix} = \begin{pmatrix} 1 & 0 \\ \dfrac{t_1'}{n_1'} & 1 \end{pmatrix}\begin{pmatrix} 1 & -k_1 \\ 0 & 1 \end{pmatrix}\begin{pmatrix} n_1\alpha_1 \\ x_1 \end{pmatrix} \tag{1-26}$$

or

$$\begin{pmatrix} n_2\alpha_2 \\ x_2 \end{pmatrix} = T_{21}R_1\begin{pmatrix} n_1\alpha_1 \\ x_1 \end{pmatrix} \tag{1-27}$$

Thus we may combine the results of the refraction and the translation by simply combining the appropriate matrices in the correct order (right to left). Further, we would expect that the refraction at the point R on surface 2 may be specified by a second refraction matrix R_2, so that we can get from Q all the way to T by the equation

$$\begin{pmatrix} n_2'\alpha_2' \\ x_2' \end{pmatrix} = R_2T_{21}R_1\begin{pmatrix} n_1\alpha_1 \\ x_1 \end{pmatrix} \tag{1-28}$$

where n_2', α_2', and x_2' are the appropriate quantities measured to the right of surface 2. This process may be continued through any number of successive refractions and translations; we can trace the behavior of a ray through any combination of lenses—no matter how involved—by multiplying the appropriate matrices in the proper order.

Finally, to show the logic behind the dual designation for parameters such as the index of refraction of the lens, labeled n_1' to the right of surface 1 and n_2 to the left of surface 2, let us introduce the conventions listed in Table 1-1, taken mostly from O'Neill.[1-1]

Table 1-1

Lens Conventions

(a) Light travels from left to right.

(b) Distances are measured from lens vertices, the vertex V_i being the intersection of lens surface i with the symmetry axis.

(c) A horizontal distance is positive if measured from left to right; a vertical distance is positive if measured up from OZ, the symmetry axis. Angles are positive when measured up from the axis.

(d) Interfaces between two different media (such as air and glass) are numbered from left to right.

(e) Quantities associated with an incident ray are unprimed; those associated with a refracted ray are primed.

(f) A subscript on a quantity indicates the refracting surface with which it is associated.

(g) If the center of curvature is to the right of an interface, the radius r_i is positive, and vice versa.

(h) External indices on matrices indicate the reference points or planes with which they are associated: P denotes the object, P' denotes the image, and $V_1, V_2, \ldots$ are the interfaces.

Although this list appears formidable, most of the conventions are either familiar from analytic geometry or have already been introduced in Fig. 1-6. However, (h) requires that we write Eq. (1-26) in the form

$$\begin{pmatrix} n_2\alpha_2 \\ x_2 \end{pmatrix}_{V_2} = \begin{pmatrix} 1 & 0 \\ \frac{t'_1}{n'_1} & 1 \end{pmatrix}_{V_2}{}_{V_1} \begin{pmatrix} 1 & -k_1 \\ 0 & 1 \end{pmatrix}_{V_1} \begin{pmatrix} n_1\alpha_1 \\ x_1 \end{pmatrix}_{V_1} \tag{1-29}$$

that is, the indices V_1, V_1 around

$$R_1 = {}_{V_1}\begin{pmatrix} 1 & -k_1 \\ 0 & 1 \end{pmatrix}_{V_1} \tag{1-30a}$$

denote a refraction associated solely with surface 1, and the indices V_1, V_2 on

$$T_{21} = {}_{V_2}\begin{pmatrix} 1 & 0 \\ \frac{t'_1}{n'_1} & 1 \end{pmatrix}_{V_1} \tag{1-30b}$$

refer to a translation from surface 1 to surface 2. The index V_1 on the right-hand column matrix denotes the parameters n_1, α_1, and x_1 as specified before refraction at surface 1, and V_2 on the final column matrix refers to the corresponding quantities after the translation t'_1. Upon multiplication, the common index is absorbed. For example, the two 2×2 matrices above can be combined to give

$$\begin{pmatrix} n_2\alpha_2 \\ x_2 \end{pmatrix}_{V_2} = {}_{V_2}\begin{pmatrix} 1 & -k_1 \\ \frac{t'_1}{n'_1} & -\frac{k_1 t'_1}{n'_1} + 1 \end{pmatrix}_{V_1} \begin{pmatrix} n_1\alpha_1 \\ x_1 \end{pmatrix}_{V_1}$$

The index V_1 between R_1 and T_{21} has disappeared, and the indices V_1 and V_2 left on the product matrix tell us automatically that it refers to a combined refraction and translation starting at surface 1 and terminating at surface 2. The dual designation feature also shows up in this process. For example, just to the right of surface 1 we denote the three ray parameters as n'_1, α'_1, and x'_1, and just to the left of surface 2 they become n_2, α_2, and x_2 in accordance with (e) and (f) of Table 1-1.

Problem 1-2

A plano-concave lens has a first surface with a radius of 10.0 units, a flat second surface, and a thickness of 1.0 units. The index of refraction is 1.500. (Note that, in optics, it is customary to work in a consistent set of units, such as inches or meters, which need not be specified).

(a) Show that R_2 is the unit matrix.
(b) Show that a ray striking the lens at $+2.0$ units from the axis and with an inclination of $+0.1$ radians will emerge at $+2.0$ units with an inclination of 0.0 radians.

■

1-3 The Gaussian Constants and Their Significance

Considering just the 2×2 matrices associated with the lens of Fig. 1-6, we define their product as the *system matrix* S_{21}, so that

$$S_{21} = {\begin{pmatrix} 1 & -k_2 \\ 0 & 1 \end{pmatrix}}_{v_2} {\begin{pmatrix} 1 & 0 \\ \frac{t'_1}{n'_1} & 1 \end{pmatrix}}_{v_2} {\begin{pmatrix} 1 & -k_1 \\ 0 & 1 \end{pmatrix}}_{v_1} \tag{1-31}$$

The four terms in the product matrix are specified by the symbols

$$S_{21} = {}_{v_2}{\begin{pmatrix} b & -a \\ -d & c \end{pmatrix}}_{v_1} \tag{1-32}$$

Comparing Eqs. (1-31) and (1-32) shows that

$$a = k_1 + k_2 - \frac{k_1 k_2 t'_1}{n'_1} \tag{1-33a}$$

$$b = 1 - \frac{k_2 t'_1}{n'_1} \tag{1-33b}$$

$$c = 1 - \frac{k_1 t'_1}{n'_1} \tag{1-33c}$$

$$d = \frac{-t'_1}{n'_1} \tag{1-33d}$$

where a, b, c, and d are known as the *Gaussian constants*.

Problem 1-3

Verify Eqs. (1-33).

■

Equations (1-20) and (1-25) show that the determinant of either a translation or a refraction matrix is unity, that is

$$\det R_1 = \det T_{21} = \det R_2 = 1$$

Since the determinant of the product of any number of matrices is the product of the individual determinants, we have

$$\det S_{21} = bc - ad = 1 \tag{1-34}$$

This means that only three of the four Gaussian constants are independent, and we generally work with a, b, and c.

Returning to Fig. 1-4, let us add to it the labels for the vertices V_1 and V_2 and the specification of cardinal planes and of object and image locations with respect to these vertices (Fig. 1-7). In accordance with our conventions, the distance between P and V_1 is designated as

$$t_1 = -l \tag{1-35}$$

and between V_2 and P it is

$$t_2' = l \tag{1-36}$$

Problem 1-4

Show that the translation matrix specifying the ray leaving the object and striking surface 1 is

$$T_{1P} = \left.\begin{pmatrix} 1 & 0 \\ -\dfrac{l}{n_1} & 1 \end{pmatrix}\right._{V_1 \qquad\quad P}$$

where $+l$ is a *negative* quantity. ■

Then the complete matrix equation connecting the object with the image is

$$\begin{pmatrix} n_2'\alpha_2' \\ x' \end{pmatrix}_{P'} = \left.\begin{pmatrix} 1 & 0 \\ \dfrac{l'}{n_2'} & 1 \end{pmatrix}\right._{P\;n_2'\;V_2} \left.\begin{pmatrix} b & -a \\ -d & c \end{pmatrix}\right._{V_1} \left.\begin{pmatrix} 1 & 0 \\ -\dfrac{l}{n_1} & 1 \end{pmatrix}\right._{P} \begin{pmatrix} n_1\alpha_1 \\ x \end{pmatrix}_{P} \tag{1-37}$$

or

$$\begin{pmatrix} n_2'\alpha_2' \\ x' \end{pmatrix}_{P'} = S_{P'P} \begin{pmatrix} n_1\alpha_1 \\ x \end{pmatrix}_{P} \tag{1-38}$$

where the 2×2 matrix $S_{P'P}$ is the *object-image matrix*. Multiplying the three 2×2 matrices in Eq. (1-37), we obtain

$$S_{P'P} = \left.\begin{pmatrix} b + \dfrac{al}{n_1} & -a \\ \dfrac{bl'}{n_2'} - d + \dfrac{al'l}{n_2'n_1} - \dfrac{cl}{n_1} & c - \dfrac{al'}{n_2'} \end{pmatrix}\right._{P'} \tag{1-39}$$

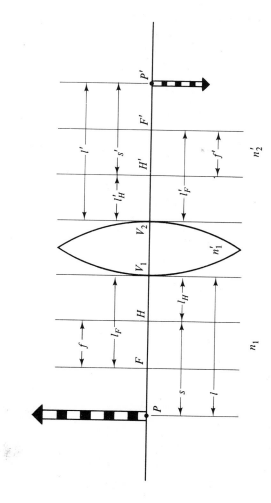

Figure 1-7

By Eq. (1-3), the magnification β is given by x'/x. Equations (1-38) and (1-39) show that the ratio x'/x will depend on α_1, the inclination of the ray leaving the object. That is, the magnification is different for each ray and this would lead to a blurry image. To eliminate this effect, we require that the lower left-hand element of the matrix $S_{P'P}$ vanish, so that β simplifies to

$$\beta = \frac{x'}{x} = c - \frac{al'}{n_2'} \tag{1-40}$$

Further, the determinant of $S_{P'P}$ is unity, giving

$$b + \frac{al}{n_1} = \frac{1}{\beta} \tag{1-41}$$

and the matrix becomes

$$S_{P'P} = \begin{pmatrix} \dfrac{1}{\beta} & -a \\ 0 & \beta \end{pmatrix}_{P'}^{} \;_P \tag{1-42}$$

Therefore Eq. (1-38) is now

$$\begin{pmatrix} n_2'\alpha_2' \\ x' \end{pmatrix}_{P'} = \begin{pmatrix} \dfrac{1}{\beta} & -a \\ 0 & \beta \end{pmatrix}_{P'}\;_P \begin{pmatrix} n_1\alpha_1 \\ x \end{pmatrix}_{P} \tag{1-43}$$

Problem 1-5

Show that

$$\frac{l'}{n_2'} = \frac{(cl/n_1) + d}{(al/n_1) + b} \tag{1-44}$$

∎

To attach a physical meaning to the Gaussian constants, consider Eq. (1-40). For unit magnification, the image must be at H' in Fig. 1-7. If $\beta = 1$ and $l = l'_{H'}$, Eq. (1-40) becomes

$$l'_H = \frac{n_2'(c - 1)}{a} \tag{1-45}$$

Further, when $l' = l'_{H'}$, then $l = l_H$, and using the fact that the determinant of $S_{P'P}$ in Eq. (1-42) is unity gives

$$b + \frac{al_H}{n_1} = 1 \tag{1-46}$$

or

$$l_H = \frac{n_1(1 - b)}{a} \tag{1-47}$$

Next, we apply Eq. (1-43) to a ray which is parallel to OZ when it leaves the object. The image is then of height $x' = 0$ at $l' = l'_{F'}$, and Eq. (1-40) shows that

$$0 = c - \frac{al'_F}{n'_2}$$

or

$$l'_F = \frac{n'_2 c}{a} \tag{1-48}$$

Similarly

$$l_F = -\frac{n_1 b}{a} \tag{1-49}$$

Combining these with Eqs. (1-45) and (1-47), we see from Fig. 1-7 that

$$f' = l'_F - l'_H = \frac{n'_2}{a} \tag{1-50}$$

and

$$f = l_F - l_H = \frac{-n_1}{a} \tag{1-51}$$

In air, where we generally assume that $n_1 = n'_2 = 1.0$, these reduce to

$$-f = f' = \frac{1}{a} \tag{1-52}$$

so that a (in air) is the reciprocal focal length. Note that even when the medium is not air, $-f = f'$, provided the same substance is on both sides of the lens.

To specifically consider b and c, we use Eq. (1-52) in Eq. (1-49), obtaining (again in air)

$$l_F = bf \tag{1-53}$$

and similarly,

$$l'_F = cf' \tag{1-54}$$

Comparing Eq. (1-54) with the right-hand portion of Fig. 1-7, we see that c, which is greater than unity, specifies the ratio l'_F/f'. This quantity, in turn, tells how far the distance $H'F' = f'$ lies beyond the lens. If H' is inside the lens (i.e., to the left of V_2), then c tells what fraction of the focal length f' actually lies outside the lens. This is the situation in a double convex lens, as we shall see in the example which follows. Hence, our consideration of the unit point in Section 1-1 was somewhat misleading, for a simple magnifier would require both object and image to actually be just inside the lens for unit magnification. Lens designers customarily use the term *back focus* to designate the distance l'_F and to emphasize that it differs from the focal length f'.

To demonstrate these remarks, consider the asymmetric double convex lens of Fig. 1-8. The conventions dictate that we denote the dimensions and

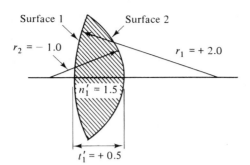

Figure 1-8

indices of refraction as

$$r_1 = +2.0$$
$$r_2 = -1.0$$
$$t'_1 = t_2 = +0.5$$
$$n_1 = 1.0$$
$$n'_1 = n_2 = 1.5$$
$$n'_2 = 1.0$$

Since $t'_i = t_{i+1}$ and $n'_i = n_{i+1}$ for the ith surface, we shall use just the first symbol, t'_1 and n'_1, for example, from now on.

By Eq. (1-19), we obtain

$$k_1 = \frac{n'_1 - n_1}{r_1} = \frac{1.5 - 1.0}{2.0} = 0.25$$

and similarly

$$k_2 = \frac{1.0 - 1.5}{-1.0} = 0.50$$

By Eq. (1-31)

$$S_{21} = {}_{V_2}\!\begin{pmatrix} 1 & -0.50 \\ 0 & 1 \end{pmatrix}_{V_2}\!\begin{pmatrix} 1 & 0 \\ 0.5/1.5 & 1 \end{pmatrix}_{V_1}\!\begin{pmatrix} 1 & -0.25 \\ 0 & 1 \end{pmatrix}_{V_1}$$

$$= {}_{V_2}\!\begin{pmatrix} 0.83 & -0.50 \\ 0.33 & 1 \end{pmatrix}_{V_1}\!\begin{pmatrix} 1 & -0.25 \\ 0 & 1 \end{pmatrix}_{V_1}$$

$$= {}_{V_2}\!\begin{pmatrix} 0.83 & -0.71 \\ 0.33 & 0.92 \end{pmatrix}_{V_1} \tag{1-55}$$

from which

$$a = 0.71 \qquad b = 0.83 \qquad c = 0.92$$

It is a good idea to check all such calculations through the use of Eq. (1-34), and we obtain

$$(0.83)(-0.92) - (-0.71)(0.33) = 0.77 + 0.23 = 1.00$$

By Eq. (1-45) and (1-47)

$$l_H = \frac{n_1(1-b)}{a} = \frac{1-0.83}{0.71} = 0.24$$

$$l'_H = \frac{n'_2(c-1)}{a} = \frac{0.92-1}{0.71} = -0.12$$

By Eqs. (1-48) and (1-49)

$$l_F = -1.18 \qquad l'_F = 1.30$$

This lens and its cardinal points are shown on the scale drawing of Fig. 1-9; H and H' are inside the lens as indicated above. This is the case for all double convex lenses for which the thickness is not negligible. Note also that the cardinal points are asymmetrically arranged, but that f and f' have the same magnitude.

There are two situations for which the lens matrices become extremely simple. One involves a flat refracting surface. Let us take the lens of Fig. 1-8, reverse it, and convert it into a doublet by adding the plano-concave lens shown in Fig. 1-10. It is customary to indicate the various constants for lens systems in the following way

r	n'	t'
1.0		
	1.500	0.5
−2.0		
	1.632	0.4
∞		

It is understood that the first entry under r is r_1, the second is r_2, and so on. The values of the indices n' and the spacings t' are placed between the appropriate values of r, with surface 2 as the common surface between the two lenses. If we regard the passage of light from the right-hand surface of the

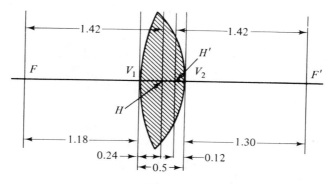

Figure 1-9

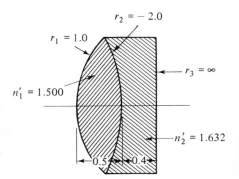

$r_2 = -2.0$

$r_1 = 1.0$

$r_3 = \infty$

$n_1' = 1.500$

$n_2' = 1.632$

0.5   0.4

Figure 1-10

first lens to the left-hand surface of the second one as a translation $t_2' = 0$, then the associated matrix is

$$T_{22} = \begin{pmatrix} 1 & 0 \\ \dfrac{t_2'}{n_2'} & 1 \end{pmatrix}_{v_2} = {}_{v_2}\begin{pmatrix} 1 & 0 \\ 0 & 1 \end{pmatrix}_{v_2} \tag{1-56a}$$

which we recognize as the unit matrix. Since multiplication by the unit matrix does not change a product, we can ignore all such translations.

For the plane surface, we have a radius $r_3 = \infty$. Hence, the value of the refracting power k is

$$k_3 = \frac{n_3' - n_3}{r_3} = 0$$

and the refraction matrix becomes

$$R = \begin{pmatrix} 1 & 0 \\ 0 & 1 \end{pmatrix} \tag{1-56b}$$

which once again is the unit matrix.

Problem 1-6

Find l_F, l_F', l_H, l_H', f, and f' for the doublet of Fig. 1-10.   ▬

Returning to Eq. (1-43), let us consider a ray leaving the z axis at an inclination α_1. Then $x = 0$, and

$$\left. \begin{array}{c} n_2'\alpha_2' = \dfrac{n_1\alpha_1}{\beta} \\[2mm] x' = 0 \end{array} \right\} \tag{1-57}$$

That is, the image point is on the axis and a ray passing through it has an inclination α_2'. We define the *angular magnification* γ as

$$\gamma = \frac{\alpha_2'}{\alpha_1} \tag{1-58}$$

If the media on each side of the lens are identical, Eq. (1-57) shows that

$$\gamma = \frac{1}{\beta} \tag{1-59}$$

In the case where $n_1 \neq n_2'$, we specify the *nodal points* N and N' as those corresponding to the condition

$$\alpha_1 = \alpha_2'$$

or

$$\gamma = 1 \tag{1-60}$$

Problem 1-7

Show that the location of the nodal points with respect to the lens vertices is given by the equations

$$\left.\begin{aligned} \frac{l_N}{n_1} &= \frac{(n_1/n_2') - b}{a} \\ \frac{l_N'}{n_2'} &= \frac{c - (n_1/n_2')}{a} \end{aligned}\right\} \tag{1-61}$$

Show also that these relations reduce to Eqs. (1-45) and (1-47) when $n_1 = n_2'$. ■

Since we rarely work with a situation in which the two media are different, Eqs. (1-61) indicate that N and N' are coincident with H and H', respectively, and we deal with four (rather than six) cardinal planes.

1-4 Approximations for Thin Lenses

Equation (1-33a), when combined with Eqs. (1-50) and (1-51), becomes

$$a = \frac{n_1}{f} = \frac{n_2'}{f'} = k_1 + k_2 - \frac{k_1 k_2 t_1'}{n_1'}$$

Using the definition, Eq. (1-19), and assuming the lens is in air ($n_1 = n_2' = 1$), we obtain

$$\frac{1}{f'} = (n_1' - 1)\left[\frac{1}{r_1} - \frac{1}{r_2} + \frac{(n_1' - 1)t_1'}{n_1' r_1 r_2}\right] \tag{1-62}$$

This result, which expresses the focal length of a lens in terms of its geometry and material, is the *lensmaker's equation*. We can specify a *thin lens* as one for which the third term in brackets on the right of Eq. (1-62) is negligible, and we then have the approximate formula

$$\frac{1}{f'} = (n_1' - 1)\left[\frac{1}{r_1} - \frac{1}{r_2}\right] \tag{1-63}$$

Since n_1' is usually about 1.5, it follows that t_1' has to be much less than r_1 or r_2 for this approximation to hold; that is, the lens is physically thin. In this case, Eqs. (1-33b) and (1-33c) reduce to

$$b = 1, \qquad c = 1 \tag{1-64}$$

and by Eqs. (1-45) and (1-47)

$$l_H = l_H' = 0 \tag{1-65}$$

Thus, the unit planes fall right at the vertices; for a thin lens used as a magnifier, the object and image just touch the two sides of the lens when $\beta = 1.0$. If we use $a = 1/f'$, the system matrix S_{21} for a thin lens takes on the form

$$S_{21} = {}_{V_2}\begin{pmatrix} 1 & -\dfrac{1}{f'} \\ 0 & 1 \end{pmatrix}_{V_1} \tag{1-66}$$

and this is convenient for calculations, as the following example shows.

Let us consider a *compound magnifier* consisting of a double convex lens (focal length $= 15.0$) and a double concave lens (focal length $= 10.0$) whose separation is 12.0 units (Fig. 1-11). For the convex lens, $f' = 15.0$ and for the concave lens, $f' = -10.0$. By Eq. (1-66) the system matrix for the two lenses is

$$S_{41} = {}_{V_4}\begin{pmatrix} 1 & 0.10 \\ 0 & 1 \end{pmatrix}_{V_3} \begin{pmatrix} 1 & 0 \\ 12.0 & 1 \end{pmatrix}_{V_2} \begin{pmatrix} 1 & -0.0667 \\ 0 & 1 \end{pmatrix}_{V_1} \tag{1-67}$$

where $t_2' = 12.0$ may be measured as shown, since the lenses are thin. Performing the multiplication gives

$$S_{41} = {}_{V_4}\begin{pmatrix} 2.200 & -0.0467 \\ 12.000 & 0.200 \end{pmatrix}_{V_1}$$

This result shows that the focal length of the combination is

$$f' = \frac{1}{a} = 21.43$$

If an object is placed 60.0 units from the first lens, we can locate the image position from Eq. (1-44). Since

$$a = 0.047, \qquad b = 2.20, \qquad c = 0.20, \qquad d = -12.0, \qquad l = -60.0$$

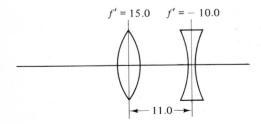

$f' = 15.0 \qquad f' = -10.0$

$\leftarrow 11.0 \rightarrow$

Figure 1-11

then

$$l' = \frac{(0.20)(-60.0) + (-12.0)}{(0.047)(-60.0) + (2.20)} = 40.0$$

The object-image matrix Eq. (1-42) is

$$S_{P'P} = \,_{P'}\begin{pmatrix} 1 & 0 \\ 40.0 & 1 \end{pmatrix}_{V_4} \begin{pmatrix} 2.20 & -0.047 \\ 12.0 & 0.20 \end{pmatrix}_{V_1} \begin{pmatrix} 1 & 0 \\ 60.0 & 1 \end{pmatrix}_P$$

$$= \,_{P'}\begin{pmatrix} -0.60 & -0.047 \\ 0 & -1.67 \end{pmatrix}_P$$

from which

$$\beta = -1.67$$

Therefore, the image is magnified and inverted.

We may obtain some useful formulas by matrix multiplication. If we express Eq. (1-67) symbolically, we have

$$S_{41} = \,_{V_4}\begin{pmatrix} 1 & -\dfrac{1}{f'_2} \\ 0 & 1 \end{pmatrix}_{V_3} \begin{pmatrix} 1 & 0 \\ t'_2 & 1 \end{pmatrix}_{V_2} \begin{pmatrix} 1 & -\dfrac{1}{f'_1} \\ 0 & 1 \end{pmatrix}_{V_1} \tag{1-68}$$

where f'_1 and f'_2 refer to the focal lengths of the first and second lenses, respectively.

Problem 1-8

(a) Show from Eq. (1-68) that the focal length f' of a pair of thin lenses is given by the formula

$$\frac{1}{f'} = \frac{1}{f'_1} + \frac{1}{f'_2} - \frac{t'_2}{f'_1 f'_2} \tag{1-69}$$

(b) Show by multiplying the exact, rather than the approximate, matrices that the focal length of a pair of thick lenses is given by the formula

$$\frac{1}{f'} = \frac{1 - k_4 t'_3/n'_3}{f'_1} + \frac{1 - k_1 t'_1/n'_1}{f'_2} - \frac{t'_2}{f'_1 f'_2} \tag{1-70}$$

(Note that this reduces to Eq. (1-69) for thin lenses.) ∎

Problem 1-9

Let D be the total distance between an object and its image in a thin lens. Show that

$$D = \frac{1}{f'}\left(2 - \frac{1}{\beta} - \beta\right) \tag{1-71}$$

(This formula is useful for making estimates of the physical size of a projection or copying instrument.) ∎

We may trace a ray of light through the compound magnifier either graphically or by matrix multiplication. For example, a ray parallel to the axis leaving a point 10.0 units above the axis will emerge as specified by Eq. (1-43), which becomes

$$\begin{pmatrix} \alpha_2' \\ x \end{pmatrix}_{P'} = \begin{pmatrix} -0.60 & 0.047 \\ 0 & -1.67 \end{pmatrix}_{P'} \begin{pmatrix} 0 \\ 10.0 \end{pmatrix}_{P} = \begin{pmatrix} -0.47 \\ -16.7 \end{pmatrix}_{P'} \tag{1-72}$$

so that

$$\alpha_2' = -0.47 \qquad x' = -16.7$$

To locate the cardinal planes, we find that

$$l_H = \frac{(1-b)}{a} = \frac{(1-2.20)}{0.047} = -25.71$$

$$l_H' = \frac{(c-1)}{a} = \frac{(0.20-1)}{0.047} = -17.14$$

$$l_F = -\frac{b}{a} = -47.14$$

$$l_F' = \frac{c}{a} = 4.29$$

and these distances are plotted to scale in Fig. 1-12. We then use the technique of Fig. 1-4 to trace both the ray specified by Eq. (1-72) and a ray through F parallel to the axis in image space.

Problem 1-10

Show that a ray passing through the geometric center of a *very* thin lens is undeviated, assuming $n_1 = n_2'$.

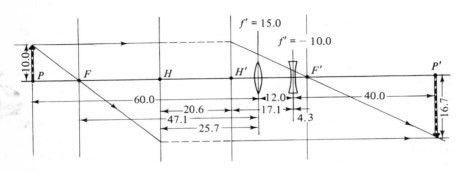

Figure 1-12

Problem 1-11

Verify that Fig. 1-12 agrees with Eq. (1-69). ∎

Problem 1-12

(a) Use Fig. 1-7 to show that the object image matrix $S_{P'P}$ may be expressed as

$$S_{P'P} = \begin{pmatrix} 1 + \dfrac{as}{n_1} & -a \\ 0 & 1 - \dfrac{as'}{n_2'} \end{pmatrix}_{P'} {}_{P} \tag{1-73}$$

(b) Use this result to show that the form of the Gauss lens equation consistent with our conventions is

$$\frac{1}{f'} = -\frac{1}{f} = \frac{n_2'}{s'} - \frac{n_1}{s} \tag{1-74}$$

(c) Show that

$$\beta = \frac{s'n_1}{sn_2'} \tag{1-75}$$

∎

Problem 1-13

(a) A *telescope* consists of an objective lens (focal length f_1') and an eyepiece (focal length f_2') separated by a distance $f_1' + f_2'$. Show that the angular magnification of the image of the objective as formed by the eyepiece is

$$\gamma = -\frac{f_2'}{f_1'}$$

(b) Consider a ray with slope α entering the objective, passing on to the eyepiece, and emerging at a new angle α'. Show that γ, above, can also be expressed as α'/α.

(c) A *Barlow* lens is a negative lens (focal length $-f_B'$) placed between V_2 and F_1' at a small distance d to the left of F_1'. Show that the effective focal length $\bar{f}_1'$ of the objective becomes

$$\bar{f}_1' = \frac{f_1'f_B'}{(f_B' - d)} \tag{1-76a}$$

(d) Show that the angular magnification of the telescope is increased by a factor M, where

$$M = \frac{f_1'}{f_1' - d}$$

and that the Barlow lens lies a distance D to the left of the new focal plane of the objective, where

$$D \sim (M - 1)f_1' \tag{1-76b}$$

∎

Problem 1-14

Four thin lenses have focal lengths -1.00, 1.33, -0.33, 1.00 and spacings 1.00, 3.00, 0.50, respectively. Show that this system is equivalent to no lens at all.   ■

1-5   Some Applications of Paraxial Matrix Methods

An optical system which really tests one's understanding of matrix methods and the associated conventions is the transparent sphere with one hemisphere silvered (Fig. 1-13), as presented by O'Neill.[1-1] Denoting the index of refraction simply as n and the radius as r, the refraction matrix at the first vertex is

$$R_1 = \underset{V_1}{\left(\begin{matrix} 1 & -\dfrac{(n-1)}{r} \\ 0 & 1 \end{matrix}\right)}_{V_1}$$

and the first translation matrix is

$$T_{21} = \underset{V_2}{\left(\begin{matrix} 1 & 0 \\ \dfrac{2r}{n} & 1 \end{matrix}\right)}_{V_1}$$

We handle the reflection at surface 2 by treating it as a refraction into a medium whose index is $-n$.

Problem 1-15

Using the refraction matrix for a plane mirror, show that reflection corresponds to the condition

$$n_1' = -n$$

and show how l and l', x and x', and α and α', respectively, are related.   ■

Hence, the refraction matrix specifying the reflection at surface 2 is

$$R_2 = \underset{V_2}{\left(\begin{matrix} 1 & -\left(\dfrac{-n-n}{-r}\right) \\ 0 & 1 \end{matrix}\right)}_{V_2}$$

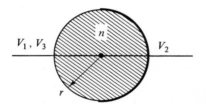

V_1, V_3 n V_2

r

Figure 1-13

where r is negative. The next translation is

$$T_{32} = \left.\begin{pmatrix} 1 & 0 \\ \dfrac{-2r}{-n} & 1 \end{pmatrix}\right._{V_2}^{} \atop {}_{V_3}$$

where the left-hand vertex is now labeled as V_3. The refraction at this surface is then

$$R_3 = \left.\begin{pmatrix} 1 & -\left(\dfrac{-1-(-n)}{r}\right) \\ 0 & 1 \end{pmatrix}\right._{V_3}^{} \atop {}_{V_3}$$

Multiplying these matrices in the proper order gives

$$S_{31} = R_3 T_{32} R_2 T_{21} R_1$$

$$= \left.\begin{pmatrix} \dfrac{n-4}{n} & -\dfrac{2(2-n)}{nr} \\ -\dfrac{4r}{n} & \dfrac{n-4}{n} \end{pmatrix}\right._{V_1}^{} \atop {}_{V_3} \tag{1-77}$$

Problem 1-16

Consider a ray parallel to the axis in Fig. 1-13 and at a distance x. Find the index of refraction needed to make the emerging ray return to the source of light. This problem indicates the usefulness of coated spheres. Particles made in this way are imbedded in highway signs and cause them to reflect the illumination from a car's headlights back to the driver. Note that this property depends only on n; the radius is immaterial, so the particle size need not be carefully controlled. ∎

    The second example which we wish to consider is the *zoom lens*, as discussed by Halbach.[1-2] This is a lens whose magnification may be varied without changing the object-to-image distance. For the paraxial approach to such an optical system, it is convenient to introduce the *focal plane matrix*, $S_{F'F}$. The matrix connecting F and F' is obtained from S_{21} for the lens in Fig. 1-14 by multiplication with translation matrices involving l_F and l'_F, giving

$$S_{F'F} = \left.\begin{pmatrix} 1 & 0 \\ l'_F & 1 \end{pmatrix}\right._{V_2} \left.\begin{pmatrix} b & -a \\ -d & c \end{pmatrix}\right._{V_1} \left.\begin{pmatrix} 1 & 0 \\ -l_F & 1 \end{pmatrix}\right._{F} \tag{1-78}$$

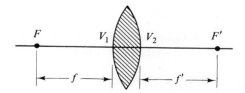

Figure 1-14

If we use Eqs. (1-48) and (1-49), this becomes

$$S_{F'F} = \begin{pmatrix} 1 & 0 \\ \dfrac{c}{a} & 1 \end{pmatrix}_{F'} \begin{pmatrix} b & -a \\ -d & c \end{pmatrix}_{V_2} \begin{pmatrix} 1 & 0 \\ \dfrac{b}{a} & 1 \end{pmatrix}_{V_1}{}_F$$

$$= \begin{pmatrix} 0 & -a \\ \dfrac{1}{a} & 0 \end{pmatrix}_{F'}{}_F = \begin{pmatrix} 0 & \dfrac{1}{f} \\ -f & 0 \end{pmatrix}_{F'}{}_F \tag{1-79}$$

Problem 1-17

(a) Show that the matrix $S_{F'F}$ predicts the behavior we would expect for rays through F and F'.

(b) How may the location of the cardinal points be determined from $S_{F'F}$? ▄

To extend these results to two lenses whose focal points F'_1 and F_2 have a separation d_1 (Fig. 1-15), we incorporate a translation matrix and get

$$S_{F_2'F_1} = \begin{pmatrix} 0 & \dfrac{1}{f_2} \\ -f_2 & 0 \end{pmatrix}_{F_2'} \begin{pmatrix} 1 & 0 \\ d_1 & 1 \end{pmatrix}_{F_2}{}_{F_1'} \begin{pmatrix} 0 & \dfrac{1}{f_1} \\ -f_1 & 0 \end{pmatrix}_{F_1}$$

$$= \begin{pmatrix} -\dfrac{f_1}{f_2} & \dfrac{d_1}{f_1 f_2} \\ 0 & \dfrac{f_2}{f_1} \end{pmatrix}_{F_2'}{}_{F_1} \tag{1-80}$$

and for three lenses (Fig. 1-16), the resulting focal plane matrix is

$$S_{F_3'F_1} = \begin{pmatrix} 0 & \dfrac{1}{f_3} \\ -f_3 & 0 \end{pmatrix}_{F_3'} \begin{pmatrix} 1 & 0 \\ d_2 & 1 \end{pmatrix}_{F_3}{}_{F_2'} \begin{pmatrix} -\dfrac{f_1}{f_2} & \dfrac{d_1}{f_1 f_2} \\ 0 & -\dfrac{f_2}{f_1} \end{pmatrix}_{F_1}$$

$$= \begin{pmatrix} -\dfrac{d_2 f_1}{f_2 f_3} & \dfrac{d_1 d_2}{f_1 f_2 f_3} - \dfrac{f_2}{f_1 f_3} \\ \dfrac{f_1 f_3}{f_2} & -\dfrac{d_1 f_3}{f_1 f_2} \end{pmatrix}_{F_3'}{}_{F_1} \tag{1-81}$$

Problem 1-18

Show that the formulas giving the positions of the cardinal points may be used with the elements of the focal-plane matrix for two lenses. ▄

The last two problems demonstrate that formulas such as Eqs. (1-48) and

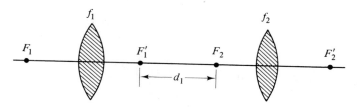

Figure 1-15

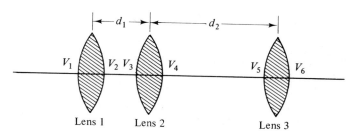

Figure 1-16

(1-49) are valid when we take the appropriate elements from $S_{F_3'F_1}$ rather than S_{61}. Hence, we find that the sum of $-l_F$ and l_F' is

$$-l_F + l_F' = \frac{(d_2 f_1/f_2 f_3) + d_1 f_3/f_2 f_3}{-(f_2/f_1 f_3) + (d_1 d_2/f_1 f_2 f_3)}$$

$$= \frac{d_1 f_3^2 + d_2 f_1^2}{-f_2^2 + d_1 d_2} \tag{1-82}$$

If we now move lens 2 but hold the other two lenses stationary, then the total distance between F and F' is a constant if we maintain the quantity $(-l_F + l_{F'})$ constant. The simplest way of doing this is to make them numerically equal, so that $-l_F + l_{F'} = 0$. Equation (1-82) shows that

and
$$\left.\begin{array}{c} f_1 = f_3 \\[2mm] d_1 = -d_2 \end{array}\right\} \tag{1-83}$$

Hence, if we make the two end lenses identical and arrange them as shown in Fig. 1-17, both conditions of Eq. (1-83) are satisfied as the middle lens moves.

Maintaining the spacing between focal planes is only the first step, however, for by Eqs. (1-51) and (1-83), the focal length is

$$f = \left[\frac{f_2}{f_1 f_3} - \frac{d_1 d_2}{f_1 f_2 f_3}\right]^{-1}$$

$$= \frac{f_1^2 f_2}{f_2^2 + d_1^2} \tag{1-84}$$

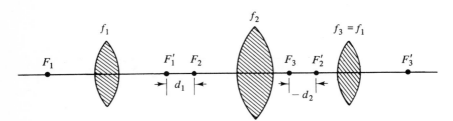

Figure 1-17

where $d_1 = -d_2$. Since f varies with d_1, then so will the positions of H and H', and the Gauss lens equation (1-74) will not be satisfied. That is, the image will be out of focus.

To study this behavior, it is convenient at this point to use a numerical approach. Let us place an object at 100.0 units to the left of the first lens and the final lens at 60.0 units to the right (Fig. 1-18). If t is the separation between the first and second lenses, then the system matrix S_{61} for the three thin lenses of focal lengths $f'_1 = 10.0$, $f'_2 = 20.0$, and $f'_3 = 10.0$, respectively, is

$$S_{61} = {}_{V_6}\begin{pmatrix} 1 & -0.1 \\ 0 & 1 \end{pmatrix}_{V_5} \begin{pmatrix} 1 & 0 \\ 60 - t & 1 \end{pmatrix}_{V_4} \begin{pmatrix} 1 & -0.05 \\ 0 & 1 \end{pmatrix}_{V_3} \begin{pmatrix} 1 & 0 \\ t & 1 \end{pmatrix}_{V_2} \begin{pmatrix} 1 & -0.1 \\ 0 & 1 \end{pmatrix}_{V_1}$$

$$(1\text{-}85)$$

The object-to-image distance D can be calculated from the elements of the system matrix S_{61} by means of the relation

$$D = 160.0 + l'$$

where, by Eq. (1-44)

$$l' = \frac{-100c + d}{-100a + b}$$

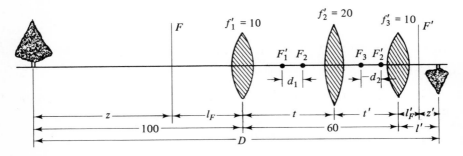

Figure 1-18

The magnification β comes from Eq. (1-41) and is

$$\beta = \frac{1}{(b - 100a)}$$

Letting t vary from 0.0 to 60.0, we can compute the variation in D (Fig. 1-19) and the corresponding variation in β, shown in Fig. 1-20. To correct for the defocusing, lens 3 can be moved a small amount at the same time that lens 2 is shifting. The amount of motion necessary is shown in Fig. 1-21. It is possible to obtain a simultaneous uniform motion of lens 2 and the nonuniform adjustment of lens 3 by constructing the cam of Fig. 1-22. This is the method actually used in the manufacture of zoom lenses.

The evaluation of the matrix product in Eq. (1-85) is quite tedious; hence such calculations should be done on a computer. The two problems which follow are examples of numerical matrix multiplication; an elementary explanation of FORTRAN programming adequate for these problems and for those which come in later chapters will be found in a previous book.[1-3]

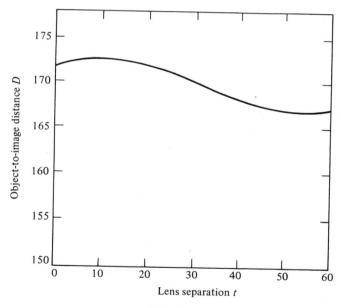

Figure 1-19

Problem 1-19

The lens shown in Fig. 1-23 is known as a *Tessar*; it is used in moderately priced cameras. The physical parameters of the lens in the figure (which is to scale) are:

r	n'	t'
1.628		
	1.6116	0.357
−27.57		
	1.0000	0.189
−3.457		
	1.6053	0.081
1.582		
	1.0000	0.325
∞		
	1.5123	0.217
1.920		
	1.6116	0.396
−2.400		

Write a computer program which will calculate the elements of the system matrix and the positions of the cardinal points.   ∎

Problem 1-20

(a) Verify numerically the graphs of Figs. 1-19 and 1-20.

(b) Starting with lens 2 at 30.0 units to the right of lens 1, write a program which will shift lens 2 toward lens 3, and calculate the corresponding motion of lens 3 necessary to hold the object-to-image distance D constant to within 0.1 unit, thus verifying Fig. 1-21.   ∎

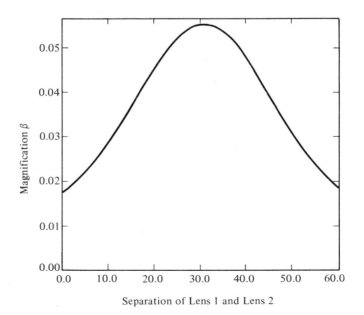

Separation of Lens 1 and Lens 2

Figure 1-20

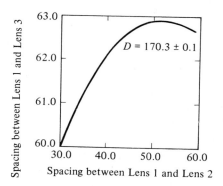

Figure 1-21

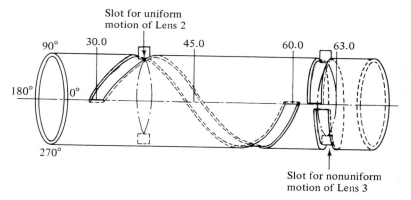

Figure 1-22

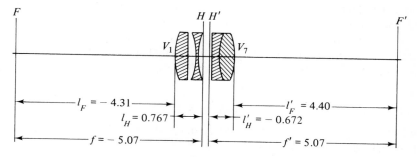

Figure 1-23

Concluding Remarks

We have seen that an optical system consisting of an arbitrary number of lenses can be characterized by its cardinal planes—a pair of focal planes and a pair of unit planes—whose positions with respect to the first and last vertices we can calculate from a knowledge of the radii of curvature, the indices of refraction, and the distance from one vertex to another. This approach is analogous to that which electrical engineers use in dealing with circuits; a complex collection of elements is replaced by a "black box" which is a much simpler circuit whose electrical behavior is equivalent to the original circuit. Figure 1-12 illustrates this very nicely. When we apply the matrix methods developed in this chapter, however, we must keep in mind that we have also assumed the validity of the paraxial approximation. Although it would appear that this imposes a severe restriction, this is not, in fact, the case. We shall see in a later chapter that a paraxial calculation is the first step in the design of a real system, for this is how we determine the approximate physical dimensions of the instrument, the diameters and focal lengths of the component lenses, and an initial estimate of the amount of light which passes through the system. Then we refine the design by making a precise calculation, using methods to be described. We shall also see that it is possible to generalize the matrix equations we have been using to the nonparaxial case without adding a great deal of complexity to the theory. This is one of the many benefits of matrix geometrical optics.

Problem 1-21

(a) A method for measuring the distance l_F of a converging lens system in the laboratory involves the use of an *autocollimator*, shown in elementary form in Fig. 1-24(a). A pinhole is placed in front of a light source, and the lens and pinhole are adjusted so that the image reflected by the mirror is in focus at a position immediately adjacent to the pinhole. Show that the lens-to-pinhole distance is then equal to l_F.

(b) To find the location of the unit point H, a *nodal slide* is added to the above arrangement. This is a lens holder which moves back and forth along OZ and also permits rotation of the lens around a vertical axis. Show by a diagram that the image described in part (a) will not move if the lens is rotated by a small amount about an axis which passes through H. (*Hint:* Use elementary ray tracing and a binomial approximation.) ■

Problem 1-22

Devise a simple way of experimentally determining the Gaussian constants of an arbitrary lens system. ■

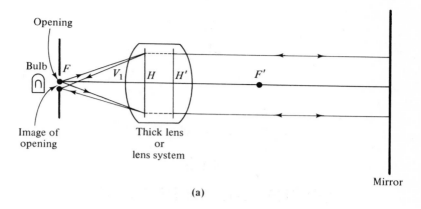

(a)

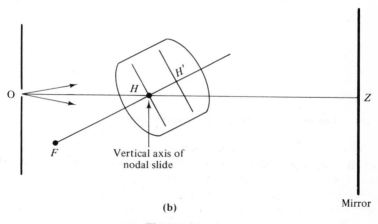

(b)

Figure 1-24

References

1-1 E. L. O'Neill, *Introduction to Statistical Optics*, Addison-Wesley (1963).
1-2 K. Halbach, *Am. J. Phys.* **32,** 90 (1964).
1-3 A. Nussbaum, *Geometric Optics*, Addison-Wesley (1968).

General References

J. W. Blaker, *Geometric Optics,* Dekker, New York (1971).
A. D. Clark, *Zoom Lenses,* American Elsevier, New York (1973).

2

MERIDIONAL RAYS
AND
SPHERICAL ABERRATION

2-1 The Origin of Geometrical Aberrations

We saw in the first chapter that we may use matrices to trace a ray of light leaving a point in object space through any number of refracting surfaces to the corresponding point in image space. One feature of this approach that should be emphasized is that *all* the rays leaving the object point intersect at the image point. The image will usually be larger or smaller than the object; it may be erect or inverted, but it will otherwise be a faithful reproduction. This is a consequence of the small-angle or paraxial approximation, valid for angles less than about 5°. However, if a light ray goes through or near the edge of a lens, the angle may be large, and the focal point for such rays could be different than for rays close to the center. The result of such a *geometrical aberration* is a blurred image. In this chapter we shall study one of these aberrations.

2-2 The Non-Paraxial, or Exact, Matrices

The first problem to be considered in connection with the removal of the small-angle approximation is to determine the exact form of the translation matrix T and the refraction matrix R. Let us consider the ray QP in Fig. 2-1

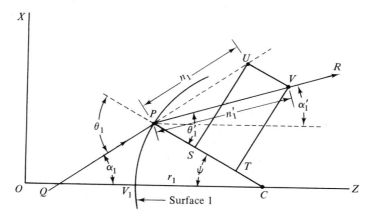

Figure 2-1

striking surface 1 of a lens system. The resulting refracted ray is labeled PR, and points U and V are chosen to satisfy the relations

$$\left.\begin{array}{l}\overline{PU} = n_1 \\ \overline{PV} = n'_1\end{array}\right\} \tag{2-1}$$

Drop perpendiculars $\overline{VT}$ and $\overline{US}$ to the line $\overline{PC}$. Then

$$\begin{aligned}\overline{US} &= \overline{PU} \sin \theta_1 = n_1 \sin \theta_1 \\ \overline{VT} &= \overline{PV} \sin \theta'_1 = n'_1 \sin \theta'_1\end{aligned} \tag{2-2}$$

and by Snell's law

$$\overline{US} = \overline{VT}$$

so that $\overline{UV}$ is parallel to $\overline{ST}$.

A portion of this figure is reproduced in Fig. 2-2. Projecting the three sides of the triangle PUV onto the x axis, we see that

$$n_1 \cos \varphi_1 = n'_1 \cos \varphi'_1 + \overline{UV} \cos (\pi - \gamma) \tag{2-3}$$

where

$$\left.\begin{array}{l}\cos \varphi_1 = \sin \alpha_1, \qquad \cos \varphi'_1 = \sin \alpha'_1 \\ \cos (\pi - \gamma) = \sin \psi\end{array}\right\} \tag{2-4}$$

But

$$\sin \psi = \frac{x_1}{r_1} \tag{2-5}$$

where (z_1, x_1) are the coordinates of P, and $r_1 = \overline{PC}$ is the interface radius. Also

$$\overline{UV} = \overline{PT} - \overline{PS} = n'_1 \cos \theta'_1 - n_1 \cos \theta_1 \tag{2-6}$$

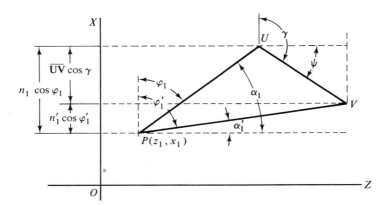

Figure 2-2

Combining Eqs. (2-3) through (2-6) gives

$$n'_1 \sin \alpha'_1 = n_1 \sin \alpha_1 - K_1 x_1 \tag{2-7}$$

where

$$K_1 = \frac{n'_1 \cos \theta'_1 - n_1 \cos \theta_1}{r_1} \tag{2-8}$$

The quantity K_1 is called the *skew power* of surface 1. For small angles it reduces to the power k_1 previously used for the paraxial approximation, and Eq. (2-7) similarly becomes

$$n_1 \alpha'_1 = n_1 \alpha_1 - k_1 x_1 \tag{1-16}$$

as before. Equation (2-7) together with

$$x'_1 = x_1 \tag{2-9}$$

gives

$$\begin{pmatrix} n'_1 \sin \alpha'_1 \\ x'_1 \end{pmatrix} = \begin{pmatrix} 1 & -K_1 \\ 0 & 1 \end{pmatrix} \begin{pmatrix} n_1 \sin \alpha_1 \\ x_1 \end{pmatrix} \tag{2-10}$$

so that the refraction matrix R_1 is now

$$R_1 = \begin{pmatrix} 1 & -K_1 \\ 0 & 1 \end{pmatrix} \tag{2-11}$$

To find the translation matrix, we use Fig. 2-3, which shows a translation T'_1, at an angle α'_1 to the z axis, corresponding to a ray traveling from surface 1 to surface 2. Then

$$x_2 = x'_1 + T'_1 \sin \alpha'_1 \tag{2-12}$$

while

$$n'_1 \sin \alpha'_1 = n_2 \sin \alpha_2$$

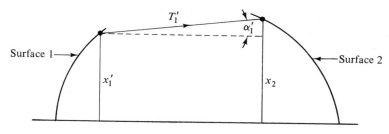

Figure 2-3

or in matrix form

$$
\begin{pmatrix} n_2 \sin \alpha_2 \\ x_2 \end{pmatrix} = \begin{pmatrix} 1 & 0 \\ \dfrac{T_1'}{n_1'} & 1 \end{pmatrix} \begin{pmatrix} n_1' \sin \alpha_1' \\ x_1' \end{pmatrix} \tag{2-13}
$$

which again reduces to the correct paraxial relation.

As before, we may combine Eqs. (2-10) and (2-13) to relate α_2, x_2 with α_1, x_1, continuing this process throughout any number of refractions and translations. However, the translation matrix T_{21} is now

$$
T_{21} = \begin{pmatrix} 1 & 0 \\ \dfrac{T_1'}{n_1'} & 1 \end{pmatrix} \tag{2-14}
$$

and the displacement T_1' must be determined; it no longer represents a spacing between two adjacent vertices. Also, we must find the angles θ_1 and θ_1' of Eq. (2-8), since the matrix R_1 is now dependent on the orientation of the light rays.

One of the delights of the paraxial matrix approach to geometrical optics is the ease with which the matrices and the few other necessary formulas may be remembered. We shall now show that this advantage is not lost when using the exact matrices. Figure 2-4 depicts an object OP of height x with a

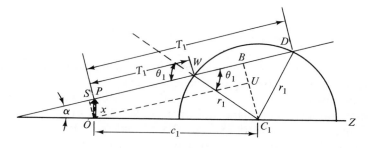

Figure 2-4

ray of inclination α leaving P and striking a circle of radius r_1 at both W and D. If the intersection point is W, then the circle is a portion of a convex surface; D therefore represents the intersection point for a concave surface. This is why two values of T_1 are possible. Let $\overline{C_1 B}$ and $\overline{OS}$ be perpendicular to the ray and $\overline{OU}$ be parallel to it. Then

$$T_1 = \overline{OU} - \overline{SP} \pm \overline{WB}$$

where the negative sign goes with the smaller of the two possible values of T_1. But

$$\overline{OU} = c_1 \cos \alpha$$
$$\overline{SP} = x \sin \alpha$$

and

$$\overline{WB}^2 = r_1^2 - \overline{BC_1^2} = \overline{BD}^2$$

where

$$\overline{BC_1} = \overline{BU} + \overline{UC_1}$$
$$= x \cos \alpha + c_1 \sin \alpha$$

Hence

$$T_1 = c_1 \cos \alpha - x \sin \alpha \pm \sqrt{r_1^2 - (x \cos \alpha + c_1 \sin \alpha)^2} \qquad (2\text{-}15)$$

Knowing the position and height of the object, the inclination of the ray, and the radius of the lens, we can find T_1 from this formula, which is easy to obtain geometrically. This is the variable quantity needed to find the incident translation matrix for surface 1, which is designated as T.

It is also a simple matter to find the variable K_1 in the refraction matrix R_1 for surface 1. By Snell's law and Eq. (2-8) we have

$$K_1 = \frac{n_1' \cos \theta_1' - n_1 \cos \theta_1}{r_1}$$
$$= \frac{\sqrt{n_1'^2 - n_1^2 \sin^2 \theta_1} - \sqrt{n_1^2 - n_1^2 \sin^2 \theta_1}}{r_1} \qquad (2\text{-}16a)$$

where

$$\sin \theta_1 = \frac{\overline{BC_1}}{r_1} = \frac{x \cos \alpha + c_1 \sin \alpha}{r_1} \qquad (2\text{-}16b)$$

Hence, K_1 may be readily remembered in the symmetrical form in Eq. (2-16a) and calculated from a term in Eq. (2-15).

We are now in a position to trace exactly (i.e., without the use of approximations) a ray lying in a plane, known as a *meridional* or *tangential plane*, which contains the symmetry axis OZ. To be confident about our extended matrix method, we shall check a calculation performed in a different way by Smith.[2-1] A ray of light leaves a point on the axis at a distance of -200.0 to the left of a lens and at an angle whose sine is 0.1. The constants of the lens are

$$r_1 = 50.0$$
$$r_2 = -50.0$$
$$t'_1 = 15.0$$
$$n'_1 = 1.50$$

Program 2-1 shows how to perform this calculation. Necessary comments are as follows:

(a) The quantities P and Q correspond to the distances $\overline{BC}_1$ and $\overline{PB}$, respectively, of Fig. 2-4. The quantity UD corresponds to $\overline{WB}$ or $\overline{BD}$.

(b) The statement

$$IF(R(J))\ 20, 10, 30$$

makes the sign in front of the radical in Eq. (2-15) opposite to that of the sign of the radius.

(c) The statement defining a new value of C shifts the origin during each passage of the DO loop. After we have traced the ray from P to W using the first translation matrix and found the inclination of the refracted ray using the matrix R_1, we essentially start over. That is, $\overline{AW}$ in Fig. 2-5 is regarded as the new object, with A as the new position of the origin. The location c_2 of the center of curvature C_2 is then given by

$$c_2 = -T_1 \cos \alpha + c_1 - r_1 + t'_1 + r_2 \tag{2-17}$$

We obtain this formula by starting at A and proceeding to C_2, following the sequence indicated by the right side. A lens with two positive surfaces was deliberately used in Fig. 2-5 to avoid confusion over signs. For the customary double convex lens with negative r_2, Eq. (2-17) automatically produces the correct value of c_2. This process of shifting the origin is not absolutely necessary, but it is very desirable. In an optical system such as a telescope, the distances along the axis are usually many times greater than the lens dimensions. Shifting the origin keeps the z distances at magnitudes comparable to the x distances, preventing errors due to calculations involving both large and small numbers.

(d) The portion of the program which follows the DO loop is used to determine the behavior of the ray after it leaves the lens. Figure 2-6 shows the geometrical meanings of the quantities involved (labeled with their FORTRAN symbols).

The result of this calculation is that the light ray crosses the axis at a distance of 45.63 units from the vertex V_2 and at an angle whose sine is -0.37.

Program 2-1

```
DIMENSION R(12), U(13), T(12)
N = 2
R(1) = 50.0 $ R(2) = -50.0
U(1) = 1.0 $ U(2) = 1.5 $ U(3) = 1.0
T(1) = 200.0 $ T(2) = 15.0
A1 = 0.1    $   A2 = 0.0
C = R(1) + T(1)
DO 10 J = 1,N
SINE = A1/U(J)
ANG = ASIN (SINE)
P = A2*COS (ANG) + C*SIN (ANG)
Q = C*COS (ANG) - A2 *SIN (ANG)
UD = SQRT (R(J)*R(J) - P*P)
IF(R(J)) 20, 10, 30
30 TT = Q - UD
GO TO 40
20 TT = Q + UD
40 POR = P/R(J)
THETA = ASIN (POR)
AK = (SQRT (U(J + 1)**2 - (U(J)*SIN (THETA))**2) - U(J)*COS (THETA
1 ))/R(J)
A3 = A1
A1 = A1*(1.0 - AK*TT/U(J)) - A2*AK
A2 = A3*TT/U(J) + A2
IF(J - N) 50, 60, 60
50 C = C - R(J) + R(J + 1) + T(J + 1) - TT*COS (ANG)
10 CONTINUE
60 ANG = ASIN (A1)
AR = ABS (R(J))
QQ = AR - SQRT (R(J)*R(J) - A2*A2)
Z = A2/TAN (ANG)
AZ = ABS (Z)
AL = AZ - QQ
PRINT 100
100 FORMAT(///////)
PRINT 150, A1, AL
150 FORMAT(10X, 2F20.6)
PRINT 100
STOP
END
```

        -.372744   45.631070

This may be compared with the paraxial result obtained from the equation

$$
\begin{pmatrix} \alpha_2' \\ x_2 \end{pmatrix} = \begin{pmatrix} 1 & -\dfrac{0.5}{50.0} \\ 0 & 1 \end{pmatrix} \begin{pmatrix} 1 & 0 \\ \dfrac{15.0}{1.5} & 1 \end{pmatrix} \begin{pmatrix} 1 & -\dfrac{0.5}{50.0} \\ 0 & 1 \end{pmatrix} \begin{pmatrix} 1 & 0 \\ 200.0 & 1 \end{pmatrix} \begin{pmatrix} 0.1 \\ 0.0 \end{pmatrix}
$$

$$
= \begin{pmatrix} -0.29 \\ 19.0 \end{pmatrix}
$$

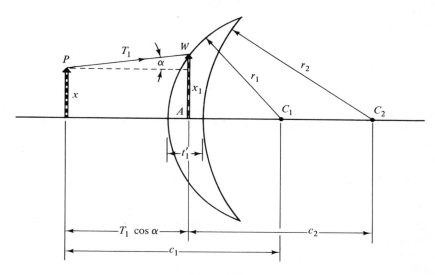

Figure 2-5

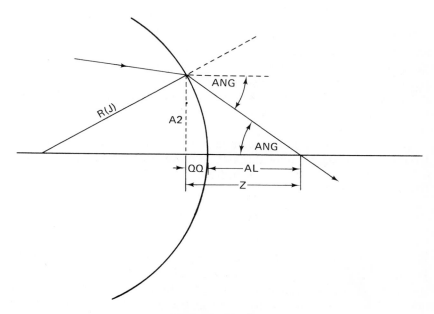

Figure 2-6

Then the intersection of the ray with the axis occurs at a distance l' from the vertex, given by

$$l' = \frac{19.0}{0.29} = 65.5$$

Problem 2-1

A straight line intersects the y axis at b and has a slope m. Derive formulas for the coordinates of the points where this line crosses a circle of radius r with center at $(c, 0)$. Use this result to obtain Eq. (2-15).   ■

2-3 Spherical Aberration

Let us consider a symmetrical convex lens and a ray of light along its axis of symmetry. This ray obviously passes through without deviation. If the ray is displaced parallel to itself a short distance x from the axis, then it goes through the paraxial focus F'. As x gets larger, this is no longer true. Although the angle with respect to the axis remains zero, the incident and refraction angles at the surfaces are getting beyond the paraxial approximations of Eqs. (1-12), and we must express these functions by the Taylor series of Eqs. (1-11); thus $\sin \theta$, for example, is

$$\sin \theta = \theta - \frac{\theta^3}{3!} + \frac{\theta^5}{5!} - \cdots \tag{2-18}$$

The presence of the second term in Eq. (2-18) means that there will be deviations from the predictions of paraxial optics which lead to defects in the image. These are the *third-order* or *Seidel aberrations*, and we shall find that there are five distinct types. The next term corresponds to *fifth-order aberrations* which are usually smaller in magnitude than the third-order aberrations (many authors refer to these as *primary* and *secondary* aberrations, respectively).

To consider aberrations quantitatively, we shall treat the situation described at the beginning of this section. Program 2-2 traces a ray of light parallel to the axis of a lens specified by

$$r_1 = 10.0, \qquad r_2 = -10.0$$
$$t'_1 = 2.0$$
$$n'_1 = 1.5000$$

This program, except for the concluding steps, is identical to Program 2-1. The ray is taken as originating at a large distance (1200.0 units) from the lens and a negligible distance ($x = 0.00001$) from the axis. This determines the position of the paraxial focus FP at the start of the calculation. The quantity XM is the maximum value of x.

Program 2-2

```
      DIMENSION R(12), U(13), T(12)
      R(1) = 10.0
      R(2) = -10.0
      N = 2.0
      XM = 4.01
      U(1) = 1.0
      U(2) = 1.5
      U(3) = 1.0
      T(1) =1200.0
      T(2) = 2.0
      A1 = 0.0
      A2 = 0.00001
      X = A2
120   C = R(1) + T(1)
          .                    .
          .                    .
          .                    .
          .                    .

60    ANG = ASIN (A1)
      TT =A2/A1
      AR = ABS (R(J))
      QQ = AR - SQRT (R(J) **2  - A2**2)
      Z = A2/TAN (ANG)
      AZ = ABS (Z)
      AL = AZ - QQ
      IF(X - 0.01) 70, 70, 80
70    PF = AL
      PRINT 500
80    ALSA = AL - PF
      TSA = -ALSA*TAN (ANG)
      PRINT 55, X, AL, TSA, ALSA
55    FORMAT(4F15.4)
      A1 = 0.0
      A2 = X + 0.1
      X = A2
      IF(X - XM) 120, 120, 110
110   STOP
      END
```

```
    .0000          9.6552              -0                0
    .1000          9.6537          -.0000           -.0015
    .2000          9.6493          -.0001           -.0058
      .                .
      .                .
      .                .
   3.9000          6.7930          -1.4535          -2.8622
   4.0000          6.5981          -1.6368          -3.0671
```

As we move the ray away from the axis, using the statement

$$A2 = X + 0.1$$

the program computes the quantities shown in Fig. 2-7. The ray for the initial value of *x* crosses the axis at the paraxial focus PF. For all other values of *x*, the ray will cross at a distance AL from the vertex, so that we may specify the deviation from paraxial conditions in one of two ways: as a *longitudinal spherical aberration* (ALSA) or a *transverse spherical aberration*

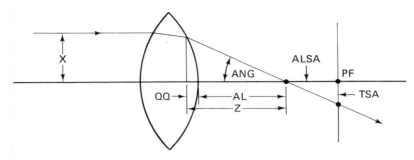

Figure 2-7

(TSA). The physical meanings of the two quantities are indicated in the figure. The longitudinal aberration is usually used by lens designers to specify lens quality, but the transverse aberration—as we shall see later—comes naturally out of the theory. We shall also see that spherical aberration is the only one which is associated with the axis of symmetry; all the others require an off-axis object.

According to the paraxial approximation, all these parallel rays through the lens meet at the paraxial focus, and the light rays define a pair of cones (Fig. 2-8). Program 2-2 shows that an exact calculation gives the situation of Fig. 2-9. Only light rays passing through a ring of very small radius (this ring is called the *paraxial zone*) meet at the paraxial focus F'. Those rays through a ring near the edge of the lens (the *marginal zone*) meet at the *marginal focus FM*, and an intermediate zone produces an intermediate focal point *FI*. The left-hand cone is then deformed into a curved surface of rotation known as the *caustic surface*. A remarkable photograph of this effect (and of the other aberrations as well) is given by Cagnet, Françon, and Thrierr.[2-2]

The caustic surface is the envelope of the set of cones generated by all the rays through each zone. The image, rather than being a point, is a circle

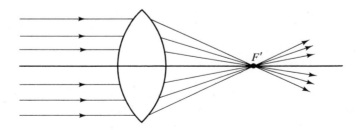

Figure 2-8

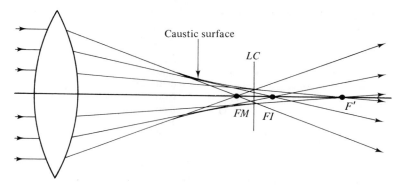

Figure 2-9

determined by the intersection of the caustic surface with a plane. Since the image plane has not yet been defined, the size of the image circle depends on the location of this plane. The minimum radius, as the figure shows, occurs when a marginal ray crosses the axis and then intersects the caustic surface. This defines the *circle of least confusion*, corresponding to the plane LC of Fig. 2-9. Hence, we may minimize transverse spherical aberration in a given lens by proper focusing.

The output of Program 2-2 also shows that TSA depends on both x^3 and x^5 as predicted in connection with Eq. (2-18). If we assume that

$$TSA = ax^3 + bx^5 \qquad (2\text{-}19)$$

and use values of TSA at $x = 1.0$ and $x = 2.0$ to determine the constants in Eq. (2-19), we find that

$$a = -0.0145$$
$$b = -0.0005 \qquad (2\text{-}20)$$

Thus—also as anticipated—the fifth-order term makes a smaller contribution to the total aberration than the third-order term. In a lens *system*, however, this may not be necessarily true.

To see what can be done if we have the size and shape of the lens at our disposal, we follow the approach of Jenkins and White.[2-3] Define the *shape factor S* of a lens as the ratio

$$S = \frac{r_2 + r_1}{r_2 - r_1} \qquad (2\text{-}21)$$

For a symmetric double convex lens ($r_2 = -r_1$), we have

$$S = 0$$

so that the shape factor measures the deviation from symmetry. We now repeat Program 2-2 for the following six lenses:

r_1	r_2	S	
−10.00	−3.33	−2.00	
∞	−5.00	−1.00	
20.00	−6.67	−0.50	
10.00	−10.00	0.00	(original lens)
6.67	−20.00	0.50	
5.00	∞	1.00	
3.33	10.00	2.00	

All seven lenses have the same thickness and index, and approximately the same focal length. Plotting the longitudinal spherical aberration for $x = 1.0$ as a function of the shape factor, we obtain the graph in Fig. 2-10. The minimum occurs at about $S = 0.7$. The process of minimizing an aberration by altering the shape of a lens while holding other properties constant is known as *bending the lens*.

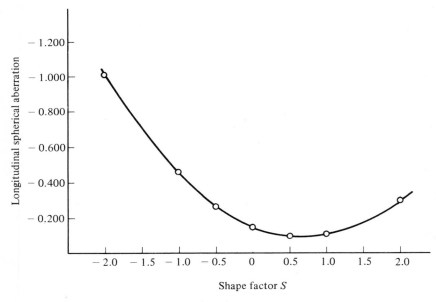

Shape factor S

Figure 2-10

Problem 2-2

(a) Find the value of S which minimizes ALSA by bending a lens for which $r_1 = 40.0$, $r_2 = -40.0$, $t_1' = 0.2$, and $n_1' = 1.523$, while maintaining f' constant.

(b) Plot ALSA vs S for $n_1' = 1.300, 1.400, \ldots, 1.900$ for the lens in Program 2-2.   ▰

All seven lenses considered in Fig. 2-10 are converging or positive. We would expect that the spherical aberration for a negative lens would have the opposite sign and it does. Hence, the next step in the reduction of this aberration is to go to a *doublet* consisting of two lenses, one element of which is positive and the other negative. To treat this quantitatively, we return to the output of Program 2-2 and show that the relation expressing the behavior of ALSA analogous to Eq. (2-19) is

$$\text{ALSA} = Ax^2 + Bx^4 \tag{2-22}$$

Let $x_{\max}$ denote the largest value of x to be used and require ALSA to vanish for this zone. Then

$$Ax_{\max}^2 + Bx_{\max}^4 = 0$$

or

$$A = -Bx_{\max}^2$$

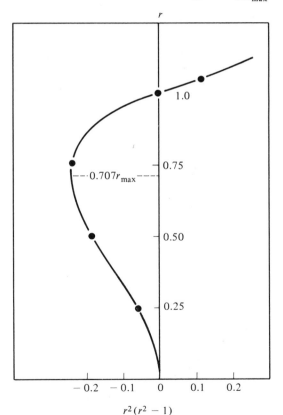

Figure 2-11

so that

$$\text{ALSA} = B(-x_{max}^2 x^2 + x^4) \qquad (2\text{-}23)$$

Differentiating this gives

$$\frac{d}{dx}(\text{ALSA}) = B(-2x_{max}^2 x + 4x^3) = 0$$

or

$$x = \frac{x_{max}}{\sqrt{2}} \qquad (2\text{-}24)$$

This result means that, if we correct spherical aberration by making ALSA zero at x_{max}, we are actually overcorrecting at larger apertures, for a plot of Eq. (2-23) has the form shown in Fig. 2-11. We obtain this graph by letting

$$r = \frac{x}{x_{max}} \qquad (2\text{-}25)$$

Then Eq. (2-23) becomes

$$\frac{\text{ALSA}}{Bx_{max}^4} = r^2(r^2 - 1) \qquad (2\text{-}26)$$

and the graph represents $r^2(r^2 - 1)$ versus r. Most lens designers specify spherical aberration with a curve of this type.

To consider a real example, let us take a cemented doublet designed by

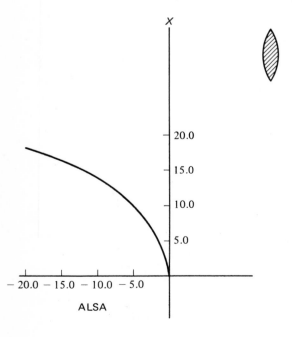

Figure 2-12

Brouwer[2-4] with the following parameters

r	t'	n'
61.070		
	4.044	1.56178
−47.107		
	2.022	1.70100
−127.098		

The front element of this lens by itself has a focal length of about 51.0 and a longitudinal spherical aberration which varies as shown in Fig. 2-12. Figure 2-13 shows the tremendous reduction in ALSA brought about by the addition of the negative element (note the relative horizontal scales of the two graphs). Both of these curves were computed by making obvious modifications in Program 2-2.

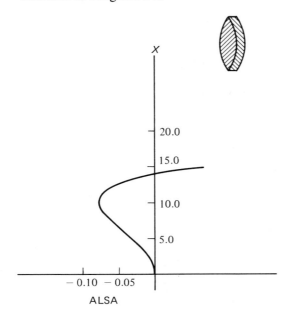

Figure 2-13

Problem 2-3

It has been suggested by Fulcher[2-5] that a lens system consisting of a series of *meniscus* lenses would have very low spherical aberration. Such lenses are shown in Fig. 2-14. They are converging lenses with two positive radii, and they are usually quite thin. Standard spectacle lenses made of crown glass ($n_1' = 1.5230$) are of this type. The figure shows Fulcher's design; the physical basis of his ideas is that the paraxial ray suffers less refraction at the eight air-glass surfaces than the marginal ray; a ray which is very close to the axis

would effectively be refracted at only two surfaces. There is an intuitive reason why the shape factor $S = 0.7$ in Fig. 2-10 gives a minimum: this is a lens for which the amount of refraction is distributed just about equally between the two surfaces. In the present case, the distribution is divided among eight surfaces, and we would expect considerable improvement.*
The specifications of the lens combination shown in Fig. 2-14 are:

r	t'	n'
20.92		
	0.200	1.5230
∞		
	0.0	1.0
13.075		
	0.200	1.5230
34.8867		
	0.0	1.0
8.7167		
	0.200	1.5230
14.9429		
	0.0	1.0
6.9733		
	0.200	1.5230
10.46		

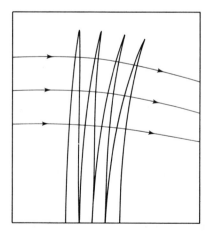

Figure 2-14

*It should be mentioned that Fulcher believed he had achieved a lens with the minimum possible spherical aberration. As we shall see in Chapter 10, this limit is set by the wavelength of visible light, and a lens system which reduces the aberration to values of about 1 μm (1 micrometer = 10^{-6} meter; this unit was formerly known as a micron) is said to be *diffraction limited*. Unfortunately, Fulcher based his design on the use of approximate formulas whose validity did not extend to the small image size he needed. His work was done before computers were readily available, when the use of exact calculations involved many hours on desk calculators and were therefore avoided if possible.

To test these ideas quantitatively, find the spherical aberration for $x = 0.1, 0.2, \ldots, 2.50$ in this lens system, and compare these results with the aberration in a single lens having the same thickness, index of refraction, and first and last radii as the system, i.e., $r_1 = 20.92$, $r_2 = 10.46$, $t'_1 = 0.8$, and $n'_1 = 1.5230$. ∎

References

2-1 W. J. Smith, *Modern Optical Engineering*, McGraw-Hill Book Co. (1966).

2-2 M. Cagnet, M. Françon, and J. C. Thrierr, *Atlas of Optical Phenomena*, Prentice-Hall (1962).

2-3 F. A. Jenkins and H. E. White, *Fundamentals of Optics*, 3rd ed., McGraw-Hill Book Co. (1957).

2-4 W. Brouwer, *Matrix Methods in Optical Instrument Design*, W. A. Benjamin, Inc. (1964).

2-5 G. S. Fulcher, *J. Opt. Soc. Am.* **37,** 47 (1947).

3

ABERRATIONS ASSOCIATED
WITH SKEW RAYS

3-1　Skew Rays and the Three-Dimensional Matrices

In the previous chapter, we saw how to extend matrix methods to non-paraxial rays lying in a meridional plane, and this capability made us realize that the failure of all the rays parallel to the axis to reach a common focus is a consequence of spherical aberration. It turns out that rays which do not lie in the plane of the diagrams of Chapter 2—these are known as *nonmeridional* or *skew rays*—lead to additional defects in the image. What we shall do here is extend nonparaxial matrix optics to three dimensions, so that we apply it to the study of skew rays.

Such a ray is shown in Fig. 3-1. It starts at point P on the object and passes through a point P_1 on the first surface of the optical system. Let x, y, z denote the coordinates of P, let x_1, y_1, z_1 be those for P_1, and let the ray PP_1 have direction-cosines γ, δ, ϵ with respect to the coordinate axes. Then the translation matrix specifying the behavior of the projection of $\overline{PP_1}$ or T_1 on the x-z plane should be like that in Eq. (2-13). However, we replace $\sin \alpha$ in this equation, where α is the angle between the ray and the z axis, with the cosine γ of the angle between the ray and the x axis. The translation

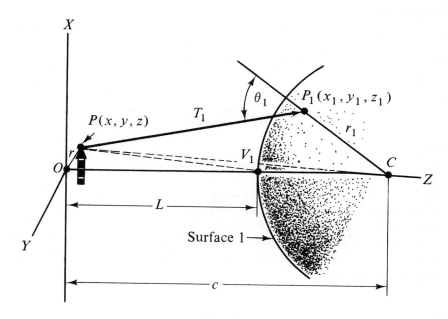

Figure 3-1

equation for the x direction then becomes

$$\begin{pmatrix} n_1\gamma_1 \\ x_1 \end{pmatrix} = \begin{pmatrix} 1 & 0 \\ \dfrac{T_1}{n_1} & 1 \end{pmatrix} \begin{pmatrix} n_1\gamma \\ x \end{pmatrix} \tag{3-1}$$

This equation is equivalent to the two individual relations

$$\left.\begin{array}{l} \gamma = \gamma_1 \\ x_1 = T_1\gamma + x \end{array}\right\} \tag{3-2}$$

The second of these we rewrite as

$$\gamma = \frac{x_1 - x}{T_1} \tag{3-3}$$

which is simply the definition of γ, thus verifying Eq. (3-1). Since skew rays are three-dimensional, we must have three matrix equations, rather than just one as in the meridional case, and the y component of the translation is specified by

$$\begin{pmatrix} n_1\delta_1 \\ y_1 \end{pmatrix} = \begin{pmatrix} 1 & 0 \\ \dfrac{T_1}{n_1} & 1 \end{pmatrix} \begin{pmatrix} n_1\delta \\ y \end{pmatrix} \tag{3-4}$$

from which

$$\delta = \frac{y_1 - y}{T_1} \tag{3-5}$$

For the third equation, we shall incorporate a change in origin from O to V_1, giving

$$\begin{pmatrix} n_1\epsilon_1 \\ z_1 + L \end{pmatrix} = \begin{pmatrix} 1 & 0 \\ \dfrac{T_1}{n_1} & 1 \end{pmatrix} \begin{pmatrix} n_1\epsilon \\ z \end{pmatrix} \tag{3-6}$$

and

$$\epsilon = \frac{z_1 - (z - L)}{T_1} \tag{3-7}$$

where z_1 is measured in the coordinate system with V_1 as its origin. This shift of the origin along the z axis is done to keep z_1 from becoming much larger than x_1 and y_1 (as in Section 2-2). It is not necessary to use all three of eqs. (3-1), (3-4), and (3-6). Two would suffice, since we could determine the third coordinate from the equation of the spherical surface, which (in the new coordinate system) is

$$x_1^2 + y_1^2 + (z_1 - r_1)^2 = r_1^2 \tag{3-8}$$

or

$$x_1^2 + y_1^2 + z_1^2 - 2z_1 r_1 = 0 \tag{3-9}$$

However, at the suggestion of Feder,[3-1] as taken from the book of Smith,[3-2] it is preferable to compute x_1, y_1, and z_1 independently and then use Eq. (3-9) as a check of possible programming errors.

The meridional calculation required auxiliary formulas for T_1 and $\cos\theta_1$. To develop the corresponding relations for skew rays, consider the geometry of Fig. 3-1. The displacement T_1 is obtained by taking the projection of $c = (L + r_1)$ along the ray and subtracting the projections of r and r_1. This process is equivalent to the relation

$$T_1 = (L + r_1)\epsilon - r \cos(r, T_1) - r_1 \cos\theta_1 \tag{3-10}$$

where $\cos(r, T_1)$ designates the cosine of the angle between $\overline{OP}$ and $\overline{PP_1}$. This angle may be expressed in terms of the direction-cosines x/r, y/r, z/r of $\overline{OP}$ and γ, δ, ϵ of T_1 through the equation

$$\cos(r, T_1) = \frac{x}{r}\gamma + \frac{y}{r}\delta + \frac{z}{r}\epsilon$$

or

$$r \cos(r, T_1) = \Sigma \tag{3-11}$$

where

$$\Sigma = x\gamma + y\delta + z\epsilon \tag{3-12}$$

To find $\cos\theta_1$, we consider the two triangles PP_1C and PV_1C, which have a common side $\overline{PC}$. By elementary trigonometry

$$\overline{PC}^2 = \overline{PP_1^2} + \overline{P_1C}^2 - 2\overline{PP_1}\,\overline{P_1C}\cos(\overline{PP_1}, \overline{P_1C}) \tag{3-13}$$

and

$$\overline{PC}^2 = \overline{PV_1^2} + \overline{V_1C}^2 - 2\overline{PV_1}\,\overline{V_1C}\cos(\overline{PV_1}, \overline{V_1C}) \tag{3-14}$$

Equating the two values of $\overline{PC}^2$ shows that

$$T_1^2 + 2T_1 r_1 \cos\theta_1 = x^2 + y^2 + (L - z)^2 + 2(L - z)r_1 \tag{3-15}$$

By Eqs. (3-10) and (3-11)

$$T_1 = (\epsilon c - \Sigma) - r_1 \cos\theta_1 \tag{3-16}$$

so that Eq. (3-15) becomes

$$(c\epsilon - \Sigma)^2 - r_1^2 \cos^2\theta_1 = x^2 + y^2 + (z - L)^2 - 2(z - L)r_1 \tag{3-17}$$

or

$$\cos\theta_1 = \pm\frac{1}{r_1}\sqrt{(c\epsilon - \Sigma)^2 - x^2 - y^2 - (z - L)^2 + 2(z - L)r_1} \tag{3-18}$$

This result expresses $\cos\theta_1$ in terms of known quantities, and the use of Eq. (3-18) in Eq. (3-16) then gives the displacement as

$$T_1 = (\epsilon c - \Sigma) \pm \sqrt{(c\epsilon - \Sigma)^2 - x^2 - y^2 - (z - L)^2 + 2(z - L)r_1} \tag{3-19}$$

which again involves only known quantities.

To resolve the sign ambiguity in Eq. (3-19), we realize that θ_1 is generally less than 90°; therefore $\cos\theta_1$ should be positive. This means that ϵc in Eq. (3-17) is diminished by the amount $(\Sigma + r_1 \cos\theta_1)$ for the situation in Fig. 3-1.

Problem 3-1

Find T_1 analytically by combining Eqs. (3-3), (3-5), and (3-7) with Eq. (3-9), and show that this expression agrees with the geometrical approach used above. ■

It is convenient to replace r_1 with the curvature c_1 of Eq. (1-19b), so that Eq. (3-16) becomes

$$T_1 = -\frac{c_1(\Sigma - \epsilon c) + \cos\theta_1}{c_1} \tag{3-20}$$

where

$$\cos\theta_1 = \sqrt{c_1^2(c\epsilon - \Sigma)^2 - c_1^2[x^2 + y^2 + (z - L)^2] + 2(z - L)c_1} \tag{3-21}$$

We note that Eq. (3-20) is indeterminate when $c_1 = 0$. In this case, however, the sphere reduces to a plane and the displacement becomes simply

$$T_1 = \frac{L - z}{\epsilon} \tag{3-22}$$

Finally, we must express the refraction process at P_1 by three matrix equations which are an extension of Eq. (2-10). These will be

$$\begin{pmatrix} n_1'\gamma_1' \\ x_1' \end{pmatrix} = \begin{pmatrix} 1 & -K_1 \\ 0 & 1 \end{pmatrix} \begin{pmatrix} n_1\gamma_1 \\ x_1 \end{pmatrix} \tag{3-23}$$

$$\begin{pmatrix} n_1'\delta_1' \\ y_1' \end{pmatrix} = \begin{pmatrix} 1 & -K_1 \\ 0 & 1 \end{pmatrix} \begin{pmatrix} n_1\delta_1 \\ y_1 \end{pmatrix} \tag{3-24}$$

$$\begin{pmatrix} n_1'\epsilon_1' - \dfrac{K_1}{c_1} \\ z_1 \end{pmatrix} = \begin{pmatrix} 1 & -K_1 \\ 0 & 1 \end{pmatrix} \begin{pmatrix} n_1\epsilon_1 \\ z_1 \end{pmatrix} \tag{3-25}$$

To see why Eq. (3-25) differs from the two equations that precede it, we use the fact that the refraction relations have the same form as Eq. (2-7). Thus in the x direction we should have

$$n_1'\gamma_1' = n_1\gamma_1 - K_1 x_1 \tag{3-26}$$

and in the y direction

$$n_1'\delta_1' = n_1\delta_1 - K_1 y_1 \tag{3-27}$$

Since the origin is now at V_1, the equation for the z direction should involve $(-z_1 + r_1)$ rather than just z_1; therefore it becomes

$$n_1'\epsilon_1' = n_1\epsilon_1 - K_1 z_1 + K_1 r_1 \tag{3-28}$$

This may be verified by the method indicated in the problem which follows.

Problem 3-2

Square and add Eqs. (3-26), (3-27), and (3-28). Then use Eq. (3-9) to show that this result is consistent with Snell's law. ■

The auxiliary relation we need to use these refraction matrices has already been derived as Eq. (3-21), and we determine $\cos\theta_1'$, needed to find K_1, by using Snell's law. As before, only two matrix equations are actually necessary, since we have available the identity

$$\gamma_1'^2 + \delta_1'^2 + \epsilon_1'^2 = 1 \tag{3-29}$$

However, it is again convenient to use an expression like this as a check on programming errors. It is also convenient to replace K_1 with a quantity G_1, defined as

$$G_1 = \frac{K_1}{n_1' c_1} \tag{3-30}$$

If we use Eq. (3-30), the refraction equations become

$$\left.\begin{array}{l} \gamma_1' = \dfrac{n_1}{n_1'}\gamma_1 - c_1 G_1 x_1 \\[2ex] \delta_1' = \dfrac{n_1}{n_1'}\delta_1 - c_1 G_1 y_1 \\[2ex] \epsilon_1' = \dfrac{n_1}{n_1'}\epsilon_1 - c_1 G_1 z_1 + G_1 \end{array}\right\} \tag{3-31}$$

Program 3-1

```
DIMENSION C(13),  U(13),  T(13)
C(1) = 0.02 $ C(2) = -0.02 $ C(3) = 0.0
U(1) = 1.0$U(2) = 1.5$U(3) = 1.0
T(1) = 10.0  $ T(2) = 15.0 $ T(3) = 65.517241
N = 3
CX = -0.1 $ CY = 0.1 $ CZ = SQRT (1.0 - CX*CX - CY*CY)
X = 10.0 $ Y = 1.0 $ Z = 2.0
DO 310 J = 1,N
BE = (C(J)*(X*CX + Y*CY + Z*CZ) - C(J)*CZ*T(J) - CZ)
CE = C(J)*(X*X + Y*Y + (Z - T(J))**2 ) -2.0*(Z - T(J))
RAD = SQRT (BE*BE - CE*C(J))
IF( C(J) ) 300, 350, 300
300 TT =(-BE - RAD)/C(J)
GO TO 340
350 TT = (T(J) - Z)/CZ
340 X = X + TT*CX
Y = Y + TT*CY
Z = Z + TT*CZ - T(J)
SPH = (X*X + Y*Y + Z*Z)*C(J) - 2.0*Z
PRINT 15, SPH
IF(J - N) 320,330,320
320 R = U(J)/ U(J + 1)
CTH = RAD
CTHP = SQRT (1.0 - (1.0 - CTH*CTH)*R*R)
G = CTHP - R*CTH
CX = R*CX- C(J)*G*X
GY = R*CY- C(J)*G*Y
CZ = R*CZ- C(J)*G*Z    +  G
SS = CX*CX + CY*CY + CZ*CZ
PRINT 15,  SS
310 CONTINUE
330 PRINT 25, X, Y, Z
PRINT 25, CX, CY, CZ
15 FORMAT(/ 10X, 1F20.6)
25 FORMAT(// 10X, 3F20.6)
STOP
END
```

```
              .000000

            1.000000

              .000000

            1.000000

              .000000


    -11.207306             6.273708              -.000000


      -.269914              .052944               .961428
```

Let us consider a ray starting at a point whose coordinates are $x = 10$, $y = 1$, $z = 2$ and with direction-cosines $\gamma = -0.1$, $\delta = 0.1$. This point is placed 8.0 units to the left of the vertex V_1 of the lens involved in Program 2-1. We wish to find the orientation of the emerging ray and the coordinates of its intersection with the paraxial focal plane, shown in Section 2-2 to be $19.0/0.29 = 65.517241$ units to the right of surface 2. Program 3-1 uses Eq. (3-19) to determine T_1, and Eq. (3-20) for $\cos \theta_1$. The changing values of the coordinates are computed as we go through the three cycles indicated, using Eqs. (3-3), (3-5), and (3-7). Similarly, the direction-cosines are determined from Eq. (3-31), but only two cycles are necessary. It should also be noted that Program 3-1 will handle meridional calculations by simply letting $\delta = 0$ and $y = 0$. This seems like a more complicated approach to the material of Chapter 2, but it has the advantage of requiring only the four arithmetical operations and square roots. Many small electronic calculators can handle this type of computation, although there is an advantage to having one with some storage capacity.

Problem 3-3

Show that the statement

$$CTH = RAD$$

which appears just below 320 in Program 3-1 may be replaced by

$$CTH = CZ - C(J) * (X * CX + Y * CY + Z * CZ) \qquad \blacksquare$$

Problem 3-4

With digital computer: Repeat Problem 2-3, using Program 3-1 with suitable modifications.

With electronic calculator: Repeat Problem 2-3, using Eqs. (3-3), (3-5), (3-7), (3-18), (3-19), and (3-31) in conjunction with an electronic calculator having a square-root capability, such as the Hewlett-Packard HP-35 or the Texas Instruments SR-10. $\blacksquare$

3-2 Coma

Our treatment of spherical aberration in Section 2-3 indicated that a ray along the symmetry axis of a lens system is undeviated, a ray parallel to the axis and quite close to it passes through the paraxial focus, and those which are farther out miss the point F'. It was this way of looking at spherical aberration that gave a quantitative way of specifying it, and we shall now extend this approach to a more complicated situation.

We start with the simple diagram of Fig. 3-2(a), which shows two rays parallel to the axis, and equidistant from it, meeting at some intermediate

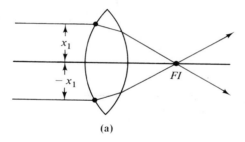

(a)

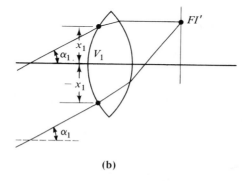

(b)

Figure 3-2

focal point *FI* (the use of multiple symbols should cause no confusion to those familiar with **FORTRAN**). If this diagram is rotated about the axis through 180°, the two rays shown sweep out a cylinder composed of an infinite number of meridional rays, all meeting at the common focus *FI*; there is no spherical aberration associated with this cylinder. Next, we incline the cylinder at an angle α_1 to the axis, as shown in Fig. 3-2(b). If the intersection with the lens is to remain a zone of radius x_1, then the cross section becomes elliptical. Furthermore, all the rays except the top and bottom (the only two shown in the figure) are now skew rays, since they no longer lie in a plane through the axis. It is highly unlikely that the skew rays will pass through the intersection *FI'* of the meridional rays, and this leads us to a new aberration called *coma*. The image is not a circle concentric with the axis as it was for spherical aberration, and we shall determine its shape by actual computation. Program 3-2 starts this process by locating the point *FI'* of Fig. 3-2(b); the plane through *FI* is then taken as the one for which we shall examine the effect of coma on the image. The part of the program down to and including statement 95 deals with the two rays of Fig. 3-2(b), where $x_1 = \pm 1.0$, and the lens is the one in Program 2-2. The subroutine ASKEW is a modification of Program 3-1, arranged so that it can be terminated with index M or N. The value M = 2 terminates with a refraction at surface 2

Program 3-2

```
CX = 0.01
CY = 0.0
CZ = SQRT (1.0 - CX*CX - CY*CY)
R1 = 10.0
X = 1.0
Y = 0.0
Z = R1 - SQRT (R1*R1 - X*X)
CALL ASKEW(X,Y,Z,CX,CY,CZ,.1,-.1,0.,1.,1.5,1.,0.,2.,0.,2,3)
B1 = X
Z1 = Z
EM1 = CX/CZ
R2 = -10.0
AR2 = ABS (R2)
QQP = AR2 - SQRT (AR2*AR2 - X*X)
X = -1.0
Y = 0.0
Z = R1 - SQRT (R1*R1 - X*X)
CX = 0.01
CY = 0.0
CZ = SQRT (1.0 - CX*CX - CY*CY)
CALL ASKEW(X,Y,Z,CX,CY,CZ,.1,-.1,0.,1.,1.5,1.,0.,2.,0.,2,3)
B2 = X
Z2 = Z
EM2 = CX/CZ
DZ = Z2 - Z1
FP = (B2 - B1 - EM2*DZ)/(EM1 - EM2)
FP = FP - QQP
PRINT 95, FP
95 FORMAT(15X, 1F25.6)
PRINT 2
2 FORMAT(//)
X = 1.0
Y = 0.0
Z = R1 - SQRT (R1*R1 - X*X)
CX = 0.01
CY = 0.0
CZ = SQRT (1.0 - CX*CX - CY*CY)
CALL ASKEW(X,Y,Z,CX,CY,CZ,.1,-.1,0.,1.,1.5,1.,0.,2.,FP,3,3)
X = -1.0
Y = 0.0
Z = R1 - SQRT (R1*R1 - X*X)
CX = 0.01
CY = 0.0
CZ = SQRT (1.0 - CX*CX - CY*CY)
CALL ASKEW(X,Y,Z,CX,CY,CZ,.1,-.1,0.,1.,1.5,1.,0.,2.,FP,3,3)
PRINT 25, X, Y, Z
25 FORMAT(3F20.6)
STOP
END


SUBROUTINE ASKEW(X,Y,Z,CX,CY,CZ,C1,C2,C3,U1,U2,U3,T1,T2,T3,M,N)
DIMENSION C(13), U(13), T(13)
C(1) = C1
C(2) = C2
C(3) = C3
U(1) = U1
U(2) = U2
U(3) = U3
T(1) = T1
T(2) = T2
T(3) = T3
```

Program 3-2 (continued)

```
      DO 310 J = 1,N
      BE = (C(J)*(X*CX + Y*CY + Z*CZ) - C(J)*CZ*T(J) - CZ)
      CE = C(J)*(X*X + Y*Y + (Z - T(J))**2 ) -2.0*(Z - T(J))
      RAD = SQRT (BE*BE - CE*C(J))
      IF( C(J) ) 300, 350, 300
  300 TT = (-BE - RAD)/C(J)
      GO TO 340
  350 TT = (T(J) - Z)/CZ
  340 X = X + TT*CX
      Y = Y + TT*CY
      Z = Z + TT*CZ - T(J)
      IF(J - N) 320,330,330
  320 R = U(J)/ U(J +1)
      CTH = CZ - C(J)*(X*CX + Y*CY + Z*CZ)
      CTHP = SQRT (1.0 - (1.0 - CTH*CTH)*R*R)
      G = CTHP - R*CTH
      CX = R*CX- C(J)*G*X
      CY = R*CY- C(J)*G*Y
      CZ = R*CZ- C(J)*G*Z  + G
      IF(J - M) 310,330,330
  310 CONTINUE
  330 RETURN
      END
```

9.505719

.103402   0.000000   -.000000

and is used to find the intercept and direction-cosines of the emerging ray. The first **CALL ASKEW** statement finds these quantities for the ray entering at $x_1 = 1.0$; the second, for the ray at $x_1 = -1.0$. The results are used as indicated in Fig. 3-3 to determine the intersection of the two emerging rays (the labeling is in **FORTRAN** symbols). The intersection point of the two rays is determined by the solution of the simultaneous equations

$$X = EM1 * Z + B1$$
$$X = EM2 * (Z - DZ) + B2$$
(3-32)

from which

$$FP = (B2 - B1 - EM2 * DZ)/(EM1 - EM2)$$
(3-33)

where FP is the symbol used for the distance from V_2 to FI'. However, the value of FP determined by Eq. (3-33) is the distance from the base of B1 rather than the vertex V_2, and this is corrected by the use of QQP. The

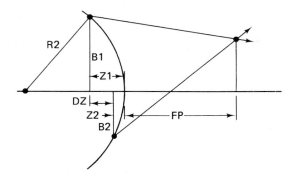

Figure 3-3

slopes **EM1** and **EM2** are given by the ratio **CX/CZ** of the emerging direction-cosines for each ray.

Having found **FP** = 9.505719, we use this value in the **CALL** statement to trace the two rays out to the focal plane. We do this by letting **T(3)** = **FP**. Note that **T(1)** = 0.0 in all four **CALL** statements, corresponding to a ray originating on a zone of radius 1.0. Also note that we find the z coordinate on surface 1 by the statement

$$Z = R1 - SQRT(R1 * R1 - X * X)$$

which we obtain from Eq. (3-9) with $y_1 = 0$.

Problem 3-5

(a) Verify this statement.

(b) Determine the form it must take for a concave surface.

(c) Show that it is unnecessary to compute **QQP** separately. (There may be other redundancies in these programs which the astute reader should stamp out.) ▄▄

The final value of x for the two rays is 0.103402, as shown in the printout. The value **M** = 3 used in the last three **CALL** statements means that the termination of the subroutine takes place just before statement 320 (as in Program 3-1), so that we are now tracing out to the image plane.

We take this program and extend it in the following two ways (Program 3-3):

(a) The statement **DO 70 JJ** = 1,10 starts with the angle γ = **CX** = 0.01, as in Program 3-2, and repeats the calculation for **CX** = 0.02, . . ., 0.1.

(b) The section starting with statement 40 moves the incident ray halfway around a circle of radius **RR** = 1.0 in steps of 15°, and this is repeated for **RR** = 0.9, . . ., 0.1.

Program 3-3

```
    PI = 3.14159
    DO 70 JJ = 1, 10
    DO 80 KK = 1, 10
    CX = 0.01*FLOAT (JJ)
    CY = 0.0
    CZ = SQRT (1.0 - CX*CX - CY*CY)
    R1=10.0    $   R2=-10.0
    X = 1.1 - 0.1*FLOAT (KK)
    Y = 0.0
    Z = R1 - SQRT (R1*R1 - X*X)
    C1 = 0.1 $ C2 = -0.1
    CALL ASKEW(X,Y,Z,CX,CY,CZ,C1,C2,0.0,1.0,1.5,1.0,0.0,2.0,0.0,2,3)
    B1 = X
    Z1 = Z
    EM1 = CX/CZ
    AR2 = ABS (R2)
    QQP = AR2 - SQRT (AR2*AR2 - X*X)
    X = 1.1 - 0.1*FLOAT (KK)
    X = -X
    Y = 0.0
    CX = 0.01*FLOAT (JJ)
    CY = 0.0
    CZ = SQRT (1.0 - CX*CX - CY*CY)
    CALL ASKEW(X,Y,Z,CX,CY,CZ,C1,C2,0.0,1.0,1.5,1.0,0.0,2.0,0.0,2,3)
    B2 = X
    Z2 = Z
    EM2 = CX/CZ
    DZ = Z2 - Z1
    FP = (B2 - B1 - EM2*DZ)/(EM 1 - EM2)
    FP = FP - QQP
    PRINT 95, FP
 95 FORMAT(15X, 1F2..6)
    RR= 1.1 - 0.1*FLOAT (KK)
    DO 85 NN = 1,13
    AF = FLOAT (NN) - 1.0
    THETA = AF*PI/12.0
 40 X =RR*COS (THETA)
    Y =RR*SIN (THETA)
    Z = R1 - SQRT (R1*R1 - X*X)
    CX = 0.01*FLOAT (JJ)
    CY = 0.0
    CZ = SQRT (1.0 - CX*CX - CY*CY)
    CALL ASKEW(X,Y,Z,CX,CY,CZ,C1,C2,0.0,1.0,1.5,1.0,0.0,2.0,FP ,3,3)
    PRINT 90, X,Y
 90 FORMAT(20X, 2F2..6)
 85 CONTINUE
 80 CONTINUE
 70 PRINT 1
  1 FORMAT(/)
    STOP
    END
```

```
      SUBROUTINE ASKEW(X,Y,Z,CX,CY,CZ,C1,C2,C3,U1,U2,U3,T1,T2,T3,M,N)
      DIMENSION C(13),  U(13),  T(13)
      C(1) = C1
      C(2) = C2
      C(3) = C3
      U(1) = U1
      U(2) = U2
      U(3) = U3
      T(1) = T1
      T(2) = T2
      T(3) = T3
      DO 310 J = 1,N
      BE =      (C(J)*(X*CX+Y*CY+Z*CZ)-C(J)*CZ*T(J)-CZ)
      CE = C(J)*(X*X + Y*Y + (Z - T(J))**2 ) -2.0*(Z - T(J))
      RAD = SQRT (BE*BE - CE*C(J))
      IF( C(J) ) 300, 350, 300
300   TT =(-BE - RAD)/C(J)
      GO TO 340
350   TT = (T(J) - Z)/CZ
340   X = X + TT*CX
      Y = Y + TT*CY
      Z = Z + TT*CZ - T(J)
      IF(J - N) 320,330,330
320   R =  U(J)/ U(J + 1)
      CTH = CZ - C(J)*(X*CX + Y*CY + Z*CZ)
      CTMP = SQRT (1.0 - (1.0 - CTH*CTH)*R*R)
      G = CTMP - R*CTH
      CX = R*CX- C(J)*G*X
      CY = R*CY- C(J)*G*Y
      CZ = R*CZ- C(J)*G*Z    +   G
      IF(J - M) 310,330,330
310   CONTINUE
330   RETURN
      END
```

```
      9.505875
              .103388            0.000000
              .103327             .000240
              .103162             .000418
              .102937             .000489
              .102715             .000436
              .102555             .000278
              .102498             .000058
              .102561            -.000167
              .102728            -.000337
              .102957            -.000409
              .103187            -.000362
              .103356            -.000212
              .103418            -.000000
      9.534255
              .103395            0.000000
              .103346             .000195

                 .                   .
                 .                   .
                 .                   .
                 .                   .
```

Figure 3-4 shows a plot for the smallest inclination (CX = 0.01) and for a few values of the zone radius. It is seen that the refracted ray strikes the focal plane in a figure which is approximately a circle for RR = 1.0, the approximation improving as RR decreases. The figure shows (and the output would demonstrate more completely) that the radii of these circles decrease as RR decreases and their centers shift towards increasing X. We shall show later that this figure is a reasonably good approximation to the situation which exists when all other aberrations except coma are absent and when fifth-order (and higher) aberrations are negligible. Then we would obtain the images of Fig. 3-5, and the comet-like shape explains why this aberration is known as coma. The angle determined by the common tangents is 60°, and the line of centers lies on the axis. By setting x_1 = constant in Fig. 3-2b, we attempted to eliminate spherical aberration. In practice, other aberrations distort the ideal pattern shown in Fig. 3-4 even for an angle as small as arc cos (0.01). As the angle increases, the approximate circles are no longer obtained.

We can visualize the generation of the ideal circles in the way shown by Fig. 3-6. The ray through the geometrical center of the lens produces a point on a plane a little to the right of FP = 9.652284; for the two rays at $x_1 = \pm 1.0$, the intersection point lies on the plane determined by FP = 9.505719. Rotating the pairs of rays for x_1 = 1.0, 0.9, . . . , 0.1 about the central ray as an axis generates the circles shown in the plan view at the far right of the figure.

We observe further that the half-circle in object space corresponding to the two meridional rays at x_1 = 1.0 and x_1 = −1.0 produces a *full* circle in image space. That is, the statements

$$X = RR * COS (THETA)$$

$$Y = RR * SIN (THETA)$$

for THETA = 0.0, PI/12.0, . . . , PI are sufficient to generate the comatic circle. Hence, a second semicircle ($\theta = \pi$ to 2π) would simply duplicate the original image. Each image circle corresponds to a double object circle, as shown in Fig. 3-7, where we have assumed an ideal image with a radius of approximately 0.000450.

The comatic patterns shown in Fig. 3-5 or Fig. 3-7 represent ideal conditions; that is, they show an image produced by a lens for which coma is the only aberration. Figure 3-4, on the other hand, is more realistic. It shows the comatic images for a single thin lens in which spherical aberration has been eliminated (since the object is a single zone), but the remaining three third-order aberrations are still present. This figure presents only half of the comatic pattern; the situation will actually appear as shown in Fig. 3-8(a). Each circle will have a mirror image with respect to the x axis, and the combined pattern is in Fig. 3-8(b). These cardioid-shaped figures are what we

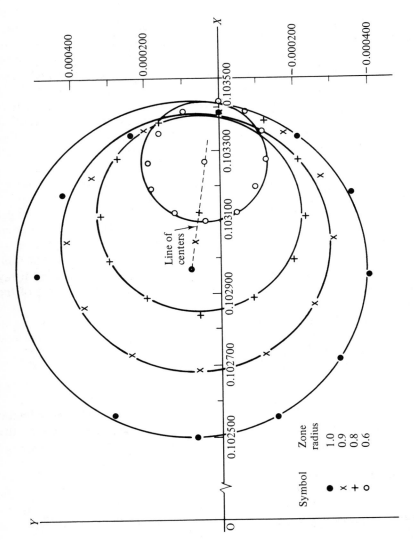

Figure 3-4

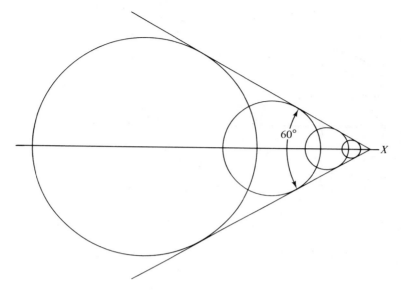

Figure 3-5

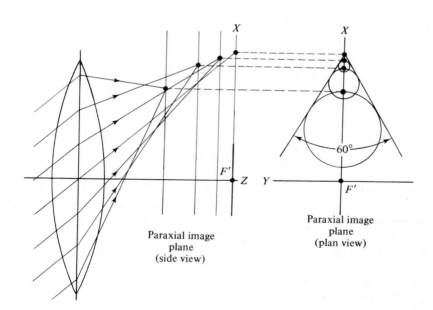

Paraxial image
plane
(side view)

Paraxial image
plane
(plan view)

Figure 3-6

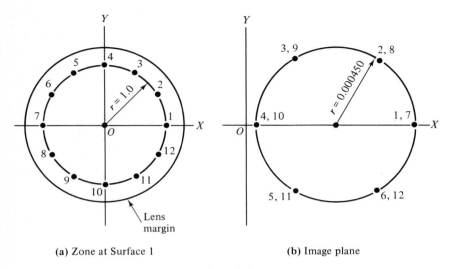

(a) Zone at Surface 1   (b) Image plane

Figure 3-7

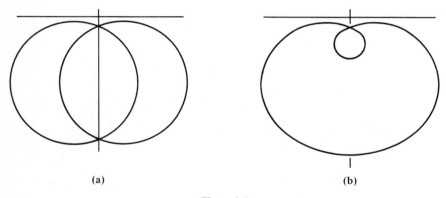

(a)   (b)

Figure 3-8

would expect to see when we attempt to find the "best" focus for this simple lens. Sketches of this phenomenon, as well as the classical textbook pattern of Fig. 3-5 have been given by Sherman[3-3].

It is customary (see Jenkins and White[3-4]) to specify the magnitude of the coma as indicated in Fig. 3-9. The length of the comatic pattern along the meridional or tangential direction is called the *tangential coma* C_T and its half-width (equal to R) is the *sagittal coma* C_S. For the ideal pattern with a half-angle $\theta = 30°$, we have

$$C_T = 3C_S \tag{3-34}$$

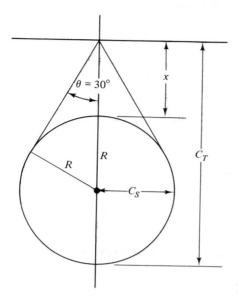

Figure 3-9

Since

$$\sin 30° = 0.5 = \frac{R}{R + x}$$

or

$$x = R$$

and

$$C_S = R, \qquad C_T = 3R$$

specifying C_S would automatically give the area of the aberration figure. That is, the light patches in the image plane corresponding to object points are larger for a poorer lens. However, in the nonideal case revealed by our calculations—due to fifth-order contributions and the third-order aberrations we have not yet considered—it seems more reasonable to actually compute the area of the comatic figure, which we take as very closely specified by

$$A = C_S C_T \tag{3-35}$$

Program 3-4 performs this calculation for the 100 combinations of **CX** (0.001 to 0.01) and **X** (0.1 to 1.0) of Program 3-3. We start with **CX** = 0.001 and **X** = 0.1 to obtain the limiting value (**XL**) of **X**; this is the location of the "tail" of the comet. We also find the edges (**XMIN** and **XMAX**) of the largest circle for each pass of the program. We then calculate C_T as

$$H = XMIN - XL$$

Program 3-4

```
      DO 70 JJ = 1,10
      DO 30 KK = 1, 10
      CX = 0.001*FLOAT (JJ)
      CY = 0.0
      CZ = SQRT (1.0 - CX*CX - CY*CY)
      R1 = 10.0 $ C1 = 1.0/R1 $ R2 = -10.0 $ C2 = 1.0/R2
      X = 0.1*FLOAT (KK)
      Y = 0.0
      Z = R1 - SQRT (R1*R1 - X*X)
      CALL ASKEW(X,Y,Z,CX,CY,CZ,C1,C2,0.0,1.0,1.5,1.0,0.0,2.0,0.0,2,3)
      B1 = X
      Z1 = Z
      EM1 = CX/CZ
      AR2 = ABS (R2)
      QQP = AR2 - SQRT (AR2*AR2 - X*X)
      X = 0.1*FLOAT (KK)
      X = -X
      Y = 0.0
      Z = R1 - SQRT (R1*R1 - X*X)
      CX = 0.001*FLOAT (JJ)
      CY = 0.0
      CZ = SQRT (1.0 - CX*CX - CY*CY)
      CALL ASKEW(X,Y,Z,CX,CY,CZ,C1,C2,0.0,1.0,1.5,1.0,0.0,2.0,0.0,2,3)
      B2 = X
      Z2 = Z
      EM2 = CX/CZ
      DZ = Z2 - Z1
      FP = (B2 - B1 - EM2*DZ)/(EM 1 - EM2)
      FP = FP - QQP
      PRINT 95, FP
95 FORMAT(15X, 1F25.6)
      X = 0.1 * FLOAT (KK)
      Y = 0.0
      Z = R1 - SQRT (R1*R1 - X*X)
      CX = 0.001*FLOAT (JJ)
      CY = 0.0
      CZ = SQRT (1.0 - CX*CX - CY*CY)
      CALL ASKEW(X,Y,Z,CX,CY,CZ,C1,C2,0.0,1.0,1.5,1.0,0.0,2.0,FP ,3,3)
      XMAX = X*1 000 000.0
      X = 0.0
      Y = 0.1 * FLOAT (KK)
      Z = R1 - SQRT (R1*R1 - X*X)
      CX = 0.001*FLOAT (JJ)
      CY = 0.0
      CZ = SQRT (1.0 - CX*CX - CY*CY)
      CALL ASKEW(X,Y,Z,CX,CY,CZ,C1,C2,0.0,1.0,1.5,1.0,0.0,2.0,FP ,3,3)
      XMIN = X*1 000 000.0
      V = (XMAX - XMIN)/2.0
      X = 0.1*FLOAT (KK)
      Y = 0.0
      Z = R1 - SQRT (R1*R1 - X*X)
      CX = 0.001*FLOAT (JJ)
      CY = 0.0
      CZ = SQRT (1.0 - CX*CX - CY*CY)
      CALL ASKEW(X,Y,Z,CX,CY,CZ,C1,C2,0.0,1.0,1.5,1.0,0.0,2.0,FP ,3,3)
      IF(KK - 1) 60,60,65
60 XL = X*1 000 000.0
65 H =  XMIN - XL
      A = H*V
```

72

```
      S = V/H
      PRINT 25, XMAX, XMIN, XL, H, V, A, S
   25 FORMAT(7F15.3)
   80 CONTINUE
      PRINT 1
   70 CONTINUE
      PRINT 3
    1 FORMAT(/)
    2 FORMAT(//)
    3 FORMAT(///)
      STOP
      END


      SUBROUTINE ASKEW(X,Y,Z,CX,CY,CZ,C1,C2,C3,U1,U2,U3,T1,T2,T3,M,N)
      DIMENSION C(13), U(13), T(13)
      C(1) = C1
      C(2) = C2
      C(3) = C3
      U(1) = J1
      U(2) = J2
      U(3) = J3
      T(1) = T1
      T(2) = T2
      T(3) = T3
      DO 310 J = 1,N
      BE =    (C(J)*(X*CX+Y*CY+Z*CZ)-C(J)*CZ*T(J)-CZ)
      CE = C(J)*(X*X + Y*Y + (Z - T(J))**2 ) -2.0*(Z - T(J))
      RAD = SQRT (BE*BE - CE*C(J))
      IF( C(J) ) 300, 350, 300
  300 TT =(-BE - RAD)/C(J)
      GO TO 340
  350 TT = (T(J) - Z)/CZ
  340 X = X + TT*CX
      Y = Y + TT*CY
      Z = Z + TT*CZ - T(J)
      IF(J - N) 320,330,330
  320 R = U(J)/ U(J + 1)
      CTH = CZ - C(J)*(X*CX + Y*CY + Z*CZ)
      CTHP = SQRT (1.0 - (1.0 - CTH*CTH)*R*R)
      G = CTHP - R*CTH
      CX = R*CX- C(J)*G*X
      CY = R*CY- C(J)*G*Y
      CZ = R*CZ- C(J)*G*Z   +   G
      IF(J - M) 310,330,330
  310 CONTINUE
  330 RETURN
      END
```

10344.749	10343.886	9.653697 / 10344.749	-.863	.432	-.372	-.500
10344.545	10341.088	9.649311 / 10344.749	-3.661	1.729	-6.329	-.472
10344.217	10336.420	9.641994 / 10344.749	-8.329	3.898	-32.469	-.468
10343.794	10329.878	9.631734 / 10344.749	-14.871	6.953	-103.402	-.468
10343.274	10321.453	9.618516 / 10344.749	-23.296	10.911	-254.171	-.468
10342.722	10311.136	9.602319 / 10344.749	-33.613	15.793	-530.855	-.470
10342.171	10298.913	9.583117 / 10344.749	-45.836	21.629	-991.369	-.472
10341.674	10284.771	9.560882 / 10344.749	-59.978	28.452	-1706.484	-.474
10341.295	10268.691	9.535576 / 10344.749	-76.058	36.302	-2761.083	-.477
10341.168	10250.653	9.507161 / 10344.749	-94.096	45.227	-4255.685	-.481
. .	. .	. .	. .	. .	. .	. .
20689.442	20687.716	9.653654 / 20689.442	-1.726	.863	-1.490	-.500
20689.033	20582.119	9.649268 / 20689.442	-7.323	3.457	-25.315	-.472
20688.377	20572.784	9.641951 / 20689.442	-16.658	7.797	-129.877	-.468
. .	. .	. .	. .	. .	. .	. .

74

Program 3-4 (continued)

and C_S as

$$V = R = (XMAX - XMIN)/2.0$$

Note that XMAX, XMIN, and XL are taken as 1,000,000 times their actual values in order to work with numbers greater than unity. The results show the increase in A with both zone size and angle of inclination, as we expect from the data of Program 3-4. The quantity S, a measure of sin θ, also shows the ideal value of 0.5 only for very small values of CX, again as expected.

Program 3-5 does the same calculation for the lens with radii $r_1 = 6.67$, $r_2 = -20.0$ considered in connection with Fig. 2-10. This is the lens with minimum spherical aberration. We see that for $CX = 0.001$, $X = 1.0$ of Program 3-4, the value of the area is

$$A = -4255.7$$

whereas for Program 3-5, A has dropped to

$$A = -1122.1$$

As Jenkins and White[3-4] show, the lens with minimum spherical aberration has a coma with the smallest absolute value. Lenses to the left of the one for S = 0.5 in Fig. 2-10 have larger negative values of coma; those to the right have positive coma, which means that the "tail" is at a smaller value of X. It should be mentioned that Jenkins and White calculate coma from approximate third-order formulas developed many years ago by Coddington.[3-5] As we have seen, these ideal calculations do not always give a true picture of aberrations in even a simple lens. For comparison with the first lens considered, we show in Fig. 3-10 the nature of the comatic figure for the lens with minimum spherical aberration. Here, the half-angle θ is considerably less than 30°, but the circles are very close to ideal and their centers lie accurately on the x axis.

The design of a lens with very low spherical aberration was considered in Problem 2-3. Diffraction-limited lenses have in fact been achieved by multi-element combinations of positive and negative lenses. Such a lens would be useful in laser machining, since a large amount of energy could be concentrated in a very small image spot. One complication, however, is that it is possible for the spherical aberration to be minimized and the coma may still be quite high. Unless the laser and the lens axis are very accurately aligned, we are not justified in regarding all incident rays as meridional, as we did in Problem 2-3.

The converse relation between coma and spherical aberration—minimizing the former automatically minimizes the latter—should be expected to hold, however, since meridional rays are a special case of skew rays. We shall consider this phenomenom from a different viewpoint in Section 3-5, which deals with the Abbe sine condition, but first we shall finish our study of the five third-order aberrations.

```
DO 70 JJ = 1,10
DO 80 KK = 1, 10
CX = 0.001*FLOAT (JJ)
CY = 0.0
CZ = SQRT (1.0 - CX*CX - CY*CY)
R1 = 6.67 $ C1 = 1.0/R1 $ R2 = -20.0 $ C2 = 1.0/R2
X = 0.1*FLOAT (KK)
Y = 0.0
Z = R1 - SQRT (R1*R1 - X*X)
        .
        .
60 XL = X**1 000 000.0
65 H = XMIN - XL
A = H*V
S = V/H
PRINT 25, XMAX, XMIN, XL, H, V, A, S
25 FORMAT(7F15.3)
80 CONTINUE
PRINT 1
70 CONTINUE
```

10259.605	10259.358	9.233686	10259.605	-.247	.124	-.031	-.500
10257.787	10256.796	9.230652	10259.605	-2.809	.495	-1.392	-.176
10254.760	10252.523	9.225590	10259.605	-7.082	1.118	-7.918	-.158
10250.530	10246.538	9.218495	10259.605	-13.068	1.996	-26.089	-.153
10245.108	10238.833	9.209357	10259.605	-20.772	3.137	-65.165	-.151
10238.503	10229.405	9.198165	10259.605	-30.200	4.549	-137.380	-.151
10230.731	10218.245	9.184904	10259.605	-41.360	6.243	-258.196	-.151
10221.808	10205.345	9.169558	10259.605	-54.260	8.231	-446.626	-.152
10211.755	10190.695	9.152107	10259.605	-68.910	10.530	-725.628	-.153
10200.596	10174.282	9.132527	10259.605	-85.323	13.157	-1122.587	-.154
.	.	.		.	.	.	.
.	.	.		.	.	.	.

Program 3-5

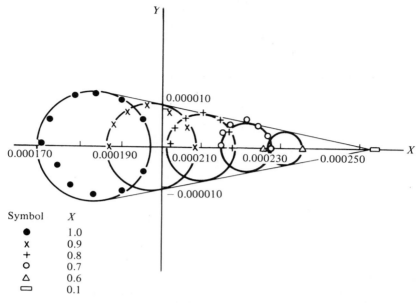

Symbol	X
●	1.0
X	0.9
+	0.8
o	0.7
△	0.6
▭	0.1

Figure 3-10

3-3 Astigmatism, Curvature of Field, and Distortion

The theory developed in Chapter 2 shows that spherical aberration can be described and quantitatively specified in terms of the behavior of rays parallel to the axis of an optical system. We started the present chapter by going on to skew rays and then using three-dimensional matrix methods to express coma in terms of parallel sets of skew rays. The approach that we have been using, while satisfactory for numerical work, is really not very systematic, for what we have done is to show how spherical aberration and coma manifest themselves under very special conditions. What we really need is a precise set of definitions for the five third-order aberrations that is independent of the manner in which they are produced. This is possible when we go to the wave description of light in Chapter 10; in the meantime, we must be content with basing the study of aberrations on geometric optics, which, although helpful in visualizing their causes and consequences, involves the use of a fair amount of intuition.

Let us consider the possibility of a point object. If the point lies on the optical axis, then all the rays leaving it will automatically be meridional. The set of rays striking a given zone on the first surface of the lens system will meet at a common image point which is also on the axis, and there will be a different image point for each zone as shown in Fig. 3-11(a). On the other

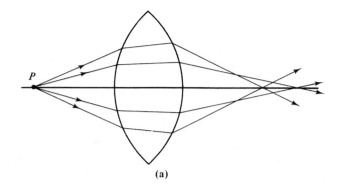

(a)

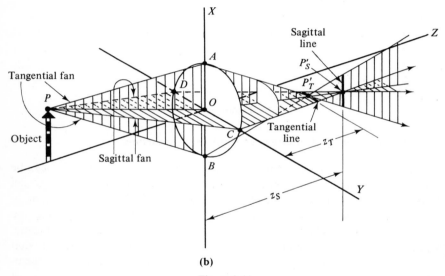

(b)

Figure 3-11

hand, it was shown at the beginning of this chapter that coma is associated with the failure of skew rays to conform to the paraxial approximation. We thus see that spherical aberration is the only defect which can exist for an axial object point. If the point P in Fig. 3-11(a) is moved to an off-axis location, there will still be spherical aberration at the image due to the spread in the meridional rays. In addition, the skew rays from P will introduce coma, which causes a further blurring of the image point; coma is a point aberration which can be produced only by nonaxial object points.

Let us next consider the point P in its nonaxial position (Fig. 3-11(b)) and form an image with just two rays, one passing through the top and the other through the bottom of a given zone on the first surface. If P is in the XOZ plane, then the two rays (designated as PA and PB) are tangential or meridional and they determine a meridional plane coincident with the XOZ coordinate plane. The image distance, as measured from a plane passing through the XOY plane (for a thin lens, this plane is arbitrary), is designated as z_T. If there were no spherical aberration, all the rays between these two— forming what is called the *tangential fan*—would meet at a common image point P'_T. On the other hand, a pair of sagittal rays (PC and PD) going through the horizontal edges of this zone, meet at a different distance z_S. This is what we would expect from our study of coma, since these are skew rays. It is found, in fact, that $z_S > z_T$ for a converging lens. If the coma is reduced as well, we may expect that the skew rays in the sagittal fan will come to a common focus P'_S, lying beyond P'_T, but unfortunately these two points in general do not merge. If we were dealing with a cylinder of skew rays, some of which were actually meridional, then there should be a single focus. For the present situation involving a point object at a finite distance from the lens and from the axis, the two fans at right angles should in general result in two distinct image points, and this defect is call *astigmatism*. This is the third and last of the point aberrations—the other two require objects which have a measurable length.

We see from the figure that the tangential fan when extended out to the sagittal image plane produces a line image, the *sagittal image*. Similarly, the sagittal fan produces a *tangential image* as it crosses the tangential image plane (which, for skew rays, is not generally a straight line). Thus, the image of the point can be one of two lines, depending on where the image plane is located. Further, if the image plane is in a position between these two extremes, then both fans contribute to the image, and we expect a two-dimensional figure due to the astigmatism of the lens. In the ideal case (all other aberrations vanishing and third-order contributions, only), the intermediate images are ellipses. For an image plane halfway between the sagittal and tangential plane, the figure becomes a circle (an ellipse of unit eccentricity), and at the tangential and sagittal image planes, the ellipse degenerates into a line. These transitions are illustrated in Fig. 3-12. The existence of two distinct image planes for tangential and sagittal rays is responsible for the fact that an object with spokes (Fig. 3-13) may have its vertical arms in focus, while

Figure 3-12

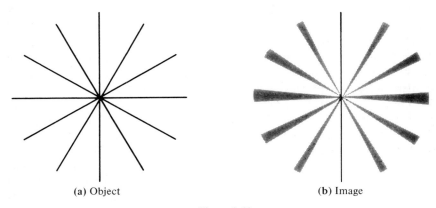

(a) Object   (b) Image

Figure 3-13

the other arms appear fuzzier as we move toward the horizontal direction. This is how an eye examination reveals this defect: the vertical rays (for example) appear quite black, but the others appear gray. This also explains the origin of the word sagittal, which comes from the Latin *sagitta*, arrow.

Returning to Fig. 3-11b, if we move the object point along the x direction, we would expect the positions of both tangential and sagittal focal planes to shift, for it is only under the paraxial approximation that the focal plane is independent of x. Hence, we obtain the two curves of Fig. 3-14 for the positions of the two image planes as a function of object height (the third curve we shall consider just below). Now we have seen that spherical aberration and coma can be minimized in a single lens if its shape is varied; they can be further reduced by the addition of a second, negative element to form a cemented doublet. Such a lens has little effect on astigmatism. This defect may be eliminated, however, by shifting the position of the second element, by placing a stop between the two lenses, or both. When this is done, the

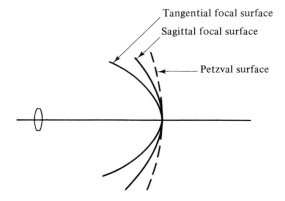

Tangential focal surface

Sagittal focal surface

Petzval surface

Figure 3-14

tangential and sagittal curves of Fig. 3-14 merge into the single curve at the right. While the astigmatism has been eliminated, the position of the focal plane is dependent on the object height, and the defect is called *curvature of the field* or *Petzval curvature* (hence the labeling on the third curve). For a single lens, a stop in the proper position will reduce the Petzval curvature, and this is usually done in inexpensive box cameras (but the astigmatism is quite high).

If we next take the object of Fig. 3-11b and shift it parallel to itself away from the z axis, then all rays leaving it are skew and the fifth (and last) aberration is introduced. As shown in Fig. 3-15, an object formed of horizontal

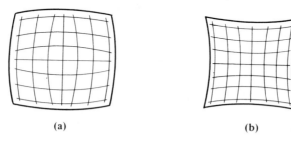

(a)   (b)

Figure 3-15

and vertical lines will either bulge outward (*barrel distortion*) or inward (*pincushion distortion*). A pinhole camera will have no distortion, since there are no skew rays, and a single thin lens of small aperture will have very little. Placing a stop near the lens to reduce astigmatism and Petzval curvature introduces distortion because, as shown in Fig. 3-16(a), the rays for large values of x and y are limited to an off-center portion of the lens. The situation of the figure corresponds to barrel distortion; placing the stop at an equal distance on the other side of the lens then introduces an equal amount of pincushion distortion. The percentage D of distortion is defined as

$$D = \frac{I' - I}{I} \times 100 \qquad (3-36)$$

where: I = paraxial image size
I' = distorted image size

It is therefore clear why highly corrected camera lenses often consist of two identical lens groupings with an iris diaphragm in the center, as shown in Fig. 3-16(b); the distortion of the front half cancels that of the back half.

Remember that spherical aberration, coma, and astigmatism are point defects and that the considerations above show that the other two aberrations are associated with objects of a finite extent. We have seen, however, that there is a close connection between curvature of the field and astigmatism.

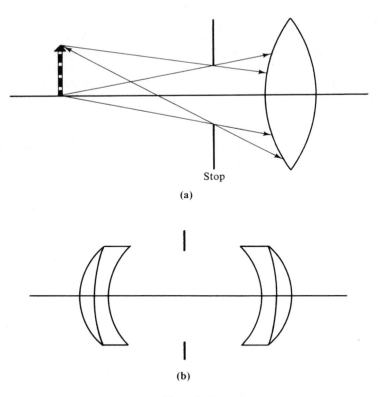

(a)

(b)

Figure 3-16

The theory to be developed in Chapter 10 will show in a more analytic way why these two aberrations are coupled. For the present, we shall continue to look at these effects from a ray-tracing point of view, as described in the following problem.

Problem 3-6

When we studied spherical aberration and coma, we performed the following calculations:

1. We began with a cylinder of rays concentric with the axis of symmetry and parallel to it. These rays all focus at a single point in image space, and the variation of the focal point with the diameter of the cylinder is a measure of the longitudinal spherical aberration.

2. We next tipped a given cylinder at an angle to the symmetry axis and observed the intersection of the skew rays with the focal plane determined by the two tangential rays. This gave us a graphical approach to coma.

We will now treat astigmatism and curvature of field in a quantitative way. We do this by a slight extension of the procedure for coma. We again start with the tipped cylinder which strikes the front face of the lens along a circle of radius R. If the axis of this cylinder lies in the XOZ plane, then the two rays which strike the lens at $x = R$, $y = 0$ and $x = -R$, $y = 0$ are tangential rays, and the two rays which strike the lens at $x = 0$, $y = R$ and $x = 0$, $y = -R$ are sagittal. Hence, we find the intersection of each pair in image space. The z coordinate of these two image points with respect to the paraxial image plane is a measure of the tangential and sagittal field curvatures, respectively, while the difference specifies the astigmatism.

In tracing the intersection of emerging rays in connection with the coma programs, let us recall that we found the image plane by considering two meridional rays. For the present situation, we must be able to determine the intersection of two skew rays. We do this in the following way:

(a) Consider a line segment with end points (x_1, y_1, z_1) and (x_2, y_2, z_2), with length d, and whose direction-cosines are γ, δ, and ϵ. Let it intersect a second line segment with corresponding parameters (x_1', y_1', z_1'), (x_2', y_2', z_2'), $(\gamma', \delta', \epsilon')$, and d'. Let these two line segments be a meridional pair of rays whose initial points lie on the last surface of an optical system and which intersect at the common point (x_2, y_2, z_2) or (x_2', y_2', z_2'). Show that

$$d = \frac{(z_1' - z_1)\gamma' + (x_1' - x_1)\epsilon'}{\epsilon\gamma' - \gamma\epsilon'} \tag{3-37}$$

Show that Eq. (3-37) is *not* valid, however, for skew rays but is replaced by

$$d = \frac{(z_1' - z_1)\delta' - (y_1' - y_1)\epsilon'}{\epsilon\delta' - \delta\epsilon'} \tag{3-38}$$

Then show that z_2 can be obtained from y by using

$$z_2 = z_1 + \epsilon d \tag{3-39}$$

(b) Compute and plot the distance between the paraxial plane and the sagittal and tangential foci for the following lens (U.S. Patent No. 2,453,260) as a function of image height:

$r_1 = 40.94$		
$r_2 = \infty$	$t_1' = 8.74$	
$r_3 = 55.65$	$t_2' = 11.05$	$n_2 = 1.617$
$r_4 = 39.75$	$t_3' = 2.78$	$n_4 = 1.649$
$r_5 = 107.56$	$t_4' = 7.63$	$n_6 = 1.617$
$r_6 = -43.33$	$t_5' = 9.54$	

as taken from Smith,[3-2] Chapter 10. Compare your results with his, noting

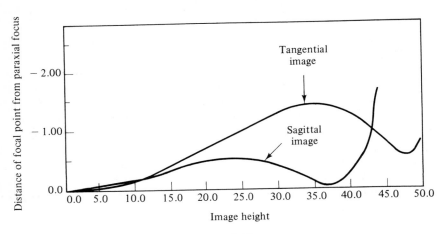

Figure 3-17

that the designer has provided for zero astigmatism at the lens margin. (Fig. 3-17.)

(c) Plot the distortion in the paraxial plane. ▬

3-4 The Petzval Condition

Prior to the time when computers were widely available, ray tracing was a tedious job. Experienced lens designers used all sorts of approximations and clever tricks to handle aberrations, but third-order theory still involved an enormous amount of algebra. To get some notion of how complex the problem was, it is suggested that the reader take a look at the classic two-volume treatise of Conrady.[3-6] By using extended paraxial methods, these early designers were able to describe the third-order and fifth-order aberrations quite accurately.

Since most books on lens theory use this approximate approach, it will be found that the aberration curves at the lens margin do not agree with results obtained by our programs (in particular, the curves obtained in Problem 3-6 will not match precisely those of Smith). To show the reasons for this and to give some idea of how to study aberrations in this approximate fashion, we shall treat Petzval curvature as an example.

Let us start with a single refracting surface of radius r separating two media of indices n and n', respectively (Fig. 3-18). The tree at P, a distance s from the vertex V, will have an image at P' at a distance s', related to s by

$$\frac{n}{s} - \frac{n'}{s'} = \frac{1}{f} = -\frac{1}{f'} \qquad (1\text{-}74)$$

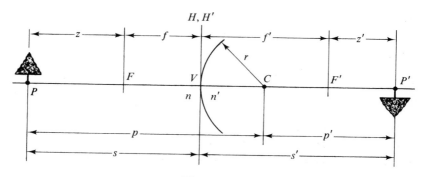

Figure 3-18

However, for a single refracting surface, the system matrix S is simply the refraction matrix R, so that

$$\begin{pmatrix} b & -a \\ -d & c \end{pmatrix} = \begin{pmatrix} 1 & -k \\ 0 & 1 \end{pmatrix}$$

or

$$a = k, \qquad b = c = 1, \qquad d = 0 \tag{3-40}$$

By Eqs. (1-45) and (1-47)

$$l_H = l'_H = 0$$

thus s and s' are now the distances from the vertex as we have previously found for thin lenses.

It is convenient at this point to find an expression like Eq. (1-74) involving the distances p and p' from object and image, respectively, to the center of curvature C. To accomplish this, we see from the figure that

$$\left. \begin{array}{l} z = p - f - r \\ z' = p' + r - f' \end{array} \right\} \tag{3-41}$$

where we shall work with magnitudes for simplicity. By Eqs. (1-50), (1-51), and (3-40), the quantities f and f' are (ignoring signs)

$$f = \frac{n}{a} = \frac{n}{k}$$

$$f' = \frac{n'}{a} = \frac{n'}{k}$$

so that Eq. (3-41) becomes

$$z = p - \frac{nr}{n'-n} - r = p - \frac{n'r}{n'-n} = p - f'$$

$$z' = p' - \frac{n'r}{n'-n} + r = p' - \frac{nr}{n'-n} = p' - f \tag{3-42}$$

Using

$$zz' = ff' \tag{1-2}$$

gives

$$pp' - pf - p'f' = 0$$

or

$$\frac{n}{p'} + \frac{n'}{p} = k = \frac{n' - n}{r} \tag{3-43}$$

which is the desired relation.

Next, let us suppose that the object and image are curved, with radius of curvature ρ or ρ', respectively (Fig. 3-19). Let a ray pass through the center of

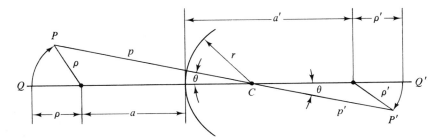

Figure 3-19

curvature, C, so that it is undeviated. Then Eq. (3-43) connects $\overline{PC} = p$ with $\overline{P'C} = p'$, since $\overline{PCP'}$ is a symmetry axis in the figure just as it was in Fig. 3-18. Further, Eq. (3-43) also applies to $\overline{QC}$ and $\overline{Q'C}$, so that we have

$$\frac{n}{a' + \rho'} + \frac{n'}{a + \rho} = k \tag{3-44}$$

where a and a' are indicated in Fig. 3-19. Subtracting Eq. (3-43) from Eq. (3-44), we obtain

$$n\left(\frac{1}{a' + \rho'} - \frac{1}{p'}\right) + n'\left(\frac{1}{a + \rho} - \frac{1}{p}\right) = 0 \tag{3-45}$$

By Fig. 3-19, we see that

$$p^2 = p^2 + a^2 - 2ap \cos \theta$$

Since θ is small, $\cos \theta$ may be expressed by a Taylor series as

$$\cos \theta = 1 - \frac{\theta^2}{2}$$

giving

$$p^2 = (p - a)^2 + pa\theta^2$$

or

$$\left(\frac{p}{p - a}\right)^2 = 1 + \frac{pa\theta^2}{(p - a)^2}$$

Using a binomial expansion, this becomes

$$\frac{p}{p-a} = 1 + \frac{pa\theta^2}{2(p-a)^2}$$

or

$$p + a = p + \frac{pa\theta^2}{2(p-a)} = p\left[1 + \frac{a\theta^2}{2(p-a)}\right]$$

Again using the binomial theorem, we obtain

$$\frac{1}{p+a} = \frac{1}{p}\left[1 - \frac{a\theta^2}{2(p-a)}\right]$$

or

$$\frac{1}{p+a} - \frac{1}{p} = \frac{a\theta^2}{2p(p-a)} \qquad (3\text{-}46)$$

In the same way, we obtain

$$\frac{1}{p'+a'} - \frac{1}{p'} = \frac{a'\theta^2}{2p'(p'-a')} \qquad (3\text{-}47)$$

Putting Eqs. (3-46) and (3-47) into Eq. (3-45) gives

$$\frac{na'}{p'(p'-a')} + \frac{n'a}{p(p-a)} = 0 \qquad (3\text{-}48)$$

But the figure shows that the relations

$$\rho = p - a, \qquad \rho' = p' - a'$$

are approximately true. Hence Eq. (3-48) becomes

$$\frac{n(p'-\rho')}{p'\rho'} + \frac{n'(p-\rho)}{p\rho} = 0$$

or

$$\frac{n'}{\rho} + \frac{n}{\rho'} = \frac{n'}{p} + \frac{n}{p'} = k \qquad (3\text{-}49)$$

using Eq. (3-43). The equation

$$\frac{n'}{\rho} + \frac{n}{\rho'} = k \qquad (3\text{-}50)$$

is the *Petzval equation*.

To extend this to an actual lens, we let the image of the first surface be the object of the second surface. If we write Eq. (3-50) as

$$\frac{1}{n_1\rho_1} + \frac{1}{n_1'\rho_1'} = \frac{k_1}{n_1 n_1'}$$

then for surface 2, this equation becomes

$$-\frac{1}{n_2\rho_2} + \frac{1}{n_2'\rho_2'} = \frac{k_2}{n_2 n_2'}$$

since the image of surface 1 is the object for surface 2. Adding, we obtain

$$\frac{1}{n_1 \rho_1} + \frac{1}{n_2' \rho_2'} = \frac{1}{n_1' f'}$$

since the focal length of a thin lens is given by

$$\frac{1}{f'} = (n_1' - n_1)\left(\frac{1}{r_1} - \frac{1}{r_2}\right)$$

Summing in the same way over an arbitrary number of thin lenses, we obtain

$$\frac{1}{n\rho} + \frac{1}{n'\rho'} = \sum \frac{1}{n_i f_i} \tag{3-51}$$

which is the famous *Petzval condition* connecting object and image field curvatures. This equation states that if the object is planar ($\rho = \infty$), there will be no Petzval curvature ($\rho' = \infty$) if the Petzval sum vanishes. Its validity depends on approximations which are essentially second-order.

Problem 3-7

Examine the validity of the Petzval condition for the lens system of Problem 3-6.   ▬

3-5 The Abbe Sine Condition

Although the Petzval condition discussed in Section 3-4 is not often used by lens designers, the Abbe* sine condition is considered to be a very necessary tool. As we shall see, it gives a very simple way of predicting whether any particular alteration in lens design will lead to a reduction in spherical aberration and coma. The programs we have developed for computing the area of a comatic pattern are rather elaborate, whereas the sine condition provides a test which can easily be handled on a calculator. Before discussing it, however, let us consider a simple example. It is possible to eliminate the spherical aberration associated with an object point on the axis by placing this point *P inside* a lens which is a complete sphere. The location of the object point is chosen to be a distance r/n to the left of the center C (Fig. 3-20(a)), where n is the index of the glass. Then

$$s = -\frac{r}{n} - r = -r\left(\frac{n+1}{n}\right) \tag{3-52}$$

To locate the image P', we start with the law of sines, which becomes

$$\frac{\sin \theta}{\sin \alpha} = \frac{r/n}{r} \tag{3-53}$$

---

*Ernst Abbe was *German*, so his name is pronounced "ah-buh". Most people (incorrectly) say "ah-bay".

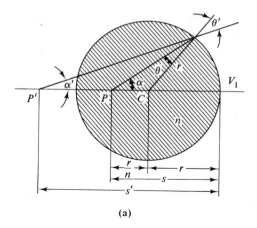

(a)

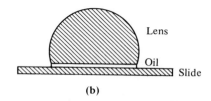

(b)   **Figure 3-20**

and by Snell's law

$$\sin \theta' = n \sin \theta \qquad (3\text{-}54)$$

Therefore

$$\sin \theta' = \sin \alpha$$

or

$$\theta' = \alpha \qquad (3\text{-}55)$$

The figure shows that

$$\theta + \alpha = \theta' + \alpha'$$

Hence

$$\theta = \alpha'$$

Finally, in the large triangle

$$\frac{s' - r}{r} = \frac{\sin \theta'}{\sin \alpha'} \qquad (3\text{-}56)$$

or

$$s' = \frac{r \sin \theta'}{\sin \theta} + r = nr + r \qquad (3\text{-}57)$$

Hence s' is independent of the angle α, so that any ray leaving P passes through P'. Note that we did not use the approximation of small angles, and our results are therefore valid in general.

Since it is not possible to place an object inside a real lens, we grind away a sector and fill the space between the lens and the object with an oil of the same index n as the glass (Fig. 3-20(b)). This is the principle of the *oil-immersion objectives* used in microscopes.

We may also use these calculations to prove the Abbe principle. By Eqs. (3-52), (3-53), and (3-56), we have

$$\frac{s + r}{s' - r} = \frac{-r \sin \theta / \sin \alpha}{r \sin \theta' / \sin \alpha'}$$

$$= -\frac{n' \sin \alpha'}{n \sin \alpha} \tag{3-58}$$

Then we consider an object of height x (Fig. 3-21) and a ray through the center of curvature C. Since this ray will be undeviated, we see that

$$\frac{x'}{x} = \frac{s' - r}{-(s + r)} \tag{3-59}$$

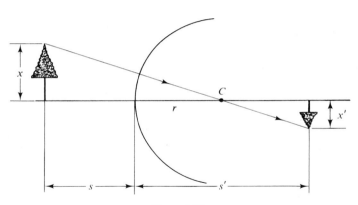

Figure 3-21

Comparing Eqs. (3-58) and (3-59) gives

$$xn \sin \alpha = x'n' \sin \alpha' \tag{3-60}$$

This is the Abbe sine condition. Since Eq. (3-58) is a consequence of the absence of spherical aberration, then so is Eq. (3-60).

To see the connection of Eq. (3-60) with coma, return to Fig. 3-7. All twelve rays which strike surface 1 are in a common zone and hence contribute nothing to the spherical aberration. Rays 1 and 7, which are meridional, correspond to zero coma. We note, however, that the sine condition was derived for meridional rays; so these two rays obey Eq. (3-60). Hence, it is the presence of skew rays which produces the comatic pattern and also violates the sine condition. We must realize that x in Eq. (3-60) refers to the distance from the top of the tree of Fig. 3-21 to the axis; thus, x should be replaced by the object height h, which is the same as x only if the object point lies in the *x-z* plane. Therefore the most general statement of the sine condition is that *hn* sin α is invariant. This quantity is called the *optical invariant*. In fact, in paraxial optics, it reduces to *hn*α, which is the *Lagrange invariant*.

As we go around the zone for which *r* = 1.0 in Fig. 3-7, we notice that ray 4 strikes the intersection of the meridional and focal planes (although it is *not* a meridional ray). We can say, therefore, that the ray with maximum "skewness" is one which falls between ray 2 and ray 3, and such a ray is a measure of the sagittal coma, as Fig. 3-9 indicates. This is why lens designers have traditionally used the failure of *hn* sin α to remain constant as a measure of sagittal coma. This property is referred to as the *offense against the sine condition (OSC)*. A lens for which *nh* sin α is constant or invariant from surface to surface is then free of both spherical aberration and coma. Two points *P* and *P'* which have an object-image relation that satisfies the Abbe condition are said to be *aplanatic*.

Problem 3-8

(a) Use the meridional matrix equation for a refraction at surface 1 of a lens followed by a translation to surface 2 to show that the magnification for this process is

$$\beta = \frac{n_1 \sin \alpha_1}{n_1 \sin \alpha_1 - K_1 x_1} \tag{3-61}$$

provided the Abbe sine condition is obeyed.

(b) Using approximations less severe than paraxial conditions, show that Eq. (3-61) indicates that the magnification is independent of the angle α_1. This is an important consequence of the Abbe condition. ■

Conclusion

We have pointed out several times that spherical aberration and coma are point aberrations, the former associated with meridional rays and the latter with skew rays. An off-axis point will produce both meridional and skew fans; even if they are separately corrected, the difference between them will result in *astigmatism*. When we consider objects of finite dimensions, the failure to obey paraxial conditions shows up as *curvature of field* and the same difficulty in the *xy* plane results in *distortion*. These are the five third-order aberrations.

The question which then arises is whether some aberrations are more important than others. No general answer can be given; the particular aberrations which are significant depend upon the application. For example, consider a photographic enlarger. If the greatly magnified image on the easel is to produce a sharp print, then coma is critical as far as each point is concerned, and so is Petzval curvature with regard to the image as a whole. That is, unless the image is absolutely flat, either the center or the edges will be slightly out of focus. In the problem which follows, this particular feature of an enlarging lens is explored.

Problem 3-9

(a) Determine the curvature of field and astigmatism of the following enlarger lens (U.S. Patent No. 2,507,164) by finding the location of the sagittal and tangential image points as the object varies in height from $x = 0$ to $x = 15$:

$r_1 = 109.12$	$t_1' = 10.60$	$n_1' = 1.517$
$r_2 = -43.74$	$t_2' = 3.26$	$n_2' = 1.617$
$r_3 = -116.12$	$t_3' = 197.88$	$n_3' = 1.000$
$r_4 = -41.82$	$t_4' = 3.50$	$n_4' = 1.517$
$r_5 = 69.72$	$t_5' = 13.12$	$n_5' = 1.734$
$r_6 = -188.92$		

The object and image distances specified are

$$l = 224.1 \qquad l' = 231.5$$

The nominal focal length is 100 mm (thus all units above are actually mm), and the nominal magnification is 3. Let the maximum opening at surface 1 have a radius of 15 mm.

(b) Compare this lens with a high-quality camera lens, that is, with a Tessar of the same focal length and with surface curvatures, indices, and axial dimensions as follows:

$c_1 = 0.0312$	$n_1' = 1.6116$	$t_1' = 7.029$
$c_2 = -0.001842$	$n_2' = 1.000$	$t_2' = 3.721$
$c_3 = -0.0147$	$n_3' = 1.6053$	$t_3' = 1.595$
$c_4 = 0.03210$	$n_4' = 1.000$	$t_4' = 6.40$
$c_5 = 0.0$	$n_5' = 1.5123$	$t_5' = 4.2726$
$c_6 = 0.02645$	$n_6' = 1.6116$	$t_6' = 7.797$
$c_7 = -0.02116$		

(c) Also compare the enlarging lens with a single lens for which $r_1 = 98.3$, $r_2 = -98.3$, $t_1' = 10.0$, $n_1' = 1.500$. This simple lens has a focal length of 100 mm. Your results should show that the tangential and sagittal field curvatures have the same sign and are identical only on the axis, whereas the lens in (a) has a more complicated behavior because of the corrections incorporated into the design.

(d) Compute the longitudinal spherical aberration of the three lenses for $x_1 = 0.1, 1.0, 3.0, 5.0, 10.0, 15.0,$ and 20.0. ■

Problem 3-10

The design of a lens which could conceivably give very low spherical aberration was discussed in Problem 2-3, and it was pointed out later that obtaining a small image spot from an incident parallel beam depended on accurate alignment of the beam with the lens axis; otherwise, coma would be present. To eliminate this difficulty, Thomson[3-7] modified Fulcher's design by adding a fifth element, as shown in Fig. 3-22. The parameters he used are as follows:

$r_1 = 117.20$	$t'_1 = 1.35$	$n'_1 = 1.9536$
$r_2 = -2020.00$	$t'_2 = 0.00$	
$r_3 = 60.25$	$t'_3 = 1.35$	$n'_3 = 1.5536$
$r_4 = 132.10$	$t'_4 = 0.00$	
$r_5 = 39.50$	$t'_5 = 1.45$	$n'_5 = 1.5536$
$r_6 = 61.50$	$t'_6 = 0.00$	
$r_7 = 29.15$	$t'_7 = 1.60$	$n'_7 = 1.5536$
$r_8 = 38.68$	$t'_8 = 0.00$	
$r_9 = 16.55$	$t'_9 = 3.50$	$n'_9 = 1.7190$
$r_{10} = 22.75$		

(a) Using a direction-cosine with respect to the x axis of 0.019 and letting $x = 3.7$, show that the coma is minimized when we change the first lens so that

$$r_1 = 113.05 \qquad r_2 = -2010.5$$

(b) Show that another way to minimize the coma is to change t'_8 to a value of about 1.0. Find this value. ∎

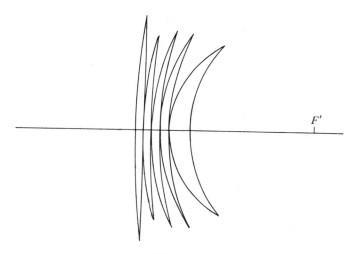

Figure 3-22

Problem 3-11

A zoom lens (U.S. Patent No. 3,045,546) has radii, spacings, and indices as follows:

r	t'	n'
+ 7.5129		
	0.1693	1.7618
+ 3.2319		
	0.7902	1.651
−20.4139		
	0.0056	
+ 2.9246		
	0.3612	1.651
+ 4.7196		
	0.0564 to 1.9995	
+ 1.7611		
	0.1129	1.62344
+ 1.0395		
	0.2484	
+ 2.9246		
	0.1129	1.62344
+ 1.6104		
	0.5757	
− 1.7367		
	0.0903	1.51507
+ 1.3767		
	0.2145	1.7618
+ 7.2263		
	2.1124 to 0.1693	
+ 1.1519		
	0.2484	1.717
− 5.9011		
	0.2484	
− 1.1519		
	0.2484	1.723
− 0.6037		
	0.1242	1.64793
+ 1.1519		
	0.3048	
+ 5.9011		
	0.2484	1.6935
− 1.5956		
	0.3951	
+ 1.4933		
	0.5080	1.6968
− 0.9178		
	0.1129	1.70035
∞		

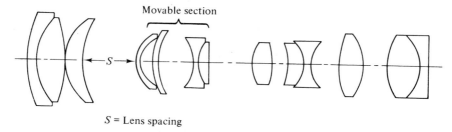

S = Lens spacing

Figure 3-23

The moving central section is shown in an intermediate position in Fig. 3-23.

(a) Calculate the variation in l_F, l'_F, l_H, l'_H, and f' as the moving section goes from one extreme position to the other.

(b) Calculate the longitudinal spherical aberration as a function of zone radius from 0.02 to 0.30 for a lens spacing S of 0.0573, 0.8345, and 1.8061.

(c) Calculate the longitudinal spherical aberration as a function of lens spacing for zones of radius 0.05, 0.10, 0.15, 0.20, and 0.25.

(d) Determine the nature of the comatic patterns for an incident angle of 0.573 using first a variable zone radius and then a variable lens spacing. Show that there is a correlation between minimum coma and minimum spherical aberration.

References

3-1   D. Feder, *J. Opt. Soc. Amer.* **41,** 630 (1951).

3-2   W. J. Smith, *Modern Optical Engineering,* McGraw-Hill Book Co. (1966).

3-3   B. Sherman, *Modern Photography* **32,** 118 (1968).

3-4   F. A. Jenkins and H. E. White, *Fundamentals of Optics,* 3rd ed., McGraw-Hill Book Co. (1957).

3-5   H. D. Taylor, *A System of Applied Optics*, Macmillan (N. Y.), 1906.

3-6   A. E. Conrady, *Applied Optics and Optical Design*, Oxford University Press, London (1929).

3-7   A. F. H. Thomson, *Serv. Elec. Res. Lab. Tech. J.* **14,** 54 (1964).

4

OPTICAL INSTRUMENTS

4-1 Introduction

An understanding of geometric optics is crucial to the design of optical instruments. In this chapter, we shall consider two specific examples from the vast range of optical systems which are commercially available. In both cases, we shall also extend the principles already discussed to cover new situations. The first example is the high-quality telescope, which requires special procedures to reduce aberrations to extremely low limits. The other is the projector, and this involves consideration of light as a form of energy. In other words, we are concerned not only with the quality of the image but its intensity as well.

4-2 Aspheric Surfaces

We saw in Chapter 3 that spherical aberration occurred when rays passing through different zones of a lens did not come to a common focus. As we look at this phenomenon, let us consider a new idea: if a marginal ray fails to pass through the paraxial focal point, this can be corrected if the radius of curvature of the last surface is changed at the point of emergence. That is, the radius of curvature need not be held constant, and the surface no longer be spherical. Such *aspheric* surfaces have been used in optical systems, but not too frequently; they are expensive to grind to a high degree of accuracy. Nevertheless, they are worth studying and we shall also show how they are used.

Let us start by considering a sphere (Fig. 4-1) of radius r and center at the point $(0, 0, r)$. By Eq. (3-9) this sphere has the equation

$$x^2 + y^2 + z^2 - 2rz = 0 \qquad (4\text{-}1)$$

from which

$$z = -\frac{x^2 + y^2}{z - 2r} = \frac{(x^2 + y^2)/r}{2 - (z/r)} \qquad (4\text{-}2)$$

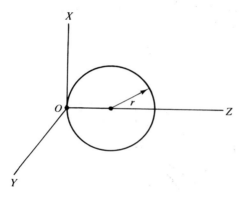

Figure 4-1

Equation (3-9) may also be written as

$$\frac{x^2 + y^2}{r^2} + \frac{(z - r)^2}{r^2} = 1$$

or

$$1 - \frac{z}{r} = \sqrt{1 - \frac{x^2 + y^2}{r^2}} \qquad (4\text{-}3)$$

Combining this with Eq. (4-2) gives

$$z = \frac{(x^2 + y^2)/r}{1 + \sqrt{1 - (x^2 + y^2)/r^2}}$$

$$= \frac{cs^2}{1 + \sqrt{1 - c^2 s^2}} \qquad (4\text{-}4)$$

where

$$c = \frac{1}{r}$$

and the quantity

$$s = \sqrt{x^2 + y^2} \qquad (4\text{-}5)$$

is the radius of a cross section of the sphere normal to the z axis. For an aspheric surface of rotation, we would expect Eq. (4-4) to involve additional terms containing $s^2, s^4, \ldots$, since such a surface should depend only on $s^2 = x^2 + y^2$ and its powers. Hence, we write

$$z_o = \frac{cs^2}{1 + \sqrt{1 - c^2 s^2}} + A_2 s^2 + A_4 s^4 + \cdots \qquad (4\text{-}6)$$

As an approximation to the intersection point of a ray and the aspherical surface, we first find the coordinates (x_0, y_0, z_0) of the point this ray pierces on the sphere of Eq. (4-4), as shown in Fig. 4-2. Let

$$s_0^2 = x_0^2 + y_0^2$$

Then the z coordinate of the point on the aspheric which has the same x and y coordinates as (x_0, y_0, z_0) may be expressed by

$$\bar{z}_0 = \frac{cs_0^2}{1 + \sqrt{1 - c^2 s_0^2}} + A_2 s_0^2 + \cdots \qquad (4\text{-}7)$$

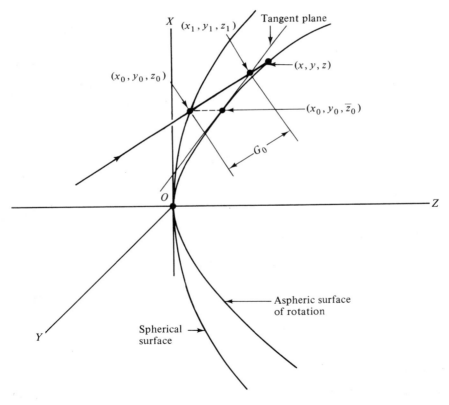

Figure 4-2

Next, we define a vector $\mathbf{t}$ joining O with an arbitrary point on the aspheric by

$$\mathbf{t} = x\mathbf{i} + y\mathbf{j} + z\mathbf{k}$$

$$= x\mathbf{i} + y\mathbf{j} + \left[\frac{cs^2}{1 + \sqrt{1 - c^2 s^2}} + A_2 s^2 + \ldots\right]\mathbf{k}$$

or

$$\mathbf{t} = x\mathbf{i} + y\mathbf{j} + \left[\frac{1}{c} - \sqrt{\frac{1}{c^2} - s^2} + A_2 s^2 + \ldots\right]\mathbf{k}$$

If $\mathbf{t}$ is changed to $\mathbf{t} + \Delta\mathbf{t}$, then $\Delta\mathbf{t}$ is a short vector lying on the aspheric surface; therefore $\partial\mathbf{t}/\partial x$ and $\partial\mathbf{t}/\partial y$ are vectors defining a tangent plane. These vectors have the form

$$\left.\begin{array}{l}\dfrac{\partial\mathbf{t}}{\partial x} = \mathbf{i} + \mathbf{k}\left[\dfrac{x}{\sqrt{\dfrac{1}{c^2} - s^2}} + 2A_2 x + 4A_4 x s^2 + \ldots\right]\\[3em]\dfrac{\partial\mathbf{t}}{\partial y} = \mathbf{j} + \mathbf{k}\left[\dfrac{y}{\sqrt{\dfrac{1}{c^2} - s^2}} + 2A_2 y + 4A_4 y s^2 + \ldots\right]\end{array}\right\} \quad (4\text{-}8)$$

Now we wish to find a vector through the point $(x_0, y_0, \bar{z}_0)$ on the aspheric surface which is normal to the tangent plane. Its direction will be specified by the vector product

$$\frac{\partial\mathbf{t}}{\partial x} \times \frac{\partial\mathbf{t}}{\partial y} = \mathbf{k} - \mathbf{j}\left[\frac{y_0 c}{n_0} + 2A_2 y_0 + 4A_4 y_0 s^2 + \ldots\right]$$

$$- \mathbf{i}\left[\frac{x_0 c}{n_0} + 2A_2 x_0 + 4A_4 x_0 s^2 + \ldots\right]$$

where

$$n_0 = \sqrt{1 - c^2 s_0^2} \quad (4\text{-}9)$$

and where we have used Eq. (4-8) with $x = x_0$, $y = y_0$, $z = \bar{z}_0$.

In terms of this normal, the coordinates x, y, z of any point on a tangent plane through $(x_0, y_0, \bar{z}_0)$ will satisfy the condition

$$[(x - x_0)\mathbf{i} + (y - y_0)\mathbf{j} + (z - \bar{z}_0)\mathbf{k}] \cdot \left[\frac{\partial\mathbf{t}}{\partial x} \times \frac{\partial\mathbf{t}}{\partial y}\right] = 0$$

or

$$(z - \bar{z}_0) - (x - x_0)f(s_0, x_0) - (y - y_0)f(s_0, y_0) = 0 \quad (4\text{-}10)$$

where

$$f(s_0, x_0) = \frac{x_0 c}{n_0} + 2A_2 x_0 + 4A_4 x_0 s_0^2 + \ldots + jA_j s_0^{j-2}$$

$$f(s_0, y_0) = \frac{y_0 c}{n_0} + 2A_2 y_0 + 4A_4 y_0 s_0^2 + \ldots + jA_j s_0^{j-2}$$

Solving Eq. (4-10) for z gives

$$z = (x - x_0)f(x_0, x_0) + (y - y_0)f(s_0, y_0) + \bar{z}_0 \quad (4\text{-}10a)$$

A vector R_t from the origin to the point (x, y, z) on the tangent plane will then have the form

$$R_t = x\mathbf{i} + y\mathbf{j} + z\mathbf{k}$$
$$= x\mathbf{i} + y\mathbf{j} + [(x - x_0)f(s_0, x_0) + (y - y_0)f(s_0, y_0) + \bar{z}_0]\mathbf{k} \quad (4\text{-}11)$$

Let us consider the section of the incident ray between the point (x_0, y_0, z_0) on the sphere and the point (x_1, y_1, z_1) on the tangent plane. Its origin is specified by the vector

$$R_i = x_0\mathbf{i} + y_0\mathbf{j} + z_0\mathbf{k}$$

and its terminal point by Eq. (4-11); therefore the ray itself is represented by the vector

$$R_t - R_i = (x_1 - x_0)\mathbf{i} + (y_1 - y_0)\mathbf{j}$$
$$+ [(\bar{z}_0 - z_0) + (x_1 - x_0)f(s_0, x_0) + (y_1 - y_0)f(s_0, y_0)]\mathbf{k} \quad (4\text{-}12)$$

It we denote the distance from (x_0, y_0, z_0) to (x_1, y_1, z_1) by G_0 and the direction-cosines of the incident ray by X, Y, Z, then

$$X = \frac{x_1 - x_0}{G_0} \quad (4\text{-}13)$$

$$Y = \frac{y_1 - y_0}{G_0} \quad (4\text{-}14)$$

and, by Eq. (4-12)

$$Z = \frac{z_1 - z_0}{G_0}$$
$$= [(\bar{z}_0 - z_0) + (x_1 - x_0)f(s_0, x_0) + (y_1 - y_0)f(s_0, y_0)]/G_0 \quad (4\text{-}15)$$

Combining these three relations gives

$$G_0 = \frac{\bar{z}_0 - z_0}{Z - Xf(s_0, x_0) - Yf(s_0, y_0)} \quad (4\text{-}16)$$

Let

$$l_0 = -x_0[c + n_0(2A_2 + \dots)]$$
$$m_0 = -y_0[c + n_0(2A_2 + \dots)]$$

so that

$$l_0 = -n_0 f(s_0, x_0) \quad (4\text{-}17)$$
$$m_0 = -n_0 f(s_0, y_0) \quad (4\text{-}18)$$

Then Eq. (4-16) becomes

$$G_0 = \frac{n_0(\bar{z}_0 - z_0)}{l_0 X + m_0 Y + n_0 Z} \quad (4\text{-}19)$$

This formula gives the distance G_0 between the point (x_0, y_0, z_0) where the incident ray crosses the spherical surface and the point (x_1, y_1, z_1) where the ray crosses a plane which is tangent to the aspheric through a point $(x_0, y_0, \bar{z}_0)$, as the figure shows. If we know x_0, y_0, z_0 and compute G_0 from Eq. (4-19), then we may find (x_1, y_1, z_1) from the relations

$$\left. \begin{aligned} x_1 &= G_0 X + x_0 \\ y_1 &= G_0 Y + y_0 \\ z_1 &= G_0 Z + z_0 \end{aligned} \right\} \tag{4-20}$$

The point (x_1, y_1, z_1) is fairly close to (x, y, z), the point where the ray actually crosses the aspheric. We can improve the approximation by using x_1, y_1, z_1 to find a new value of G_0 and put this in Eq. (4-20). We thus obtain a better value for the coordinates of the intersection point, and this is repeated until the values do not change significantly.

Having brought the ray to the point (x, y, z), we next consider the refraction process at this point. Let (x_k, y_k, z_k) be the coordinates of (x_1, y_1, z_1) after k times through the process described above. We introduce a quantity P by the relation

$$P^2 = l_k^2 + m_k^2 + n_k^2 \tag{4-21}$$

where in analogy with l_0, m_0, n_0, and s_0 we use the definitions

$$\left. \begin{aligned} l_k &= -x_k[c + n_k(2A_2 + \ldots)] \\ m_k &= -y_k[c + n_k(2A_2 + \ldots)] \\ n_k &= \sqrt{1 - c^2 s_k^2} \\ s_k^2 &= x_k^2 + y_k^2 \end{aligned} \right\} \tag{4-22}$$

Using the same reasoning as before, we obtain a vector perpendicular to the aspheric surface at (x, x_k, z_k) by first using Eq. (4-8) to find two vectors specifying the tangent plane. These will be

$$\left. \begin{aligned} \frac{\partial t}{\partial x_k} &= \left[\frac{x_k c}{n_k} + 2A_2 x_k + \ldots \right] \mathbf{k} + \mathbf{i} \\ \frac{\partial t}{\partial y_k} &= \left[\frac{y_k c}{n_k} + 2A_2 y_k + \ldots \right] \mathbf{k} + \mathbf{j} \end{aligned} \right\} \tag{4-23}$$

or

$$\begin{aligned} \frac{\partial t}{\partial x_k} &= -\frac{l_k}{n_k} \mathbf{k} + \mathbf{i} \\ \frac{\partial t}{\partial y_k} &= -\frac{m_k}{n_k} \mathbf{k} + \mathbf{j} \end{aligned} \tag{4-24}$$

so that the normal is specified by the relation

$$\frac{\partial t}{\partial x_k} \times \frac{\partial t}{\partial y_k} = \mathbf{k} + \mathbf{j}\frac{m_k}{n_k} + \mathbf{i}\frac{l_k}{n_k}$$

We can thus write this normal vector as

$$\mathbf{R}_k = l_k\mathbf{i} + m_k\mathbf{j} + n_k\mathbf{k} \tag{4-25}$$

The incident ray is specified by the vector

$$\mathbf{R}_i = X\mathbf{i} + Y\mathbf{j} + Z\mathbf{k} \tag{4-26}$$

and the incident Snell's law angle is

$$\cos \theta = \frac{Xl_k + Ym_k + Zn_k}{\sqrt{l_k^2 + m_k^2 + n_k^2}} \tag{4-27}$$

using

$$\sqrt{X^2 + Y^2 + Z^2} = 1$$

Letting

$$F = P \cos \theta \tag{4-28}$$

then

$$F = Xl_k + Ym_k + Zn_k \tag{4-29}$$

To find the value of θ after refraction, we use Snell's Law in the form

$$n' \cos \theta' = \sqrt{n'^2 + n^2(\cos^2 \theta - 1)}$$

from which

$$F' = P \cos \theta' = \frac{1}{n'}\sqrt{P^2(n'^2 - n^2) + n^2 F^2} \tag{4-30}$$

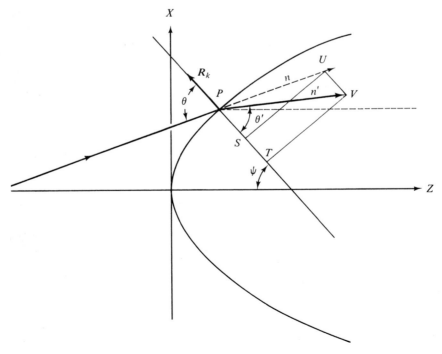

Figure 4-3

Let the normal vector $\mathbf{R}_k$ of Eq. (4-25) intersect the z axis at an angle ψ (Fig. 4-3). The magnitude of $\mathbf{R}_k$, by Eq. (4-21), is

$$R_k = \sqrt{l_k^2 + m_k^2 + n_k^2} = P \tag{4-31}$$

and its projection along the x axis is

$$\mathbf{R}_k \cdot \mathbf{i} = l_k$$

Hence

$$\sin \psi = \frac{l_k}{P} \tag{4-32}$$

Figure 4-3 is similar to Fig. 2-1. Since ψ has the same meaning in both figures (the angle between the normal at the point of refraction and the z axis) and since $\overline{UV}$ is defined in the same fashion in Figs. 2-1, 2-2, and 4-3, then Eq. (2-3) may be rewritten in the present case as

$$\overline{UV} \sin \psi = n' \cos \varphi' - n \cos \varphi \tag{4-33}$$

and Eq. (2-6) gives

$$\overline{UV} = n' \cos \theta' - n \cos \theta$$

or by Eqs. (4-28) and (4-30)

$$\overline{UV} = \frac{n'F'}{P} - \frac{nF}{P} \tag{4-34}$$

Then

$$\cos \varphi' = \frac{\overline{UV} \sin \psi + n \cos \varphi}{n'}$$

$$= \left(\frac{F'}{P} - \frac{nF}{n'P}\right)\frac{l_k}{P} + \frac{n}{n'}\cos \varphi$$

$$= \frac{n}{n'}\cos \varphi + \frac{1}{P^2}\left(F' - \frac{n}{n'}F\right)l_k \tag{4-35}$$

But $\cos \varphi'$ is the direction-cosine of the refracted ray with respect to the x axis, and $\cos \varphi$ is the incident cosine X, so that Eq. (4-35) becomes

$$X' = \frac{n}{n'}X + g_k l_k \tag{4-36}$$

and similarly

$$Y' = \frac{n}{n'}Y + g_k m_k \tag{4-37}$$

$$Z' = \frac{n}{n'}Z + g_k n_k \tag{4-38}$$

where

$$g_k = \frac{1}{P^2}\left[F' - \frac{n}{n'}F\right] \tag{4-39}$$

These are the refraction equations for an aspheric surface.

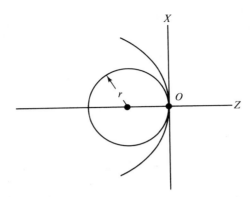

Figure 4-4

A simple example, as given by Smith,[2-1] is the paraboloidal mirror (this is a parabolic surface of revolution). Consider such a mirror with its reflecting surface facing as shown in Fig. 4-4, so that the incident light travels from left to right, as our conventions require. For small apertures, this mirror is similar to the tangent sphere shown, with the equation

$$x^2 + y^2 + (r - z)^2 = r^2$$

or

$$r - z = \sqrt{r^2 - (x^2 + y^2)}$$

$$= r\left(1 - \frac{s^2}{r^2}\right)^{1/2}$$

$$= r\left[1 - \frac{1}{2}\frac{s^2}{r^2} + \frac{1}{8}\left(\frac{s^2}{r^2}\right)^2 + \cdots\right]$$

$$= r - \frac{1}{2}\frac{s^2}{r} - \frac{s^4}{8r^3} - \cdots$$

and

$$z = \frac{1}{2}\frac{s^2}{r} + \frac{s^4}{8r^3} - \cdots \tag{4-40}$$

Equation (4-40) shows that a spherical surface which is not too large ($s^2 = x^2 + y^2$ is smaller than r^2) can be regarded as a paraboloid plus higher-order correction terms.

Problem 4-1

Use matrix methods to show that the focal length of a spherical mirror is related to the radius and the Gaussian constant a by the equation

$$f = +\frac{1}{a} = \frac{r}{2} = -f' \tag{4-41}$$

Equation (4-41) then permits us to write Eq. (4-40) as

$$z = \frac{s^2}{4f'} \qquad (4\text{-}42)$$

for the paraboloidal mirror. If f' is taken as -5.0, then Eq. (4-42) becomes

$$z = -0.05s^2 \qquad (4\text{-}43)$$

and comparison with Eq. (4-6) shows that

$$c = 0, \qquad A_2 = -0.05, \qquad A_4 = \ldots = 0$$

The use of these values is equivalent to saying that the parabolic surface of Fig. 4-4 represents a small deviation from the x-y plane.

As an example, consider a ray which travels towards the paraboloidal mirror with direction-cosines having the values

$$X = 0.1$$
$$Y = 0.0$$
$$Z = (1 - X^2 - Y^2)^{1/2}$$
$$= 0.994\,987\,4$$

and which strikes the x-y plane passing through the vertex of the mirror at the point $(0, 1, 0)$.

Problem 4-2

Using two iterations, show that the incident ray will strike the mirror at the point whose coordinates are

$$x_2 = -0.005\,025$$
$$y_2 = 1.000\,000$$
$$\bar{z}_2 = -0.050\,001$$

and that the direction-cosines after reflection are

$$X_1 = -0.100\,990$$
$$Y_1 = 0.197\,017$$
$$Z_1 = 0.975\,184$$

In performing these calculations, note that (x_0, y_0, z_0) is taken as $(0, 1, 0)$, since the "sphere" which approximately represents the paraboloid is actually a plane which lies to the right of the mirror. Hence the ray never actually passes through the point $(0, 1, 0)$, but is reflected just before it reaches it. However, the method developed in this section does not depend on the specific relation of the aspheric to the approximating spherical surface. Note also that $n' = -1.0$. ∎

Problem 4-3

By a slight extension of the program used in the previous problem, show that all rays parallel to the axis of revolution of a paraboloidal mirror meet at the focal point. In this case, there is no spherical aberration.   ∎

4-3   The Schmidt System

One of the most remarkable optical systems ever designed is the *Schmidt camera*. Astronomical telescopes use either a large-diameter objective lens or a spherical mirror. The latter is not too difficult to grind but suffers from spherical aberration and coma. Schmidt proposed to correct these defects by placing a glass plate of rather peculiar shape in front of the mirror, as shown in Fig. 4-5. Note that this *corrector plate* has very little effect on rays near the

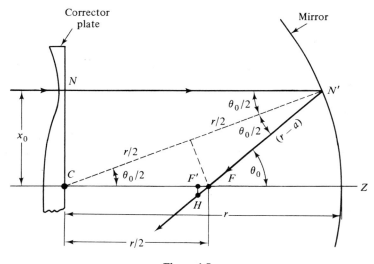

Figure 4-5

axis, but marginal rays will be altered in direction. Miller[4-1] states that Schmidt[4-2] ground his corrector plates by supporting a glass blank at its circumference, applying a vacuum to the underneath, grinding the top surface optically flat, and releasing the vacuum. How he arrived at this procedure is not known, but the method nearly went to the grave with him.

To see why the plate has the shape indicated, consider the ray NN' parallel to the z axis in Fig. 4-5. By Problem 4-1, the paraxial focus is at F', where $\overline{CF'} = r/2$. The ray NN', however, will be reflected and will pass through F, which lies to the right of F'. This follows from the fact that the triangle $FN'C$ is isosceles, so that $\overline{CF}$ must be larger than $r/2$. In fact

$$\overline{CF} = \frac{r/2}{\cos(\theta_0/2)} = \frac{r}{2\sqrt{1 - (x_0/r)^2}}$$

so that

$$\overline{FF} = \frac{r}{2} - \frac{r}{2}\left\{1 + \frac{1}{2}\left(\frac{x_0}{r}\right)^2 + \cdots\right\}$$

$$= -\frac{x_0^2}{4r} = \frac{x_0^2}{8f'} \tag{4-44}$$

The slope of the ray $N'F$ is approximately given by $x_0/(r/2) = x_0/f'$, and the transverse spherical aberration as measured at the paraxial image plane is then

$$\overline{F'H} = \frac{x_0^2}{8f'}\frac{x_0}{f'} = \frac{x_0^3}{8f'^2}$$

Shifting the image plane by an amount Δ towards F reduces the aberration to a value given by the relation

$$\delta = \left(\frac{x_0^2}{8f'} - \Delta\right)\frac{x_0}{f'} \tag{4-45}$$

If we now arrange for the corrector plate to deflect the ray through an angle $\alpha = \delta/f'$, approximately, then it will pass through F' rather than F. This angle is

$$\alpha = \frac{\delta}{f'} = \left(\frac{x_0^2}{8f'} - \Delta\right)\frac{x_0}{f'^2} \tag{4-46}$$

To obtain the angle of deviation specified by this expression, consider the wedge of Fig. 4-6, with a light ray striking it normally. The emerging ray suffers an angular deviation

$$\alpha = \theta - \theta'$$

By Snell's law in paraxial form

$$n\theta' = \theta$$

so that

$$\alpha = \theta'(n - 1)$$

For the triangle ABC

$$\tan\theta' = \frac{\overline{AB}}{\overline{BC}}$$

and again for small angles

$$\theta' = \frac{\overline{AB}}{\overline{BC}} = \frac{t}{h}$$

Then

$$h\alpha = (n - 1)t$$

or

$$\alpha\,dh = (n - 1)\,dt \tag{4-47}$$

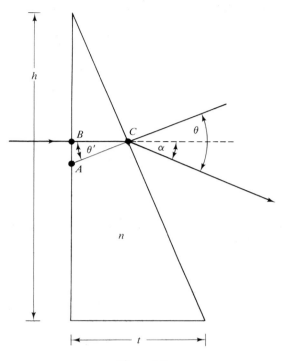

Figure 4-6

The Schmidt corrector plate can be regarded as a wedge whose thickness t varies with height $h = x_0$, resulting in a deviation α which is a function of x_0. Equating the values of α in Eqs. (4-46) and (4-47) gives

$$\frac{dt}{dx_0} = \left(\frac{x_0^3}{8f'^3} - \frac{x_0\Delta}{f'^2}\right)\left(\frac{1}{n-1}\right) \tag{4-48}$$

Integrating

$$t = \frac{(x_0^4/32f'^3) - (x_0^2\Delta/2f'^2)}{(n-1)} + t_0 \tag{4-49}$$

where

$$t = t_0 \quad \text{at} \quad x_0 = 0$$

That the corrector plate has a shape specified by a fourth-degree polynomial should come as no surprise; Eq. (4-49) agrees with Eq. (2-22), a computed expression describing the longitudinal spherical aberration in a simple lens. The preceding result contains some approximations, but it is valid well beyond the paraxial region.

It is customary to put Eq. (4-49) in a different form by introducing the new variables

$$u = \frac{2x_0}{D}$$

$$a = \frac{64f'\Delta}{D}$$

where D is diameter of the Schmidt plate and, therefore, the aperture of the telescope. Then Eq. (4-49) becomes

$$t = \frac{(u^4 - au^2)D^4}{512(n-1)f'^3} - t_0 \qquad (4\text{-}50)$$

Values of a which are usually used lie in the range $1.0 \leq a \leq 2.0$; Fig. 4-7 shows the shape of a plate for which $a = 1.5$.

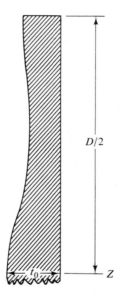

Figure 4-7

Problem 4-4

A Schmidt system uses a mirror with a radius of 8.0 cm. The corrector plate has an index of refraction $n = 1.500$, a thickness $t_0 = 1.0$ cm, and a diameter $D = 10.0$ cm. Verify the shape illustrated in Fig. 4-7 by computing $t - t_0$ for $x_0 = 1.0, 2.0, 3.0, 4.0,$ and 5.0 cm. ∎

The above discussion is concerned solely with the effect of the corrector plate on spherical aberration. This is only part of the story, however. The reason for locating one side of the plate at the center of curvature C in Fig. 4-5 is to create an optical system which—to a first approximation—has no unique axis of symmetry. That is, the light-gathering aperture can pivot by a small amount about *C* without affecting the validity of our analysis. Hence, coma and astigmatism are automatically reduced as well, since these are point aberrations associated with nonmeridional rays. Schmidt's design has been widely used by astronomers. The largest system to date is the 48-inch-diameter camera at Mt Palomar; this achieves a very sharp central area by fitting the film to a concave holder, thus eliminating field curvature also.

A different type of corrector plate, using only spherical surfaces, was proposed independently by Bouwers[4-3] and Maksutov.[4-4] The Bouwers system, shown in Fig. 4-8, is constructed so that the two surfaces of the corrector plate and the surface of the mirror all have a common center. Thus, a ray through this point from any direction is axial. The Maksutov system (Fig. 4-9) is not concentric, and the corrector plate—rather than being a portion of

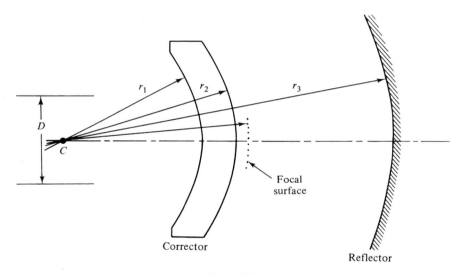

Figure 4-8

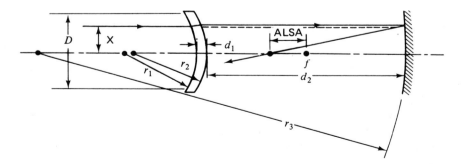

Figure 4-9

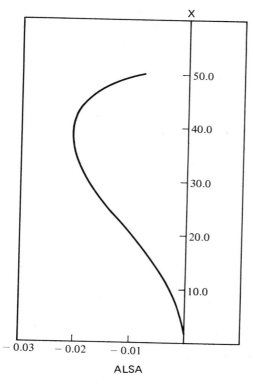

ALSA

Figure 4-10

a spherical shell—has a meniscus shape. Figure 4-10 shows the longitudinal spherical aberration for a Maksutov telescope having a mirror 100 cm in diameter and with a focal length of 400 cm.

Problem 4-5

A Maksutov system has the following parameters

$$r_1 = -152.80 \qquad n = 1.5225$$
$$r_2 = -158.62 \qquad d_1 = 10.00$$
$$r_3 = -823.17 \qquad d_2 = 539.10$$
$$D = 100.0$$

Verify Fig. 4-10.

4-4 Light as a Form of Energy

The principal topic we have considered so far has been the formation of an image of acceptable quality. There are optical systems, however, where a criterion of equal importance is the efficiency with which the light energy is used. Examples of such systems are slide projectors and microfilm readers. To understand the design principles involved, we must first consider how to specify the amount of energy associated with a light source and falling on a detector or a screen.

Let us consider an element of area dS which emits electromagnetic radiation (Fig. 4-11). Suppose that there is a receiving element of area dS' at a

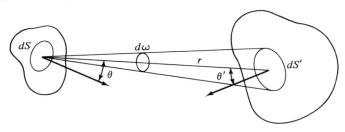

Figure 4-11

distance r and the two elements have normals making angles θ and θ', respectively, with the line joining them. The energy dU being radiated is measured in *joules*. The rate of emission is a power dP expressed in *watts* or joules/sec. In optics, this power is called the *radiant flux*. The surface dS radiates dP/dS watts/meter^2, where

$$W = \frac{dP}{dS} \qquad (4\text{-}51)$$

is the *radiant emittance*. Since we shall introduce a number of other similar quantities (plus a confusing collection of units), a summary of this information will be found in Table 4-1, in which the name, symbol, and unit is shown for each quantity; in some cases, a definition and conversion units are given. The last entry in the table is the *irradiance H*, which is the power per unit surface area being received by dS', or

$$H = \frac{dP}{dS'} \qquad (4\text{-}52)$$

If r in Fig. 4-11 is large, then dS is effectively a point source, and the energy received by dS' is inversely proportional to r^2.

Let us consider the irradiance H_n falling on the projection dS'_n of dS' normal to r. This irradiance decreases as the square of the distance; the proportionality constant J is called the *radiant intensity* and Eq. (4-52) may be written as

$$H_n = \frac{dP}{dS'_n} = \frac{J}{r^2}$$

or

$$J = \frac{dP}{dS'_n/r^2} \qquad (4\text{-}53)$$

But dS'_n/r^2 will be recognized as the solid angle $d\omega$ subtended by dS' at dS. Hence

$$J = \frac{dP}{d\omega} \qquad (4\text{-}54)$$

and we see that the intensity J is the flux per unit solid angle radiated by a source. Solid angle is specified in *steradians* or *spherical radians*.

When the area of the source is not negligible, then we should also be able to specify the intensity per unit area. This is known as the *radiance N* and is defined by the relation

$$N = \frac{J}{dS_n} = \frac{J}{dS \cos \theta}$$

$$= \frac{dP}{d\omega \, dS \cos \theta} \qquad (4\text{-}55)$$

where the normal component of dS is used since the energy received at dS' will depend both on the size and the orientation of dS.

Problem 4-6

To review the concept of solid angle ω, consider a point P at a distance a from the center of a disc of radius R, where a is measured along a line normal to the disc.

(a) Show that

$$\omega = 2\pi a \left[\frac{1}{a} - \frac{1}{\sqrt{a^2 + R^2}} \right] \qquad (4\text{-}56)$$

(b) Show that when point P is very far from the disc, Eq. (4-56) reduces to

$$\omega = 0$$

(c) Show that when P is very close to the disc, we obtain

$$\omega = 2\pi.$$

(d) Show that when P is far enough away so that $R/a < 1$, then Eq. (4-56) becomes

$$\omega = \frac{\pi R^2}{a^2}$$

(e) Explain the physical significance of (b), (c), and (d). ■

Returning to Eq. (4-55), we should realize that the radiant intensity J refers to the power per unit solid angle radiated in a particular direction as specified by θ; that is, we really should write J as $J(\theta)$. If the energy supplied by dS is being received from another source and if dS is retransmitting it isotropically—that is, dS is a *perfect diffuser*—then $J(\theta)$ is related to the inten-

Table 4-1
Summary of SI (and other) Units for Radiant Energy

Radiometric	Photometric
Radiant energy U joule	Luminous energy Q talbot
Radiant flux P watt	Luminous flux F lumen

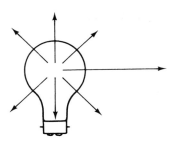

Flux is total output power ("light").
1 W = 680 lm at 555 nm

Radiant emittance W watt meter^{-2}	Luminous emittance L lumen meter^{-2}

Emittance is power per unit area or flux density
from a source.

(See Illuminance for conversion factors.)

Table 4-1 *cont.*

Radiant intensity J watt steradian^{-1}	Luminous intensity I candela (lumen steradian^{-1}, candle)

Intensity is power per unit solid angle ("candlepower").

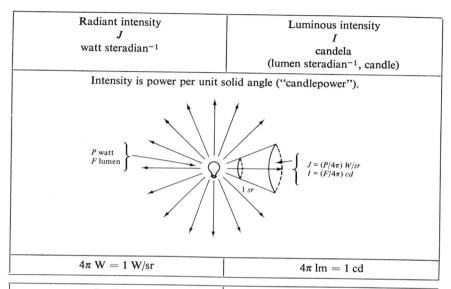

$$J = (P/4\pi) \ W/sr$$
$$I = (F/4\pi) \ cd$$

4π W $= 1$ W/sr	4π lm $= 1$ cd

Radiance N watt steradian^{-1} meter^{-1}	Luminance B candela meter^{-2} (lumen steradian^{-1} meter^{-2}, lambert)

Radiance or luminance is power per unit solid angle per unit projected area emitted or scattered ("brightness").

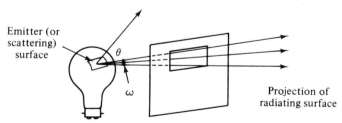

Emitter (or scattering) surface

Projection of radiating surface

For a Lambertian source only:

$$B \ cd/m^2 = \pi L \ lm/m^2$$
$$1 \ L = 1 \ lm/cm^2$$
$$1 \ m\text{-}L = 1 \ lm/m^2$$

Also:

$$1 \ m\text{-}L = (0.0929 \ m^2/ft^2) \ ft\text{-}L$$

Table 4-1 *cont.*

Irradiance H watt meter^{-2}	Illuminance E lux (lumen meter^{-2}, meter-candle)
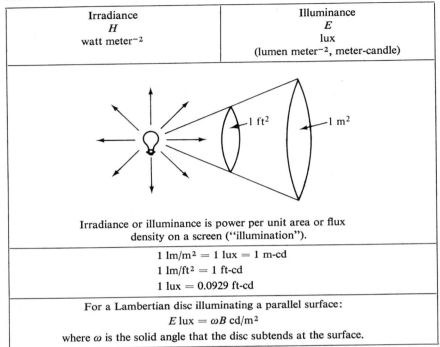 Irradiance or illuminance is power per unit area or flux density on a screen ("illumination").	
1 lm/m^2 = 1 lux = 1 m-cd 1 lm/ft^2 = 1 ft-cd 1 lux = 0.0929 ft-cd	
For a Lambertian disc illuminating a parallel surface: E lux = ωB cd/m^2 where ω is the solid angle that the disc subtends at the surface.	

sity J_n leaving in the normal direction by the equation

$$J(\theta) = J_n \cos \theta \tag{4-57}$$

This is known as *Lambert's law*, and a surface which satisfies Eq. (4-57) is said to be *Lambertian*. (Such a surface can be regarded as a collection of point sources all radiating isotropically.) Then Eq. (4-55) becomes

$$N = \frac{J_n}{dS} \tag{4-58}$$

which indicates that the apparent brightness of a Lambertian surface is everywhere constant. For example, the sun gives the appearance of being a disc of constant brightness.

For a Lambertian surface only, there is a useful relation between radiant emittance W and radiance N. Consider a plane light source radiating dP watts per unit area into a solid angle $d\omega$ (Fig. 4-12). By Eqs. (4-54) and (4-57)

$$dP = J(\theta) \, d\omega$$

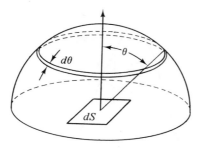

Figure 4-12

or

$$P = \int J(\theta)\, d\omega = J_n \int \cos\theta\, d\omega$$

The solid angle $d\omega$ is the surface enclosed by the intersection of the two cones of Fig. 4-12 with the surface of a unit sphere, or

$$d\omega = 2\pi \sin\theta\, d\theta$$

so that the flux is

$$P = 2\pi J_n \int_0^{\pi/2} \sin\theta \cos\theta\, d\theta$$
$$= \pi J_n$$

Dividing both sides by the surface area normal to the flux gives

$$W = \pi N_n \qquad\qquad (4\text{-}59)$$

which is what we wished to demonstrate.

The quantities that have been introduced so far deal with radiated energy of any wavelength and are called *radiometric*. If we take the same set of definitions, but restrict ourselves to the visible spectrum, then the corresponding items are *photometric*. Every radiometric quantity has a photometric mate, as listed in the right-hand half of Table 4-1. For example, the *luminance* B is the photometric equivalent of the radiance N (either one of these may be regarded as having the common meaning of brightness, although—strictly speaking—brightness is one of those sensations which should be expressed in logarithmic units like the decibel). The luminous flux F (simply called *light* by illuminating engineers) is measured in a unit known as the *lumen*. Its radiometric equivalent, the radiant flux P, is the total energy radiated by a given source over the full frequency range of the electromagnetic spectrum. However, a source in the visible range is detected in accordance with the sensitivity of the human eye, as shown in Fig. 4-13. The maximum at 555 nm (0.555 micrometer or 5550 A) lies in the green region of the visible spectrum. It has been internationally agreed that 1 watt of green light with this wavelength is equivalent to 680 lumens.

The quantity corresponding to J is the *luminous intensity* I, expressed in lumens per steradian (lm/sr). This unit, called the *candela*, used to be the

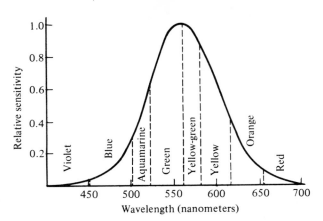

Figure 4-13

candle, and I was known as "candlepower." Then B is measured in candela per meter² (cd/m²). The *illuminance E* corresponds to H and has units of lm/m². This is the quantity we think of as the "illumination" on a screen. An illuminance of 1 lm/m² is also known as 1 *lux* or 1 *meter-candle*. The present universal system of units, called *SI units* (from the French *Système Internationale*) uses the meter, kilogram, second, ampere, kelvin, mole, and candela as its fundamental units, and the radian and steradian are supplementary units. The lumen and lux are derived units for optics.

Equation (4-59) is valid for photometric quantities as well; thus a plane Lambertian source with a luminance B of 1 candela/meter² has a luminous emittance of π lumens/meter². To eliminate this factor of π, a unit called the *lambert* was introduced. When $L = 1$ lumen/cm², then $B = 1$ lambert. In addition, we have the *meter-lambert* (and foot-lambert) for $L = 1$ lm/m² (or ft²). It is hoped that all of these non-SI units (and many more that we have not mentioned) will soon disappear.

To determine how to compute the energy of a receiving surface due to a specific source, consider a disc of radius R receiving light of intensity I (Fig. 4-14). We wish to find the relation between the luminance B of the disc and the illuminance E on an element of area dS' normal to the z axis, that is, the disc is receiving light from a source and reradiating it to a small screen of area S'. Writing the photometric analogue of Eq. (4-55) gives the power received on the screen as

$$dF = B \, d\omega \cos \theta \, dS \tag{4-60}$$

where the solid angle subtended by dS' at the radiating point P is

$$d\omega = \frac{dS' \cos \theta}{r^2} \tag{4-61}$$

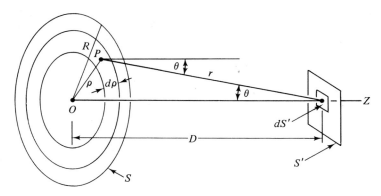

Figure 4-14

and the radiation comes from an element of area given by

$$dS = 2\pi\rho \, d\rho \tag{4-62}$$

Then Eq. (4-60) becomes

$$F = 2\pi B \int dS' \int_0^R \frac{\cos^2 \theta}{r^2} \rho \, d\rho$$

$$= 2\pi BS' \int_0^R \frac{\cos^2 \theta}{\rho^2 + D^2} \rho \, d\rho \tag{4-63}$$

assuming that B is constant across the disc. But

$$\cos^2 \theta = \frac{D^2}{D^2 + \rho^2} \tag{4-64}$$

and Eq. (4-63) may be integrated to give

$$F = \pi BS' \frac{R^2}{R^2 + D^2} \tag{4-65}$$

Using Eq. (4-52) as applied to illuminance, we have

$$E = \frac{dF}{dS'} = \frac{\pi BR^2}{(R^2 + D^2)} = \frac{BS}{r^2} \tag{4-66}$$

where S is the area of the disc. By Problem 4-6

$$E = \omega B \tag{4-67}$$

which is the desired relation.

We should also note that a source which is radiating 4π lumens of flux uniformly in all directions will be providing 1 lumen/steradian or 1 candela. Hence, 12.57 lumens of radiant flux from an isotropic source means that $I = 1$ cd. For example, a common 100-watt light bulb will radiate about 1700 lumens of visible energy. If the distribution were uniform, the luminous intensity I would be about 135 candela, and it used to be customary to say that the light bulb has an output of 135 candlepower. This may be compared to a flashlight lamp (1 candlepower) or an arc lamp (10,000 candlepower).

Problem 4-7

A lens whose diameter is 1.25 in. and whose focal length is 2.0 in. projects the image of a lamp capable of producing 3000 cd/cm². Find the illuminance E on a screen 20 ft from the lens in lm/ft² (footcandles). ∎

Problem 4-8

(a) A lamp of 100 cd illuminates a book 4 m directly below it. Find the illuminance in SI units.

(b) A small-aperture spherical mirror is placed above the lamp at its focal distance of 3 m. The reflection coefficient of the mirror is 0.9. Find the increase of illuminance at the book. ∎

4-5 Projection Systems

A transparent object such as a color slide or a microfilm frame is used in a projection system. The viewer wants to see a large, bright, and aberration-free image, and he wants this accomplished with minimum cost, both initially and in terms of energy consumption. Figure 4-15 shows the essential parts of a typical microfilm reader, which uses 16-mm motion picture film and projects it from the rear onto a translucent screen. The frame is approximately 0.5 in. × 0.5 in. and the screen is 12 in. × 12 in., so that the magnification is customarily specified as 24X. The light comes from a lamp described by the maker as having an output of 4500 lumen, consuming 150 W at 120 V. This amount of energy is being radiated over most of a full sphere from a plane filament whose dimensions are 3/8 in. × 3/8 in.

For the geometry shown, the solid angle subtended by a frame whose area is 0.25 in² at a distance of 3.5 in. from the bulb is less than 0.1 steradian. Since light is emitted into $4\pi = 12.5$ sr, only about 1% of the light leaving the bulb can reach the screen. We need to increase the efficiency of the system and can do so in two ways. First, a spherical mirror is placed behind the bulb at a distance equal to its radius. If the bulb were a true point source and located at the center of the sphere, half of the rays leaving the bulb would strike the mirror, be reflected back on themselves, and simply

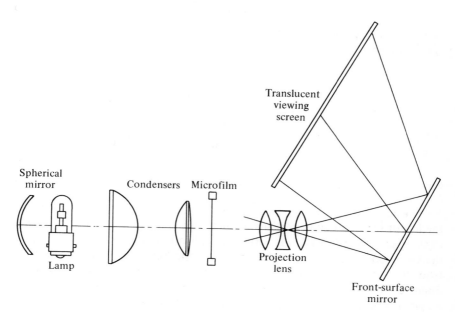

Figure 4-15

double the effective output being radiated towards the microfilm. For a plane filament, we obtain an approximation to a plane image. However, it is found that if the image of the filament coincides with the filament itself, this large amount of energy in a small region raises the temperature of the filament so much that it seriously shortens its life. The image is therefore arranged to fall adjacent to the filament, and the two together serve as a planar light source of twice the original size. When this is enlarged by a factor of 24, it also results in a more uniform light distribution on the viewing screen.

The second method used to improve the efficiency is to take the light from the filament and its image, radiated over a region of about 2π spherical radians, and project it to just fit within the area determined by the size of the microfilm. This is accomplished with a pair of *condenser lenses*. The double condenser serves the same function as the lens in a lighthouse beacon. It decreases the solid angle of the beam and therefore increases the energy per square meter being received at some distant surface. Most condensers consist of a pair of plano-convex lenses with the curved surfaces facing each other, as shown in the figure. The reason for this is connected to an effect which we have not previously considered: the variation of the index of refraction with wavelength. We shall examine this topic in more detail at the end of the present section.

Problem 4-9

The condenser system used in the microfilm reader of Fig. 4-15 has the following specifications:

r	t'	n'
∞		
	0.624	1.518
-1.087		
	0.985	1.000
1.102		
	0.370	1.523
-6.693		

Find the size and position of the image of the filament. ■

The solution to this problem will show that a filament located 1.0 in. to the left of the first condenser lens will have an inverted image about 2.25 in. to the right of the second lens with a magnification of about -1.4. We must therefore position the microfilm frame so that the cone of light emerging from the second condenser lens and coming to a focus 2.25 in. away has a diameter of about 5/8 in. when it strikes the object (corresponding to the diagonal length of the frame). This ensures that the object is properly illuminated but that no light is wasted.

Locating the projection lens involves a number of considerations. First, the image of the bulb should fall somewhere between the first and last surfaces of the projection optical system to ensure that an image of this image does not appear on the screen of the reader. An enlarged image of the filament would have a disastrous effect on legibility. A more fundamental consideration, however, is based on the obvious notion that the projection lens should pass to the screen all the light energy it receives. Every optical system has some limiting aperture. For example, a telescope in its simplest form consists of a large objective lens at one end of a tube and a small eyepiece lens at the other end. If we insure that the diameter of the tube is large enough so that it offers no hindrance to the light passing through it, then it seems rather obvious that the more expensive objective lens would be the limiting feature. That is, we would like it to be large in order to gather as much light as possible, but cost and weight considerations force some sort of compromise. On the other hand, we may at times want to limit the amount of light collected; for example, this is what we do when we adjust the diaphragm of a good camera. This adjustable opening, or even a simpler one like a hole in a metal plate, is called a *stop*.

Let us consider the effect of a stop on some point of the object. For convenience, we shall take a point that lies on the lens axis—the base of the tree in Fig. 4-16. This choice simplifies the geometrical arguments we shall use. The limiting rays—those which just pass through the lens—are *AB* and *AC*.

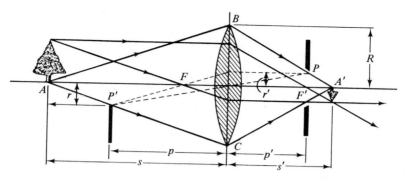

Figure 4-16

Let us assume that rays pass from the top of the tree through F and F'; A' is then the image of A. Now we introduce a stop which just permits the passage of rays BA' and CA'; P is the top edge of this stop. We shall show that the image P' of this point has the same effect in object space, that is, P' must lie on AC. If we suppose that r' is the radius of the stop, r is the radius of the image of the stop, and R is the radius of the lens, then we must prove that

$$\frac{r}{s-p} = \frac{R}{s} \tag{4-68}$$

From the Gaussian lens equation, we have

$$\frac{1}{f'} = \frac{1}{s'} - \frac{1}{s} = \frac{1}{p'} - \frac{1}{p}$$

or

$$\frac{s-p}{ps} = \frac{s'-p'}{s'p'} \tag{4-69}$$

Also,

$$\frac{r'}{s'-p'} = \frac{R}{s'} \tag{4-70}$$

and

$$\frac{r}{r'} = \frac{p}{p'} \tag{4-71}$$

By combining Eq. (4-69) through (4-71) we can prove Eq. (4-68). Hence the image of the stop in object space does exactly the same thing as the stop itself in image space—either the stop or its image represents the limit of the light-gathering power of the lens. If we decrease the diameter of the stop somewhat, its image is correspondingly decreased, and now the image of the stop rather than the lens determines the amount of light passed through the lens. The same effect occurs if we consider an object to the left of A; the point P' again cuts off more rays than the lens itself.

Thus, if we have a lens and a stop to its right, for some rays the lens will be the limiting factor, and for others it will be the image of the stop, depending on the distance from A to the unit plane. Whatever establishes this limit—lens or stop—is called the *entrance pupil*. For a more complex system having several lenses and stops, we must find the image of each optical element in object space, and then determine which one acts as the entrance pupil.

In a similar way, we may define an *exit pupil*. For an object to the left of A, the rays forming the image will be limited by the stop. That is, the stop and its image compose the exit pupil and the entrance pupil, respectively, for all rays originating to the left of A. Conversely, the lens is both an entrance and an exit pupil for rays starting to the right of A. This general principle will be found to hold in more complicated systems.

Problem 4-10

A telescope has the following parameters:

	Focal Length	Diameter
Objective	100.0 cm	10.0 cm
Eyepiece	5.0 cm	2.0 cm

If the eyepiece and objective are separated by a distance equal to the sum of the focal lengths (105.0 cm), show that the objective serves as the entrance pupil and that its image is the exit pupil. Find the position and size of the exit pupil. ■

Since light is a form of energy, it is conserved (ignoring absorption by lenses and mirrors and reflections at transparent surfaces) as it passes through a system. This principle is of great practical importance, as we shall show. Consider an area, like dS in Fig. 4-11 but taken normal to the axis, which is radiating a total of dF lumens into a solid angle $d\omega$. Let this beam strike the entrance pupil of a lens at an angle α to the axis, as shown in Fig. 4-17, intercepting an elementary area given by (when r is large and $d\alpha$ is small)

$$dA = (r\, d\alpha)(r \sin \alpha\, d\varphi)$$

By Eq. (4-60)

$$dF = B\, dS \cos \alpha\, d\omega$$

or

$$dF = B\, dS \cos \alpha \left(\frac{dA}{r^2}\right)$$

$$= B\, dS \cos \alpha \sin \alpha\, d\alpha\, d\varphi$$

$$= B\, dS\, d\varphi\, d\left(\frac{\sin^2 \alpha}{2}\right)$$

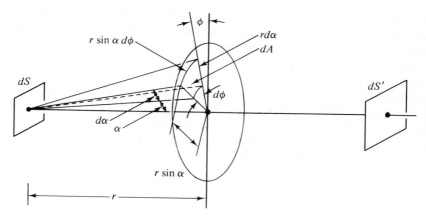

Figure 4-17

If this beam then forms an image of area dS' to the right of the lens, conservation of energy requires that

$$B \, dS \, d\varphi \, d(\sin^2 \alpha) = B' \, dS' \, d\varphi' \, d(\sin^2 \alpha') \tag{4-72}$$

where the primed quantities refer to parameters measured at the image.

For perfect imaging of off-axis points, spherical aberration and coma must vanish; that is, the Abbe sine condition as expressed by Eq. (3-60) must be satisfied. It then follows that

$$(xn \sin \alpha)^2 = (x'n' \sin \alpha')^2 \tag{4-73}$$

But the magnification of the area is simply

$$\frac{dS'}{dS} = \left(\frac{x'}{x}\right)^2$$

Hence, Eq. (4-73) becomes

$$dS \, n^2 \sin^2 \alpha = dS' n'^2 \sin^2 \alpha' \tag{4-74}$$

Inserting this into Eq. (4-72) gives

$$\frac{B}{n^2} \, d\varphi = \frac{B'}{n'^2} \, d\varphi' \tag{4-75}$$

For a system with axial symmetry, the angle φ remains unchanged, so that $d\varphi = d\varphi'$. If object and image are in the same medium, then Eq. (4-75) shows that

$$B = B' \tag{4-76}$$

Thus, conservation of energy implies conservation of brightness when the restrictions stated above are obeyed.

At first sight this is surprising, for it means that reducing the image size does not increase the brightness, and this appears to be contrary to experience. If we take the image of a bulb, look at it on a screen, and then reduce the magnification, the eye sees a brighter image. What we really should do is observe the image *directly* with the eye (e.g., use zoom binoculars). Then when we reduce the magnification, we are simultaneously reducing the area dS' and increasing the solid angle $d\omega'$ as subtended at the eye. Figure 4-18 shows what is happening: if the image size is decreased from dS'_1 to dS'_2, the solid angle is increased from $d\omega'_1$ to $d\omega'_2$. However, as indicated by Eq. (4-72), the product $dS'\, d\omega'$ remains invariant. These arguments do not apply, on the other hand, when the image is observed on the screen, for then the light is scattered in all directions. Thus, the apparent paradox is resolved.

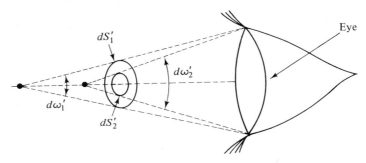

Figure 4-18

After these digressions, we return to the microfilm reader of Fig. 4-15, realizing that the image of the filament must fall somewhere within the immediate vicinity of the entrance pupil of the projection lens and be of the same size. Although there is a sharp image of the filament formed within the body of the projection lens, back at the microfilm frame this light is not focused; thus the object is illuminated in a fairly uniform fashion. In fact, Cox[4-5] states that it has been empirically found that best results are obtained when the image of the filament is brought to a focus near the last surface of the projection lens. This gives a blurrier image at the microfilm plane without making the incoming cone of light too large to pass through the entrance of the pupil. (This latter requirement will be recognized as the optical counterpart of the principle of impedance matching used by engineers for mechanical and electrical systems.) Since the filament plus its image as produced by the mirror is a 3/8 in. × 3/4 in. rectangle with a diagonal of about 7/8 in., the image has a diagonal of about 1.2 in., which requires a fairly large (and expensive) lens to pass all the light. The lens that was actually used had a diameter of only 7.4 mm or 0.29 in., which is too small to accept all of the energy in the image of the filament. A larger diameter for the projection lens was impractical in a product which must be sold at a competitive price. The use of a smaller than optimum projection lens resulted in a mismatch between the projection lens and the condensing lenses which led to a considerable loss in image illuminance.

To find the illuminance E at the viewing screen, we must determine the brightness B' at the entrance pupil, and this quantity—by Eq. (4-76)—is equal to B at the bulb. Although we do not know the effective area of the filament, we can compute B' by first finding the fraction of the 4500 lumens radiated by the bulb which strikes the condenser system. This quantity is found from the ratio of the solid angle subtended by the condenser at the filament to the solid angle for a full sphere, or

$$4500\frac{\pi(2.4)^2/(3.0)^2}{4\pi} = 720 \text{ lumens}$$

where the dimensions are as indicated in Fig. 4-19. Although the 720 lumens is being radiated from the second condenser lens to the entrance pupil, if we

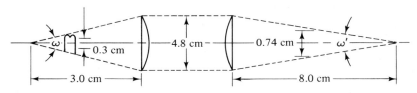

Figure 4-19

regard the direction of travel as reversed, then we can compute B' from its definition as

$$B' = \frac{720}{[\pi(0.37)^2][\pi(2.4)^2/(8.0)^2]}$$
$$= 6350 \text{ lm/cm}^2/\text{sr}$$

The use of the spherical mirror doubles this figure; so by Eq. (4-67)

$$E = B'\omega = (6530)\left[\pi\left(\frac{0.29}{40.0}\right)^2\right]$$
$$= 1.03 \text{ lm/cm}^2$$

This represents the illuminance on the screen. Converting this to units more commonly used by designers, we obtain

$$(1.03 \text{ lm/cm}^2)(929 \text{ cm}^2/\text{ft}^2) = 957 \text{ lm/ft}^2 = 957 \text{ ft-candles}$$

Measurements gave a value of 23 ft-candles at the center of the screen and 19 ft-candles at the corners. Hence, the losses due to the mismatch between the projection lens and the condensers are significant in this example of an inexpensive projection system.

The fall-off of the illuminance from center to edge should also be considered. It is known that the human eye can tolerate a variation of about 50% without serious objection. The calculation above is valid only for the center of the screen. For a region off the axis, we must take into consideration a number of effects, as indicated in Fig. 4-20. The solid angle at A may be computed from the formula in Eq. (4-56). At B, however, the solid angle is due to a disc of diameter $\overline{FG}$, so that the area subtending this solid angle is decreased by a factor of cos ψ. The distance to the screen is reduced by cos ψ as we go out to B and contributes a factor of cos² ψ. Finally, we have already seen that the illumination falls off as cos ψ; therefore the resultant dependence of E on the angle of inclination is expressible as a cos⁴ ψ law. The microfilm reader in Fig. 4-15 measured values that obey this relation.

A factor that we should discuss is the quality of the two systems—the condenser and the projection lens—used in this instrument. The projection lens clearly should be as good as possible, consistent with cost limitations. A practical problem which turned up on the production line is the failure of the lens manufacturer to meet specifications set for geometrical aberrations because of poor assembly. All of our calculations in Chapters 2 and 3 assumed that the lens system was perfectly symmetric about its axis, but this is not an easy criterion to meet. Lenses are mounted in metal tubes by threaded rings which clamp the edges, and it is difficult to avoid slight misalignments. These mechanical flaws result in an uneven distribution of light with respect to the axis, and a phenomenon known as *flare* shows up on

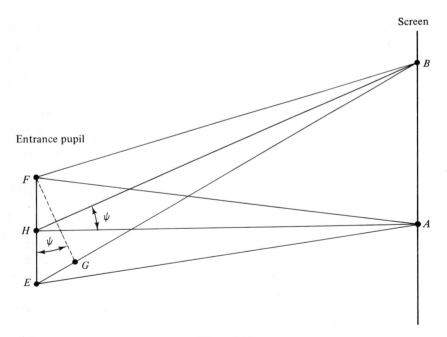

Figure 4-20

the viewing screen. Portions of the document being read have bright patches, so that the letters are smeared and difficult to read.

The condensers, on the other hand, do not demand rigid control over their dimensions and assembly, for they do not produce the image on the viewing screen. For this reason, it is possible to use inexpensive lenses; they are generally molded glass or plastic. In fact, the availability of the molding process puts aspheric lenses in a reasonable price range. The use of aspherics reduces coma (with spherical aberration as a special case), and this is important when we want a uniform light distribution over the screen. Streaks of bright and dark patches would be very distracting. Hence, we need to worry only about the point aberrations of the condensers.

4-6 Chromatic Aberration

Another aspect of lens system design which we have not yet mentioned is associated with the fact that white light (such as sunlight) has a range of wavelengths or colors, and the index of refraction n for a given medium depends on the color. We can see this experimentally by passing white light through a prism. The emerging beam will be spread, each color emerging at a different angle. Since the refraction angle depends on n by Snell's law, we see that the index n varies with wavelength λ. Some typical values of n for two types of glass, crown and flint, are given in Table 4-2.

Table 4-2

Fraunhofer line	Color	Wavelength (nanometer)	Index (crown glass)	Index (flint glass)
C	Red	656.28	1.51418	1.69427
D	Yellow	589.59	1.51666	1.70100
F	Blue	486.13	1.52225	1.71748

The first column of the table identifies the wavelength in terms of its position in the *Fraunhofer spectrum* of the sun. Fraunhofer arbitrarily labeled a number of bright lines in the solar spectrum A through H; the three we have shown correspond to hydrogen (C), sodium (D), and calcium (F). These provide convenient sources for making index of refraction measurements, and most tables specify n in terms of Fraunhofer wavelengths.

It is possible to express the dependence of the index of refraction n of a given sample of glass on the wavelength by defining a quantity called the *dispersive power V* as

$$V = \frac{n_B - n_R}{n_Y - 1} \tag{4-77}$$

where n_B, n_Y, and n_R are the indices for blue, yellow, and red light, respectively.

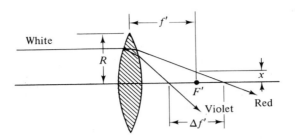

Figure 4-21

To obtain an idea of the physical significance of V, consider the lens of Fig. 4-21 with radius R. Because of the different colors, there is a spread $\Delta f'$ in the focal point f'. The corresponding spread in the image point has a radius x. By similar triangles, we have

$$\frac{R}{f'} = \frac{x}{\Delta f'/2}$$

or

$$x = \frac{R}{2}\frac{\Delta f'}{f'} \tag{4-78}$$

We may write the definition Eq. (4-77) in the form

$$V = \frac{dn}{n-1} \tag{4-79}$$

where we say that $dn = n_B - n_R$, since the two indices are quite close, and where $n = n_Y$ represents a sort of average index for the lens. For a thin lens, from Eq. (1-63), the Gaussian constant a is given by

$$a = \frac{n-1}{r_1} + \frac{1-n}{r_2}$$

and if we want to determine the change in a (or f') due to the variation of n with wavelength, we have

$$da = dn\left(\frac{1}{r_1} - \frac{1}{r_2}\right)$$

However,

$$a = \frac{1}{f'} = (n-1)\left(\frac{1}{r_1} - \frac{1}{r_2}\right)$$

Therefore

$$da = \frac{dn}{(n-1)f'} = d\left(\frac{1}{f'}\right) = \frac{-df'}{f'^2} \tag{4-80}$$

or

$$-\frac{df'}{f'} = \frac{dn}{(n-1)} = V \tag{4-81}$$

Hence, Eq. (4-78) may be written as

$$x = -\left(\frac{R}{2}\right)V \tag{4-82}$$

This means that the dispersive power V is a measure of the spread in the image due to the change in the index n with wavelength. We call this effect *chromatic aberration*. We correct for this aberration by designing *achromatic* lenses or lens systems; that is, we adjust the focal lengths for two different colors until they both have the same length. These colors are usually chosen as the red and blue wavelengths in Table 4-2. Figure 4-22 shows the hydrogen

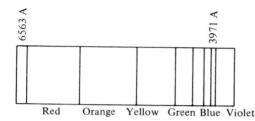

Red   Orange   Yellow   Green Blue   Violet   **Figure 4-22**

spectrum, with the red and blue lines lying near each end. If the lens is corrected at these points, it is approximately corrected for other wavelengths as well. Occasionally, lenses are corrected for all three colors. Such lenses are said to be *apochromatic*. These lenses are expensive and are used in critical applications such as color printing.

It is possible to incorporate provisions for color correction in the condenser system of Fig. 4-15. Using Eq. (1-69) for two thin lenses, we express the condition for constant focal length by the equation:

$$d\left(\frac{1}{f'}\right) = 0 = d\left(\frac{1}{f'_1}\right) + d\left(\frac{1}{f'_2}\right) - td\left(\frac{1}{f'_1 f'_2}\right)$$

However, by using Eqs. (4-80) and (4-81)

$$d\left(\frac{1}{f'}\right) = -\frac{df'}{f'^2} = \frac{V}{f'}$$

for each lens. Hence

$$0 = \frac{V_1}{f'_1} + \frac{V_2}{f'_2} - \left(\frac{V_1}{f'_1 f'_2} + \frac{V_2}{f'_2 f'_1}\right)$$

or

$$t = \frac{V_1 f'_2 + V_2 f'_1}{V_1 + V_2}$$

If we now specify that the two lenses are made of the same glass, then $V_1 = V_2$, and

$$t = \frac{f'_1 + f'_2}{2} \tag{4-83}$$

Hence, by separating the two lenses by the numerical sum of the focal lengths, we achieve an achromatized condenser.

Problem 4-11

Show that by the proper choice of glass, it is possible to achromatize a single thick lens, where the index of refraction of the glass selected is related to the radii and thickness of the lens by the equation

$$n_1'^2 = \frac{t_1'}{r_1 - r_2 + t_1'} \tag{4-84}$$

■

The indices of commercially available optical glasses range from about 1.446 to 1.961. The most complete recent tabulation is that of McClenahan and Christensen,[4-6] who give the index n_D, dispersive power V, supplier, and catalog number of 600 different glasses.

Returning to the microfilm reader for the last time, we now have an appreciation for the fact that the condenser lens system is deceptively simple; there are many factors to be balanced in the specification of even this pair of simple, inexpensive lenses.

References

4-1 E. E. Miller, Unpublished notes for Physics 625, University of Wisconsin (1970).

4-2 B. Schmidt, *Mitt. Hamburg* **7**, 15 (1932).

4-3 A. Bouwers, *Achievements in Optics*, Elsevier, Amsterdam (1946).

4-4 D. D. Maksutov, *J.O.S.A.* **34**, 270 (1944).

4-5 A. Cox, *Photographic Optics*, Focal Press, London (1966).

4-6 R. E. McClenahan and V. Christensen, *Optical Spectra* **3**, 63 (1969).

PART

II

THE WAVE PROPERTIES
OF LIGHT

We can learn a great deal about the behavior of optical systems by regarding light as a ray. This approach (geometrical optics) is really an approximation in which the effects of the finite size of lens systems are ignored. In other words, the ratio of the wavelength of light to the size of the system is taken to be zero in geometrical optics. In the next five chapters, we shall consider the wave nature of light in a fairly elementary way and then discuss interference and its practical applications. We shall see that the ultimate capability of an optical device can be determined only when we utilize the wave description of light. From this point on, we shall be primarily concerned with the subject known as physical optics. The material in this part is important in its own right because the wave nature of light is one of the fundamental characteristics of nature. Chapter 5 also serves as preparation for Part III, where we return to the subject of lenses and optical systems.

5

ELECTROMAGNETIC WAVES

5-1 The Nature of Physical Optics

So far we have regarded light as a form of energy propagated in straight lines or rays, and this approach is the foundation of *geometrical optics*. Now we are ready to consider the wave nature of light, or *physical optics*. Physical optics encompasses the phenomena of interference, diffraction, and polarization, all of which we shall consider in great detail. Physical optics is also used in describing the emission and detection of light.

Recalling the material covered in Chapters 1 through 4, we realize that about the only physics involved is Snell's law as it applies to both reflection and refraction, plus the use of the principle of conservation of energy. With this chapter, physical principles are introduced and will be used throughout the rest of the book.

5-2 Waves and the Classical Wave Equation

One approach to the subject of physical optics would be to first establish that light is electromagnetic in nature and therefore is described by Maxwell's equations. The fundamental properties of light are then deduced from these equations. This approach is largely mathematical and deductive in character and is taken in Born and Wolfe's[5-1] definitive treatise on optics. We feel that an applied subject such as optics is better treated using the inductive or historical approach. The experiments and observations which lead to the equations will be reviewed and the reader will think in terms of phenomena rather than equations.

We shall treat waves in a very elementary way and review some of their important characteristics. We start by considering a long piece of wire bent into sinusoidal shape and oriented as shown in Fig. 5-1(a). This curve will represent a wave. It has the equation

$$x = a \cos kz \qquad (5\text{-}1)$$

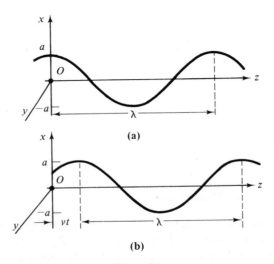

Figure 5-1

where x is the *amplitude* with maximum value a, and k is a quantity to be determined. The distance between two maxima on the curve is called the *wavelength* λ, so that when $z = 0$ or λ, then $x = a$. Therefore

$$k\lambda = 2\pi$$

or

$$k = \frac{2\pi}{\lambda} \qquad (5\text{-}2)$$

where k is called the *propagation constant* or *wave number*. By Eq. (5-2), k is a measure of the number of waves per unit length.

Now let the rigid curve move to the right with velocity v (Fig. 5-1(b)). At time t, the entire curve will have moved a distance vt, and Eq. (5-1) must be written

$$x = a \cos \left[\frac{2\pi}{\lambda} (z - vt) \right] \qquad (5\text{-}3)$$

Problem 5-1

(a) Verify that Eq. (5-3) describes a cosine wave moving to the right with velocity v.

(b) Show that Eq. (5-3) satisfies the partial differential equation

$$v^2 \frac{\partial^2 x}{\partial z^2} = \frac{\partial^2 x}{\partial t^2} \tag{5-4}$$

(c) Show that Eq. (5-3) may also be written as

$$x = a \cos (kz - \omega t)$$

where

$$\omega = vk \tag{5-5}$$

is the *angular frequency* of the wave.

(d) Show that Eq. (5-5) is also satisfied by

$$x = ae^{\pm i(kz - \omega t)} \tag{5-6}$$

∎

In three dimensions, we specify both the direction and the reciprocal wavelength by the *propagation vector* **k**. In place of the cosine form of the solution, an exponential function is often used. This simplifies the evaluations of products of waves by eliminating the need for trigonometric identities. A wave propagating along the direction of **k** may be expressed as

$$\mathbf{A} = \mathbf{A}_{max} e^{\pm i(\mathbf{k} \cdot \mathbf{r} - \omega t)} \tag{5-7}$$

which we recognize as a generalization of Eq. (5-6). The quantity $\mathbf{r} = x\mathbf{i} + y\mathbf{j} + z\mathbf{k}$ is the position coordinate, and the magnitude $(\mathbf{k} \cdot \mathbf{r} - \omega t)$ of the exponent is called the *phase*. The variable **A** satisfies the three-dimensional wave equation

$$\nabla^2 \mathbf{A} = \frac{1}{v^2} \frac{\partial^2 \mathbf{A}}{\partial t^2} \tag{5-8}$$

Problem 5-2

Verify that Eq. (5-7) is a solution to Eq. (5-8).　　　　∎

Problem 5-3

Sound travels with velocity of 330 meters/s in air at sea level. The audio range is approximately 50–15,000 hertz. Find the wavelength range.　∎

Problem 5-4

The center of the commercial radio broadcast band is approximately 1000 kHz. What is the corresponding wavelength?　　　∎

Problem 5-5

From your experience at the beach or in a boat, estimate the wavelength and frequency of a water wave. Then calculate its velocity. Does it agree with the velocity that you estimate from experience on the water? (In shallow water, the velocity depends on the depth of the bottom, and in deep water the velocity depends on wavelength.) ■

A wave moving with velocity v is said to be a *traveling wave*. Figure 5-1(b) represents a very simple form of such a wave, since it is a pure sinusoid.

Many waves encountered in practice can have very complicated shapes, and to deal with them mathematically, we use Fourier methods. This subject is reviewed in the Appendix, but for the time being it is not necessary to bring in the specific details of this approach. It should be mentioned, however, that very involved waves can be decomposed into separate components, each one of which is a sine or a cosine whose frequency and maximum amplitude generally bear a simple relation to one another. In the case of a nonperiodic wave, the "sum" of the contributions is actually an integral. Any wave can be expressed as either a Fourier series or a Fourier integral depending on whether it is periodic or nonperiodic.

On the other hand, one aspect of wave theory we can treat analytically without a large amount of detailed computation is the concept of *intensity I*, widely used in optics. Both the term and the symbol may be the source of some confusion, which we shall attempt to clarify here. First, it is shown in electromagnetic field theory[5-2] that a wave whose electric field strength is **E** will have associated with it an energy $E_V = \epsilon_0 E^2/2$, per unit volume where ϵ_0 is the permittivity of free space (assuming the wave is being propagated in a vacuum or nothing denser than air). There is a corresponding expression for the magnetic term, and the total energy density is the sum of the two contributions. We shall see in Chapter 10, however, that we are mainly interested in the variations of energy density with respect to some reference value; therefore we need not worry about the magnitude and units of ϵ_0. However, the use of complex expressions like Eq. (5-7) leads to difficulties if we are not careful. If we want to relate **A** to either a sine or a cosine, we should take just the real or the imaginary part. Squaring the amplitude to find the intensity involves scalar products. It is important to note that

$$\text{Re}\,(\mathbf{A}e^{-i\omega t}) \cdot \text{Re}\,(\mathbf{B}e^{-i\omega t}) \neq \text{Re}\,[(\mathbf{A}e^{-i\omega t}) \cdot (\mathbf{B}e^{-i\omega t})] \qquad (5\text{-}9)$$

where Re is the standard symbol for "real part of," and we have considered only the time-dependent part of the vectors **A** and **B** because we are going to relate intensity to time averages very shortly. The way to handle calculations of products of waves is to realize that any complex quantity may be written as

$$\text{Re}\,(\mathbf{A}) = \frac{\mathbf{A} + \mathbf{A}^*}{2} \qquad (5\text{-}10)$$

Hence

$$\text{Re}\,(\mathbf{A}_{max}e^{-i\omega t}) \cdot \text{Re}\,(\mathbf{B}_{max}e^{-i\omega t})$$

$$= \frac{\mathbf{A}_{max}e^{-i\omega t} + \mathbf{A}_{max}^*e^{i\omega t}}{2} \cdot \frac{\mathbf{B}_{max}e^{-i\omega t} + \mathbf{B}_{max}^*e^{i\omega t}}{2}$$

$$= \frac{\mathbf{A}_{max} \cdot \mathbf{B}_{max}^* + \mathbf{A}_{max}^* \cdot \mathbf{B}_{max}}{4} + \frac{\mathbf{A}_{max} \cdot \mathbf{B}_{max}e^{-2i\omega t} + \mathbf{A}_{max}^* \cdot \mathbf{B}_{max}^*e^{2i\omega t}}{4}$$

$$= \frac{1}{2}\,\text{Re}\,(\mathbf{A}_{max} \cdot \mathbf{B}_{max}^* + \mathbf{A}_{max} \cdot \mathbf{B}_{max}e^{-2i\omega t}) \tag{5-11}$$

where the last step follows from the fact that

$$\text{Re}\,(\mathbf{A} \cdot \mathbf{B}^*) = \text{Re}\,(\mathbf{A}^* \cdot \mathbf{B})$$

which the reader should verify for himself. The *time average* of a function $f(t)$ is defined as

$$\langle f(t) \rangle = \frac{1}{\tau} \int_0^\tau f(t)\, dt \tag{5-12}$$

where $\tau = 2\pi/\omega$ is the period. Applying this definition to Eq. (5-11), we obtain

$$\langle \text{Re}\,(\mathbf{A}_{max}e^{-i\omega t}) \cdot \text{Re}\,(\mathbf{B}_{max}e^{-i\omega t}) \rangle = \tfrac{1}{2}\,\text{Re}\,(\mathbf{A}_{max} \cdot \mathbf{B}_{max}^*) \tag{5-13}$$

Since the average of the exponential term, whose real part is $-\cos 2\omega t$, will vanish over one period, the positive and negative areas of this function are equal in magnitude. Taking the electric field E as an example of a wave, we would normally expect the energy density to be proportional to $\mathbf{E} \cdot \mathbf{E}^* = E^2$, and we will call this quantity the intensity. However, for a time-varying field, it is the average value rather than the instantaneous energy which is the significant physical quantity (this is what the eye detects, for example); thus we define intensity by the relation

$$I = \langle \text{Re}\,(\mathbf{E}) \cdot \text{Re}\,(\mathbf{E}^*) \rangle \tag{5-14}$$

Then the use of Eq. (5-13) shows that

$$I = \tfrac{1}{2}\,\text{Re}\,(\mathbf{E}_{max} \cdot \mathbf{E}_{max}^*) = \tfrac{1}{2}E_{max}^2 \tag{5-15}$$

In the situation where the electric field is not a function of time, we simply define I as

$$I = \mathbf{E} \cdot \mathbf{E}^* = E_{max}^2 \tag{5-16}$$

The factor $\tfrac{1}{2}$ which appears in Eq. (5-15) comes about because the average value of a time-varying quantity is not as large as the maximum value.

The second possible source of confusion is the relation of our usage here with the radiant or luminous intensity of Table 4-1; these latter quantities are power per unit solid angle. Although we are using the same symbol for two different purposes, we shall not have occasion to employ them in a single calculation.

5-3 Transverse and Longitudinal Waves

As will be explained later in greater detail, the observation of double refraction in a calcite crystal implies that light is "double-sided." This can hold true only for a *transverse* wave. In a transverse wave, the amplitude is perpendicular to the direction of propagation. A convenient way to visualize such a wave is to think of a line of cheerleaders with pompoms held out to the side. In succession, each cheerleader raises her pompom over her head. This creates a wave which, when viewed from the side, appears to travel down the line of girls. The motion of a single pompom is vertical, but the wave front travels in the horizontal direction.

A transverse wave such as light can have either of two perpendicular orientations or *directions of polarization*. Arago and Fresnel demonstrated that these two polarization states are perpendicular to one another by observing the interference pattern produced by combining two waves of the same polarization vanished when one wave had its direction of polarization rotated by 90°.

A wave described by an equation such as Eq. (5-3) or (5-7) is said to be a *plane wave*. The magnitude of the exponent in Eq. (5-7) is the phase, and it is easy to show[5-2] that, at a given instant in time, all points with the same phase lie in a plane normal to the direction of propagation. This plane then moves along its normal with a velocity v; in free space, $v = c = 3 \times 10^8$ m/s.

Light waves, deep water waves, and string waves are examples of transverse waves. The other type is a *longitudinal* wave. An example of a longitudinal wave is sound which compresses and rarefies the gas which carries it. The displacement of the individual atoms of the gas takes place along the direction of propagation. The associated compression and rarefaction can be regarded as an oscillation about the individual equilibrium positions.

5-4 The Principle of Superposition

If light waves from two different sources are radiating through a common medium, their amplitudes will add if the medium is linear. This fact represents a statement of the *principle of superposition*, which is also used in the study of mechanical and electrical systems. The ideal behavior is implied by the wave equation, Eq. (5-8), which contains only first-power terms in the amplitude. When superposition holds, we can take two waves such as in Eq.

(5-3) and determine the total amplitude as

$$E_T = E_1 + E_2$$
$$= E_{1m} \cos(k_1 x - \omega_1 t + \varphi_1) + E_{2m} \cos(k_2 x - \omega_2 t + \varphi_2) \qquad (5\text{-}17)$$

where the quantities φ_1 and φ_2 are used to specify *relative phase*, that is, they locate the maxima and minima of the amplitude at any desired reference position. Another term that is often used is *phase shift*, but it is quite common to refer to this quantity simply as the *phase* of the wave since what we are generally interested in is the phase of one wave with respect to another.

Problem 5-6

Use Eq. (5-17) to show that the superposition of two cosine waves with the same amplitudes, frequencies, and wavelengths is itself a cosine wave whose amplitude depends on the relative phase between the two waves. ■

We also note that the principle of superposition implies that individual waves do not affect one another; another way of saying this is that the waves are *uncoupled*.

5-5 Standing Waves

Waves which do not travel but remain stationary are *standing waves*. One way to produce such a wave is to take a traveling wave, reflect it back on itself, and add the incident and reflected parts. For example, the wave

$$E_1 = E_m \sin(kz - \omega t)$$

travels to the right along the z axis. The wave traveling in the negative direction is then (assuming there are no losses on reflection)

$$E_2 = E_m \sin(-kz - \omega t)$$

and their sum is

$$E_T = E_1 + E_2$$
$$= -2E_m \sin(\omega t) \cos(kz)$$
$$= A(z) \sin \omega t \qquad (5\text{-}18)$$

This is a wave whose amplitude has a maximum

$$A(z) = -2E_m \cos kz \qquad (5\text{-}19)$$

which itself depends on position. The time-dependent part, $\sin \omega t$, does not contain the kz term of the traveling waves we have previously considered. Such a wave then appears as indicated in Fig. 5-2. We note that there are

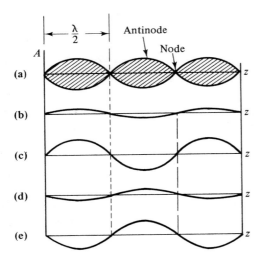

Figure 5-2 (a) The time-averaged view of a plucked string. (b)–(e) A sequence of instantaneous views.

points for which $A(z) = 0$. Since these points or *nodes* occur when $\cos kz$ vanishes, they must be separated by a distance of one half of the wavelength.

Problem 5-7

Prove that the nodes of a standing wave are spaced at a distance of $\lambda/2$. ■

5-6   The Superposition of Waves with Random Phases

An important question which frequently arises in the superposition of waves is: What happens when a large number of waves with random phases are superimposed? As an example, this problem comes up when we treat the light emitted by atoms in a gas whose atoms act independently of each other. Whenever the waves are emitted independently, their phases are random by definition.

For simplicity, let us consider a collection of waves of equal amplitude E_m having the same frequency ω. Our point of observation will be taken as $z = 0$, and the phase of the αth wave will be designated φ_α. The total disturbance E_T due to N such contributions is

$$E_T = E_m[e^{i(\omega t+\varphi_1)} + e^{i(\omega t+\varphi_2)} +\ldots + e^{i(\omega t+\varphi_N)}]$$

$$= E_m e^{i\omega t} \sum_{\alpha=1}^{N} e^{i\varphi_\alpha}$$

The average value of E_T^2 is then

$$\langle E_T^2 \rangle = \langle E_T^* E_T \rangle = E_m^2 \langle \sum e^{-i\varphi_\alpha} \sum e^{i\varphi_\alpha} \rangle \tag{5-20}$$

The product of the summations becomes

$$\left\langle \sum e^{-i\varphi_\alpha} \sum e^{i\varphi_\alpha} \right\rangle = \left\langle 1 + e^{-i(\varphi_2 - \varphi_1)} + e^{i(\varphi_2 - \varphi_1)} + 1 + \dots \right\rangle$$

There will be a contribution of unity for each term of the form $e^{-i(\varphi_N - \varphi_N)}$, and for N such terms

$$\langle E_T^2 \rangle = E_m^2 \langle [N + 2 \cos(\varphi_2 - \varphi_1) + \dots] \rangle \qquad (5\text{-}21)$$

For random phases there will be as many positive contributions from the cosine terms as negative ones. Hence, Eq. (5-21) reduces to

$$\langle E_T^2 \rangle = E_m^2 N \qquad (5\text{-}22)$$

and we take the amplitude as

$$E_T = \sqrt{N}\, E_m \qquad (5\text{-}23)$$

We see that the superposition of N waves of equal amplitudes and frequencies but with random phases produces a wave whose amplitude is $\sqrt{N}$ times the amplitude of a single wave. The energy (which is proportional to E^2) is equal to N times the energy of one wave.

Now let us perform a similar calculation assuming that the waves have identical phases. Now all the exponential terms in Eq. (5-20) become unity, each summation has a value of N, and the product gives

$$E_T^2 = N^2 E_m^2$$

so that

$$E_T = N E_m \qquad (5\text{-}24)$$

Comparison of this with Eq. (5-23) shows the difference between the superposition of waves with random phases and the superposition of waves with highly correlated phases. This is quite important in treating lasers, for example, which we do in a later chapter.

5-7 The Electromagnetic Spectrum

Light waves cover a small portion of the total electromagnetic spectrum, which goes from long-wave radio to the cosmic-ray region. The visible region corresponds to wavelengths from 380 nm to 760 nm (see Fig. 4-13). The recognized SI units for lengths of this magnitude are the micrometer (1 μm = 10^{-6} m) and the nanometer (1 nm = 10^{-9} m). Formerly, the Angstrom unit A was used (1A = 10^{-10} m). The remaining SI units and the general arrangement of the electromagnetic spectrum are shown in Fig. 5-3.

The long-wavelength limit of the visible region merges with the *infrared* (IR) region. Infrared waves are also called heat waves because they can be detected by the observation of the increase of temperature of the object absorbing them. The short-wavelength limit of the visible region merges with the *ultraviolet* (UV) region. Ultraviolet light is sometimes called *black light*.

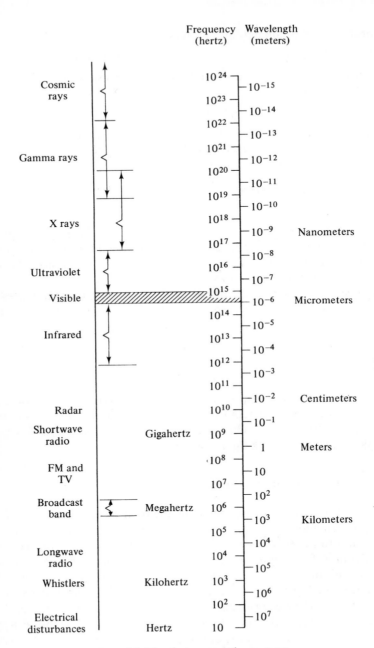

Figure 5-3 The electromagnetic spectrum.

144

It is widely used to make signs and posters fluoresce and as a germicidal agent. The *erythemal region* of ultraviolet light (0.29–0.32 μm) is responsible for sun tans. Most of the phenomena exhibited by light waves are also exhibited by UV and IR waves, and the techniques for their generation and detection are similar. One indication of the continuity of phenomena across the boundaries of the visible is the fact that the largest number of lasers that have been developed to date operate in the near IR (1–10 μm). We shall frequently extend our discussion of optics to those parts of the UV and IR spectra which lie close to the visible region.

The dominant interaction of light waves with materials is usually through the electric field, whose amplitude E is measured in volts/meter in SI units. Optical detectors respond to the square of the electric field, because the field itself changes sign at a rate of 10^{15} times per second (the frequency in the visible region), and detectors are too slow to follow such changes. In optics, it is customary to work with the intensity I of Eq. (5-14), which is defined in terms of a time-averaging process. It was also pointed out that when we work with absolute rather than relative quantities, we obtain the energy density or energy per unit volume from the relation

$$U_V = \tfrac{1}{2}\epsilon E^2 + \tfrac{1}{2}\mu H^2 \tag{5-25}$$

where ϵ is the dielectric constant, μ is the magnetic permeability of the material, and H is magnetic field intensity. In free space the energy in the magnetic field equals that in the electric field and

$$U_V = \frac{1}{2}\epsilon_0 E^2 \frac{\text{joules}}{\text{meter}^3} \tag{5-26}$$

The value ϵ_0 of ϵ in free space is 8.85×10^{-12} farad meter^{-1}. In a material medium, ϵ is $\epsilon_R\epsilon_0$, where ϵ_R is the relative value. It is related to the index of refraction[5-1] by

$$n^2 = \epsilon_R \tag{5-27}$$

Both n and ϵ_R are frequency dependent; therefore Eq. (5-27) refers to values at a specified frequency.

The *flux* is the rate of energy flow per unit time per unit area through a surface normal to the direction of propagation. This is given by the *Poynting vector* **S** as

$$\mathbf{S} = \mathbf{E} \times \mathbf{H} \tag{5-28}$$

which for plane waves has the magnitude

$$S = c\sqrt{\epsilon_R}\,E^2 \text{ joule second}^{-1}\text{ meter}^{-2} \tag{5-29}$$

The relation between the flux S and energy density E_V is

$$U_V = \frac{nS}{c} \tag{5-30}$$

Problem 5-8

An argon laser operating at 488.0 nm has an output of 500 milliwatts. The area of the beam is 1 mm². Calculate the electric field strength of the wave. ■

References

5-1 M. Born and E. Wolf, *Principles of Optics*, 4th ed., Pergamon (1970).

5-2 A. Nussbaum, *Electromagnetic Theory for Engineers and Scientists*, Prentice-Hall, Englewood Cliffs, N.J. (1966).

General References

E. Ruechardt, *Light Visible and Invisible*, University of Michigan Press, Ann Arbor, Michigan (1958).

F. Jenkins and H. White, *Fundamentals of Optics*, McGraw-Hill, New York (1957).

6

THE MICHELSON INTERFEROMETER
AND
THE VISIBILITY OF FRINGES

6-1 Introduction

Many of the properties of light can be explained if we assume that it is a wave. The unique feature of a wave is its ability to *interfere* with itself or another wave. There are two classes of interference; one is produced by *division of wave front* and the other by *division of amplitude*. Division of wave front occurs in the two-slit experiment (Fig. 6-1); two portions of a monochromatic wave front are transmitted through a double slit and combined on a screen producing an *interference pattern*. This type of interference will be described in Chapter 9, and its generalization to include diffraction will be considered in Chapter 10.

Division of amplitude occurs in the *Michelson interferometer*, the subject of this chapter. This instrument is an excellent point from which to embark on a study of light waves and interference because it is easy to understand and it has many applications. It illustrates *two-beam interference*, one of the simplest kinds. When interference with more than two beams is involved, the analysis gets complicated and will be taken up later. The Michelson interferometer can be used to make precise measurements of the wavelength of the light emitted from the source, of the length of an object, or of the refractive index of a material. As an illustration of the accuracy that is attainable,

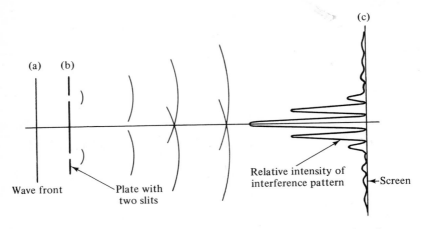

Figure 6-1. The *two slit* experiment. A wave front (a) approaches the double slit from the left. After passing through the slits, the front is divided into two portions (b). On a distant screen, the two portions combine to produce an interference pattern (c).

in a typical undergraduate laboratory it is possible to measure the wavelength of the mercury green line (546.1 nm) to an accuracy of 0.1 nm, the difference in wavelength of the pair of sodium D lines (0.59 nm) can be determined to within 0.1 nm, and the index of refraction of air is measured with six significant figures as 1.00029.

We will begin by reviewing the basic ideas of the interferometer and then progress to the more sophisticated studies of fringe visibility, coherence, and the use of the interferometer as a high-resolution spectrometer (Fourier transform spectroscopy). A. A. Michelson (1852–1931) was the first American to win the Nobel prize in physics; he received the award in 1907 for his studies in light. He was a graduate of the Naval Academy, and while an instructor there, measured the velocity of light. The superintendent of the Academy asked Michelson why he wasted his time on such useless experiments and suggested that he spend his time on the more practical applications of science. Today the Navy uses many optical devices. Several, including a laser gyroscope, are based on Michelson's work. In addition to the interferometer, which works on the principles of division of amplitude, Michelson designed the stellar interferometer which utilizes division of wave front and is used to measure the angular diameter of stars. Michelson had exceptional experimental skill; years after he performed an experiment, others with more up-to-date equipment would sometimes be unable to achieve his accurate results.

A simplified form of the Michelson interferometer is shown in Fig. 6-2. The incident wave with electric field strength E_0 strikes a glass plate at 45°. The back side of the plate is lightly coated with silver so that it serves as a beam splitter; part of the wave is transmitted to mirror M_1 and part is

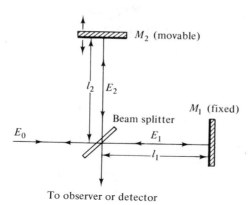

Figure 6-2. The Michelson interferometer. A light source on the left emits a wave of amplitude E_0. The beam splitter divides the wave in two. The transmitted wave amplitude is E_1 and the reflected wave amplitude is E_2. After reflection by mirrors M_1 and M_2, the two are combined.

reflected to mirror M_2. The coating on the beam splitter is made with a density which will give the two parts of the beam equal amplitudes.

First, let us consider a monochromatic beam incident along the axis of the interferometer. After reflection by mirrors M_1 and M_2, the components are recombined at the beam splitter where—on the average—half of the energy is reflected onto a screen and the other half is transmitted back toward the source. The path difference Δ for the two waves, which may be varied by changing the position of M_2, is given by $\Delta = 2(l_1 - l_2)$, where l_1 and l_2 are the lengths of each arm of the interferometer. The phase difference δ is $2\pi\Delta/\lambda$, where λ is the wavelength. For a monochromatic wave of angular frequency ω, the total field at the output is

$$E_T = E_1 + E_2 = 0.5E_0 e^{i\omega t} + 0.5E_0 e^{i(\omega t+\delta)} \tag{6-1}$$

or

$$E_T = 0.5E_0(1 + e^{i\delta})e^{i\omega t} \tag{6-2}$$

From Eq. (5-15), the intensity is

$$I = \tfrac{1}{2} \operatorname{Re}(E_T E_T^*) = \tfrac{1}{2}(0.5E_0)^2(1 + e^{i\delta} + e^{-i\delta} + 1)$$
$$= 0.25E_0^2(1 + \cos \delta) \tag{6-3}$$

Here we should note an important property of interference: when $\delta = n\pi$, where $n = 1, 3, 5, \ldots$, the two axial components will combine in such a manner that there is zero intensity or amplitude on the screen. Yet if we place a card in either arm of the interferometer, blocking one beam, we will see one quarter of the initial intensity. With no card (i.e., using both beams), we get zero intensity because one beam is 180° out of phase with the other. This is the essence of interference. To explain the phenomenon, we have to invoke the wave nature of light.

Problem 6-1

What happens to the energy when there is no card? Where does it go? ■

Next, let us consider a beam incident at some angle θ to the axis. This configuration can be analyzed using an equivalent configuration. The fixed mirror M_1 is replaced by its image M_1' as produced by the beam splitter, and an arbitrary point S on the source is also replaced by its image S_1' (Fig. 6-3). All the components, from source to observer, are treated as lying

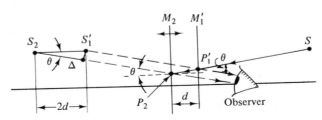

Figure 6-3

on a common axis. The point S then has a virtual image S_1' due to mirror M_1' and another at S_2 due to M_2. The spacing between S_1 and S_2 is twice that of the spacing d between M_1' and M_2, where

$$d = l_1 - l_2$$

Hence

$$\overline{S_1'S_2} = 2(l_1 - l_2)$$

and the figure further shows that the path difference Δ is now

$$\Delta = 2(l_1 - l_2)\cos\theta$$

The corresponding phase difference is then

$$\delta = \frac{2\pi}{\lambda}\Delta = \frac{4\pi}{\lambda}(l_1 - l_2)\cos\theta \qquad (6\text{-}4)$$

We also realize that if the path difference Δ is exactly equal to an integral multiple $m\lambda$ of a wavelength, where m is the *order*, the two rays received by the observer will reinforce one another. Therefore

$$2d\cos\theta = m\lambda \qquad (6\text{-}5)$$

where $m = 1, 2, \ldots$. In Fig. 6-3, we have considered a single point S' and its images. Since the figure is perfectly symmetrical about the axis shown, the two points S_1' and S_2 corresponding to reinforcement should actually lie on circles, and these circular interference patterns are shown in Fig. 6-4. There are rings of complete destructive interference between the additive ones, the position of the rings depending on the angle θ or on the path difference Δ. As the path difference goes to zero, the angle for the first ring approaches 90°, i.e., the ring is outside of the field of view and the viewer sees an area with constant illumination.

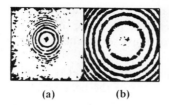

Figure 6-4. The interference pattern of a Michelson interferometer. In (a), the path difference is many wavelengths causing the rings to be closely spaced. In (b), the path difference is a few wavelengths and the rings are widely spaced.

(a)   (b)

Problem 6-2

Examination of the light emitted by a Cd lamp using a prism spectroscope reveals that there is an intense line in the red region and several weaker lines in other regions of the spectrum. A red piece of glass is placed in front of the lamp and it is found to absorb all of the other lines while transmitting the red line. A Michelson interferometer is used to measure the wavelength of this red line. The movable mirror advances a distance of 0.0322 millimeters (the change in path length is twice this amount) while 100 fringes pass the center of the screen. Find the wavelength. Look up the value in the *Handbook of Chemistry and Physics*. Do the two values agree?   ▰

Problem 6-3

A cell having an internal length of 5.0 cm is placed in one arm of a Michelson interferometer. It is evacuated to a pressure of 10^{-2} mm of Hg (atmospheric pressure is 760 mm of Hg). The ambient temperature is 23°C and the pressure is 760 mm. As air is slowly let into the cell until atmospheric pressure is reached, 44 fringes are counted. The source is a Cd lamp. Find the index of refraction of air at standard conditions. (*Hint:* This experiment measures the difference between n of air and of the vacuum, i.e., $n - 1$. The difference is measured only to two or three significant figures, but the value for n is good to five or six figures.)   ▰

6-2 Procedure for Alignment of the Michelson Interferometer

The interferometer must be aligned before it can be used. Alignment of the optical elements of a system is a task frequently encountered, and it is instructive to follow the procedure for the Michelson interferometer.

The first step is to set the length of the movable arm equal to that of the fixed arm. This is accomplished by measuring with a ruler the distances from the center of the beam splitter to the center of each mirror. The beam from the adjustable arm passes through the beam splitter three times, whereas the beam from the fixed mirror passes through the beam splitter only once. A correction plate equal in thickness to the beam splitter is set in the fixed arm at 45° with respect to the central ray. This compensates for the optical path difference.* If the interferometer does not have the compensating plate, the position of the adjustable mirror must set so that the optical paths in each arm are equal.

The second step is to insert a small nail that is standing on its head (or some similar pointed object) in front of the beam splitter (Fig. 6-5(a)). The viewer will see two images of the nail, one reflected from the mirror in each arm (Fig. 6-5(b)).

The third step is to illuminate the beam splitter with a monochromatic

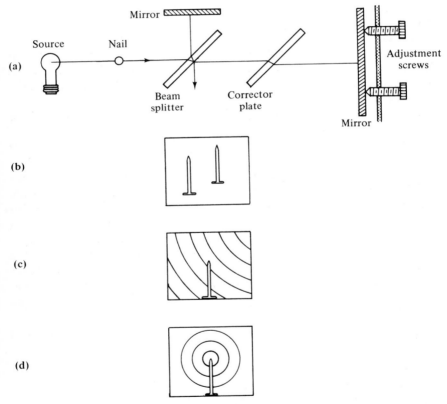

Figure 6-5. Alignment procedure for a Michelson interferometer. A nail is placed in front of the beam splitter (a). The two images (b) are brought into coincidence (c) using the coarse adjusting screws. The interference pattern is then centered (d) using the fine adjusting screws.

*With the correcting plate inserted, the position for the widest spacing of the fringes is the same as for the maximum contrast. This facilitates working with the interferometer. The contrast or visibility of fringes will be described later in this chapter.

source and to use the coarse adjustment screws on the fixed mirror to bring the two images of the nail into coincidence. When this is done, interference fringes will appear across the field of view (Fig. 6-5(c)).

The fourth step is to bring the center of the pattern into the center of the field of view by using the fine adjustment screws on the fixed mirror (Fig. 6-5(d)).

Finally the position of the movable mirror is adjusted so that the path difference is close to zero. This is accomplished by adjusting the length of the arm to maximize the ring size. Large rings are easier to observe and work with.

6-3 Fringe Visibility

Our analysis has proceeded under the assumption that the source is perfectly monochromatic. When Michelson used his interferometer to examine real sources, he found a very interesting phenomenon. As the path difference between the two arms of the interferometer was increased, the visibility of the fringes decreased and finally disappeared. A simple interpretation of this result is that light from an ordinary spectral lamp consists of wave trains of finite length. When the path difference is less than the length of the wave train, part of the train coming from one arm of the interferometer overlaps the part of the same train coming from the other arm, and interference occurs (Fig. 6-6(a)). When the path difference is greater than the length of the wave train (Fig. 6-6(b)), the light from one arm is added to the light from the other arm which consists of a different wave train. Since the phase relation between two trains is a random variable (the trains come from independent atoms) and since many such superpositions take place in an observation time, no interference will be observed, and the intensity will be a constant. The length of the wave train is related to the *temporal coherence length* of the light. Ordinary spectral-line sources have coherence lengths on the order of millimeters, whereas lasers have coherence lengths on the order of kilometers.

An alternate but equally correct explanation of the disappearance of

E_1

E_2

(a)   (b)

Figure 6-6. In (a), the wave train E_1 from one arm of the interferometer overlaps the wave train E_2 from the other arm producing an interference pattern. In (b), the trains do not overlap and there is no interference.

fringes can be made if we consider the spectral distribution of the source. If more than one wavelength is emitted, each will form its own interference pattern. At zero path difference, all wavelengths will form a maximum and the fringes will be very sharp (Fig. 6-7(a)), but at large path differences the fringe maxima for the different wavelengths will not coincide and the fringes will vanish (Fig. 6-7(b)). Since one may Fourier analyze a wave train and obtain its spectral distribution, it turns out that these two explanations are equivalent.

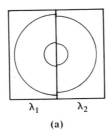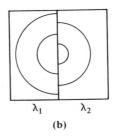

$\lambda_1 \qquad \lambda_2$ $\qquad\qquad$ $\lambda_1 \qquad \lambda_2$

(a) $\qquad\qquad\qquad\qquad$ (b)

Figure 6-7. A source emits radiation of wavelengths λ_1 and λ_2. With small path differences (a), the maxima coincide and the interference pattern is sharp. For larger path differences (b), the fringes are out of step and disappear.

After observing the loss in visibility of fringes with increasing path difference, Michelson went on to explain this effect, and he obtained an extremely important result which we shall consider next. Let the incoming wave contain a number of frequencies, rather than being monochromatic; the intensity I therefore involves a summation over frequency. As seen in connection with Eq. (5-13), when we compute intensity by squaring an amplitude which itself involves a summation, the real cross-product terms are converted into functions such as the cosine, and these average to zero over one or several cycles. This will be the case here for all the products of the form $E_1(\omega_1)E_2(\omega_2)$, since any physically realistic observation time will extend over many cycles. The total intensity I_T will then be

$$I_T = \sum_m I(\omega_m) \tag{6-6}$$

If we introduce the idea that the distribution of frequencies involved is continuous rather than discrete, this sum should be replaced by an integral. It is customary to work with wavelength rather than frequency as the variable of integration; thus the integral becomes

$$I_T = \int_{-\infty}^{\infty} I\left(\frac{1}{\lambda}\right) d\left(\frac{1}{\lambda}\right) \tag{6-7}$$

since $1/\lambda$ is proportional to the angular frequency ω. To eliminate the awkward factor of $1/\lambda$, spectroscopists define the *linear wave number σ* as

$$\sigma = \frac{1}{\lambda} \tag{6-8}$$

where the name comes from the fact that σ is equal to the number of cycles or waves per unit length and the units are usually cm^{-1}. (The quantity $k = 2\pi/\lambda$, which we have called the propagation constant, is sometimes referred to as the *circular wave number*, since it has a similar interpretation.) For our purposes here, it is convenient to specify σ_0 as the value of the wave number at the center of the spectrum produced by the source and write

$$\frac{1}{\lambda} = \sigma_0 + \sigma \tag{6-9}$$

Combining these definitions with the result given in Eq. (6-3) makes the integral in Eq. (6-7) become

$$I_T = \int_{-\infty}^{\infty} I(\sigma)[1 + \cos\{2\pi\Delta(\sigma_0 + \sigma)\}]\, d\sigma \tag{6-10}$$

This extremely important formula states that, in a Michelson interferometer, the average intensity $I_T(\Delta)$ as a function of spacing is the Fourier transform* of the intensity $I(\sigma)$ as a function of wavelength (spectrum). We may then regard this instrument as a form of computer which will perform a Fourier analysis of the spectrum of the source. (We shall see in Chapter 10 that a lens behaves in a similar way. In fact, it can instantly produce a two-dimensional Fourier transform of the spatial distribution of a source, and this is something which is extremely complicated when done numerically or by an electrical network.)

One may ask about the advantages of the interferometer as compared to a conventional grating spectrometer. In the latter, light from the source is dispersed over a wide area, and a narrow spectral band is detected. It is difficult to detect a signal for weak sources and simultaneously attain high resolution because most of the light available at a given moment is not used. On the other hand, in Fourier transform spectroscopy, only one half of the light, on the average, is discarded. This gives a better signal-to-noise ratio for a given detector. This technique is widely used for high-resolution analysis of weak sources, particularly in the infrared region. It is ironic that Eq. (6-10), derived many years ago by Michelson,[6-1] had become by 1950 only an historical curiosity. An impressive recent application is the work of Connes,[6-2] who has measured spectral lines in the atmosphere of Venus with a resolution of 0.005 cm^{-1}.

Let us return to the problem of the visibility of fringes. The cosine term of

---

*The subject of Fourier transforms is covered in the Appendix.

Eq. (6-10) can be expanded to give

$$\cos\left[2\pi\Delta(\sigma_0 + \sigma)\right] = \cos\left(2\pi\Delta\sigma_0\right)\cos\left(2\pi\Delta\sigma\right) - \sin\left(2\pi\Delta\sigma_0\right)\sin\left(2\pi\Delta\sigma\right) \tag{6-11}$$

With the substitution of

$$\left.\begin{aligned}
\theta &= 2\pi\Delta\sigma_0 \\[4pt]
P &= \int I(\sigma)\,d\sigma \\[4pt]
C &= \int I(\sigma)\cos\left(2\pi\Delta\sigma\right)\,d\sigma \\[4pt]
S &= \int I(\sigma)\sin\left(2\pi\Delta\sigma\right)\,d\sigma
\end{aligned}\right\} \tag{6-12}$$

Eq. (6-10) becomes

$$I_T = P + C\cos\theta - S\sin\theta \tag{6-13}$$

To have a quantitative measure of the nature of the fringes, we introduce Michelson's definition of *visibility* V, which is

$$V = \frac{I_{max} - I_{min}}{I_{max} + I_{min}} \tag{6-14}$$

Referring to Fig. 6-8, we see that fringes of the highest visibility are those with the greatest difference between the maximum and minimum intensity, and vice versa. This indicates the logic behind this choice of definition. We can compute I_{max} and I_{min} by utilizing the approximation that P, C, and S of Eq. (6-13) are constants for a narrow spectrum ($\sigma \ll \sigma_0$). Differentiating Eq. (6-13) with respect to θ and equating the result to zero gives

$$\tan\theta = -\frac{S}{C} \tag{6-15}$$

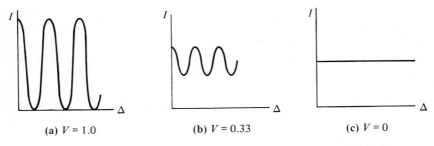

(a) $V = 1.0$ (b) $V = 0.33$ (c) $V = 0$

Figure 6-8. Michelson's definition of the visibility of fringes. In (a), the fringes having high contrast are visible ($V = 1$). In (b), the average intensity is the same but the fringes are less sharp ($V = 0.33$). In (c), no fringes are visible ($V = 0$).

Problem 6-4

Use this with Eq. (6-14) to show that

$$V = \frac{\lfloor (C^2 + S^2)^{1/2} \rfloor}{P} \tag{6-16}$$

6-3a Gaussian Lines

Our first example is the fringe system produced by a single spectral line of *Gaussian* shape, which will be specified by the equation

$$I(\sigma) = A e^{-\sigma^2/a^2} \tag{6-17}$$

This is the type of output that we get from a radiating gas whose natural shape, normally very narrow, is broadened by the thermal velocities of the atoms; the phenomenon is described as *Doppler-broadening*. Since the line width is an even function of σ, then the integral S in Eq. (6-12) vanishes and Eq. (6-16) gives

$$V = \frac{\displaystyle\int_{-\infty}^{\infty} A \exp\left[-\sigma^2/a^2\right] \cos\left(2\pi\Delta\sigma\right) d\sigma}{\displaystyle\int_{-\infty}^{\infty} A \exp\left[-\sigma^2/a^2\right] d\sigma} = e^{-(\pi\Delta a)^2} \tag{6-18}$$

which is itself Gaussian and is illustrated in Fig. 6-9. The half-width of this function is $1/(\pi a)$, and the corresponding half-width of the spectral line of Eq. (6-17) is a; their product is

$$a\frac{1}{\pi a} = \frac{1}{\pi} \tag{6-19}$$

This result is one form of the uncertainty principle, which we normally encounter as $\Delta x \times \Delta p \approx \hbar$. Equation (6-19) indicates that reducing the spec-

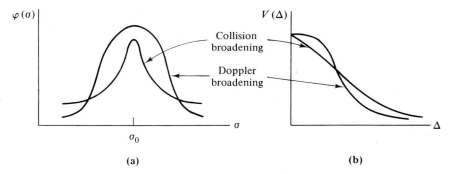

Figure 6-9. The shape of a Doppler-broadened and collision-broadened line. In (a), the integrated intensities of the lines are the same. In (b), the collision-broadened line has the straighter V curve.

tral half-width will broaden the visibility half-width, and this is a general property of variables which are related through a Fourier transform. The same thing is true for position and momentum.

Equation (6-19) provides us with a way of estimating the monochromaticity or spectral purity of the source. A typical line generated by a low-pressure electric discharge produces fringes over a distance on the order of millimeters. If we use 2 mm as the visibility half-width, we calculate the spectral half-width from Eq. (6-19) as 1.6 cm^{-1}, which corresponds to 0.1 nm at the center of the visible spectrum (550 nm). We can also tie this in with the concept of coherence if we think of the fringe visibility as a measure of the ability of a beam to interfere with itself. The greater the distance over which the amplitudes and phases of the direct and reflected portions can maintain correlated values, the larger is the distance over which the fringes are visible. Hence, we take this distance as proportional to the coherence length. This means that a laser, for example, which is a highly monochromatic source, has a long coherence length. That is, a narrow spectral width implies high visibility distance, which in turn leads to a large coherence length. This very important conclusion will be considered again in Chapter 9.

As mentioned earlier, the Gaussian shape of the lines from gases under low pressures is associated with thermal velocities. Since the average kinetic energy $\frac{1}{2}mv^2$ of a gas atom is also $\frac{3}{2}kT$, where k is Boltzmann's constant, a large mass implies a smaller velocity (at a given temperature). Michelson observed this experimentally during his studies of heavy elements. He found a spectral half-width of only 0.045 nm for the red cadmium line at 643.8 nm. Since then, mercury vapor composed of the single isotope Hg^{198} has been found to produce even narrower lines.

6-3b   Lorentzian Lines

Let us next consider a gas at high pressure, so that the atoms undergo frequent collisions. A collision interrupts the radiation process, reducing the length of the wave train. The shape of the resulting line is of the form

$$I(\sigma) = \frac{A}{\sigma^2 + a^2} \tag{6-20}$$

where σ is measured from the center σ_0 of the line. The function is said to be *Lorentzian* and the process which produces it is called *collision-broadening*. Again we have $S = 0$ and

$$C = A \int_{-\infty}^{\infty} \frac{\cos 2\pi\Delta\sigma}{\sigma^2 + a^2}\, d\sigma = A \frac{\pi}{a} e^{-|2\pi\Delta a|}$$

from which we obtain

$$V = e^{-|2\pi\Delta a|} \tag{6-21}$$

The results of this calculation are compared with the previous one in Fig. 6-9. A discussion of the relation between the properties of materials and the line shapes they produce will be found in Christy[6-3].

6-3c   The Sodium D Lines

When an electric current passes through sodium vapor, it causes a glow with a yellow-orange color. A spectral analysis of the light shows that it is predominantly concentrated into two closely spaced lines, the famous *D* lines of Fraunhofer. These were discussed in Chapter 4 in connection with chromatic aberration. One line is at 589.0 nm and the other, at 589.5 nm, is 80% as intense.

We thus have two Gaussian terms to deal with, so that

$$I(\sigma) = A\left[\exp\left\{-\left(\frac{\sigma+b}{a}\right)^2\right\} + 0.8\exp\left\{-\left(\frac{\sigma-b}{a}\right)^2\right\}\right] \qquad (6\text{-}22)$$

and Eq. (6-16) yields

$$V = \frac{e^{-(\pi\Delta a)^2}}{1.8}\sqrt{1.64 + 1.60\cos(4\pi\Delta b)} \qquad (6\text{-}23)$$

This function consists of two parts: the Gaussian term $e^{-(\pi\Delta a)^2}$ is due to the shape of the individual lines and the cosine term is due to the interference of the centers of the two components, illustrated in Fig. 6-10.

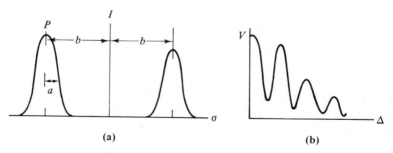

(a) **(b)**

Figure 6-10. The spectrum (a) and visibility curve (b) produced by light emitted by sodium vapor near 589 nm (5890 A).

6-3d   The Ideal Laser

A high-quality laser is monochromatic to better than 1 part in 10^{11}; therefore the spectral output can be represented as a delta function (see the Appendix for a discussion of the nature and the applications of the delta function). Hence, we write

$$I(\sigma) = A\delta(\sigma) \qquad (6\text{-}24)$$

and make use of the relations

$$\int_{-\infty}^{\infty} f(x)\, \delta(x-a)\, dx = f(a)$$

$$\int_{-\infty}^{\infty} \delta(\sigma)d\sigma = 1$$

Evaluating Eq. (6-12) leads to $S = 0$, $C = A$, and $P = A$ from which

$$V = 1 \tag{6-25}$$

This means that a perfectly monochromatic source would produce visible fringes regardless of the path difference, and a good laser should give visible fringes over a path difference exceeding 1 kilometer.

6-3e An Inexpensive Laser

The spectrum of a typical helium-neon gas laser consists of three narrow lines with equal spacing. Each can be represented by a delta function. The center component is strongest, and the outer two components have intensities equal to 0.5 of the intensity of the central component. This gives a function

$$I(\sigma) = A_1\delta(\sigma) + 0.5A_1\delta(\sigma - a) + 0.5A_1\delta(\sigma + a) \tag{6-26}$$

which leads to $S = 0$ and

$$C = A_1 + 0.5A_1 \cos 2\pi\Delta a + 0.5A_1 \cos 2\pi\Delta(-a)$$

$$= A_1[1 + \cos 2\pi\Delta a]$$

so that

$$V = \tfrac{1}{2}[1 + \cos 2\pi\Delta a] \tag{6-27}$$

Thus the visibility is unity at $\Delta = m/a$, where $m = 0, \pm 1, \pm 2, \dots$ and is zero at $\Delta = (m + \tfrac{1}{2})/a$. We see that it is a periodic function, which indicates that a laser produces sharp fringes for certain path differences and no fringes at others. This may at first sight appear surprising, since intuitively one expects a laser to be monochromatic and hence to have fringes for any path difference. But from the foregoing analysis, we see that several equally spaced spectral lines produce a periodicity in V. The narrowness of the individual components determines the maximum Δ for which fringes are visible. When we use a delta function to represent a component, we assume that the component is infinitely narrow.

One feature of the analysis that we have developed should be emphasized: V depends on the square root of $C^2 + S^2$. This means that more than one spectrum can produce a given visibility curve. Therefore, knowing V is necessary but not always sufficient to obtain $I(\sigma)$. Only if we can make a simplifying assumption (for example, that the spectrum is symmetrical about σ_0) can we unambiguously determine $I(\sigma)$. Most of the inferences that Michelson made about spectra by studying their visibility curves were correct, but he did make a number of mistakes. The proper way of obtaining the spectrum is to work with $I_T(\Delta)$, Eq. (6-10). By taking its Fourier transform we obtain $I(\sigma)$, but this requires measuring the location of the fringe as a function of Δ and is much more tedious. Computers are employed for this task.

Problem 6-5

The objective of this problem is to determine what happens to the fringe visibility as the number of laser modes increases. Let us consider a laser operating in five modes with $I(x) = A\delta(x) + 0.5A(x - a) + 0.5A\delta(x + a) + 0.5A\delta(x - 2a) + 0.5A\delta(x + 2a)$. Calculate the visibility curve and compare it to the one for three modes. As the number of modes increases beyond five, what happens to V? ▬

Problem 6-6

Calculate V for a four-mode laser having $I(x) = A\delta(x + a) + A\delta(x - a) + A\delta(x + 3a) + A\delta(x - 3a)$. ▬

6-4 Line Fringes

There are a number of variations of the Michelson interferometer that are widely employed. In one variation, line fringes are produced instead of circular fringes.

From Eq. (6-4), it is clear that for large path differences, the angular separation between fringes is small. Sometimes the separation is so small that the fringes cannot be seen. The solution is to slightly tilt one of the mirrors and use a collimated beam (i.e., a plane wave) [Eq. (6-3)]. This produces line fringes whose separation depends on the angular separation of the beams and not on the path difference. (Of course, fringes appear only in the region where the two beams overlap.)

6-5 White-Light Fringes

When white light is emitted by a source, each color sets up its own system of fringes. Observable fringes exist only when the path difference is less than a few wavelengths. At zero path difference, there will be a central white fringe and for $\Delta = \frac{1}{2}$ there will be a black region. (These two conditions can be reversed, depending on phase changes at the beam splitter.) This region will be surrounded by colored fringes for several orders. For higher orders, the maxima of one color will overlap the minima of another, and the fringes wash out.

White-light fringes are useful for determining the order of an interference fringe shift. With monochromatic light, fractional fringe shifts down to $\frac{1}{20}$ can be measured, but the order of the fringe shift cannot be determined.

Problem 6-7

An interferometer is illuminated by white light and adjusted so that a white fringe appears at the center of the field of view. A piece of transparent mica 0 millimeters thick is inserted into one arm of the interferometer. The length of the arm is shortened 0 mm and the white fringe reappears at the same position. Find the index of refraction of mica.   ■

6-6 The Mach-Zehnder Interferometer

An instrument which resembles Michelson's is the *Mach-Zehnder interferometer* shown in Fig. 6-11. Two partially silvered plates P_1 and P_2 split

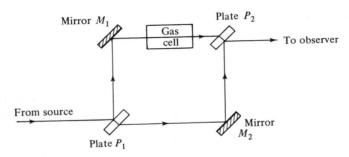

Figure 6-11. A Mach-Zehnder interferometer.

the incoming beam and then recombine the two parts again. For an ideal geometry, there is no difference in the two paths. However, when a gas is placed in the transparent cell shown, a phase shift is introduced, and it is possible to relate changes in the index of refraction to phase changes in the interference pattern. The cell can be extremely long, making it useful for applications such as wind-tunnel studies, where local changes in gas properties can be detected more easily. Fig. 6-12 shows how these density changes can be used to map temperature gradients.

6-7 The Twyman-Green Interferometer

Another variation of the Michelson interferometer is the *Twyman-Green* interferometer, an instrument widely used to test optical components (Fig. 6-13). A point monochromatic source (which can be obtained from an extended source if we focus the image of the source onto a pinhole) is placed at the focus of a lens L_1. The lens collimates the beam and provides a uniform

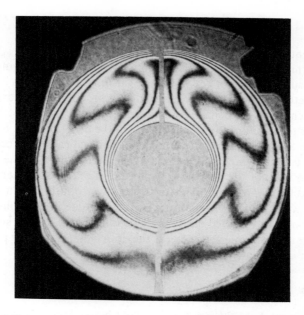

Figure 6-12. Mach-Zehnder interferogram showing convective heat flow in the annulus between two concentric cylindrical bodies. (*Photo by R.J. Goldstein and T. Kuehn, Mech. Eng. Dept., Univ. of Minnesota*)

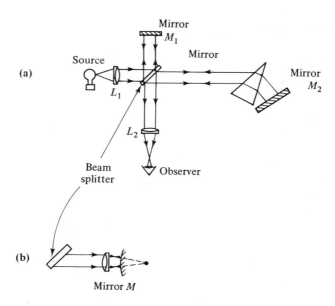

Figure 6-13. The Twyman-Green interferometer for testing a prism (a). A lens is tested (b) by replacing the plane mirror with a curved mirror *M* that matches the wave front formed by the lens or by carefully positioning a plane mirror at the focus of the lens.

and coherent wave front at the entrance to the interferometer, while another lens L_2 forms a small image of the point source. The diverging rays from the image enter the eye of the viewer. If the mirrors and beam splitter in the interferometer are polished and aligned to a fraction of a wavelength, the path difference will be constant all over the field of view, and the aperture will appear to be uniformly illuminated. On the other hand, if a transparent object such as a prism is inserted into one arm, optical imperfections on the test object will show up as variations in the illumination of the aperture.

Contour lines of the variations in optical path are drawn on the surface of the object using a small paint brush and wet polishing rouge. Then the object is reworked with the contours used as a guide in the polishing operation; the process is repeated until the element meets the desired specifications. The Twyman-Green interferometer is the device that originally permitted the production of virtually perfect lenses and prisms.

References

6-1 *Light Waves*, A. Michelson, University of Chicago Press (1927).

6-2 P. Connes, *J. Opt. Soc. Amer.* **58,** 738 (1968).

6-3 R. W. Christy, *Am. J. Phys.* **40,** 1403 (1972).

General References

Transformations in Optics, L. Mertz, Wiley (1965).

7

MULTIPLE-BEAM INTERFERENCE
AND THE FABRY-PEROT
INTERFEROMETER

7-1 The Fabry-Perot Interferometer

Interference can take place among more than two beams. An instrument that employs multiple-beam interference is the *Fabry-Perot interferometer*, developed in France in 1899 shortly after the Michelson interferometer and used today for high-resolution spectrum analysis. In the visible region it has largely superseded the Michelson interferometer (and Fourier transform spectroscopy) as a means of spectrum analysis because the spectrum can be read directly from a photographic record of the Fabry-Perot output. However, Fourier transform spectroscopy is still used in the infrared region. As we shall see, a laser is a Fabry-Perot cavity containing a light-amplifying medium.

The Fabry-Perot interferometer, shown in Fig. 7-1, consists of two pieces of glass or other transparent material. The two inner surfaces are polished flat to within $\frac{1}{20}$ to $\frac{1}{200}$ of a wavelength over the working region and are coated with high-reflectivity films. The films can be metallic (silver, gold, and aluminum are frequently used) or alternating layers of high-and low-index dielectric material. The glass plates are made with a small (0.1°) wedge angle between major surfaces so that the reflections from the outer surfaces are directed to the side and can be neglected. Interferometers are also made with

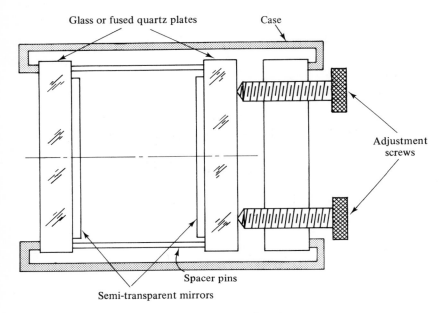

Figure 7-1. The Fabry-Perot interferometer.

one mirror on a carriage so the spacing can be adjusted, but it is difficult to keep this type of interferometer in adjustment.

The essence of the interferometer is the two highly reflecting films separated by a distance t in air (Fig. 7-2). It will be sufficient to consider only the reflecting properties of the films; the glass wedges need not be included in the calculation. The amplitude of the wave incident on the first surface is E_0. Let the amplitude of the reflected wave be ρE_0 and the amplitude of the transmitted wave be σE_0. If there is no absorption, the sum of the transmitted and reflected energies equals the incident energy, or

$$\sigma^2 + \rho^2 = 1 \tag{7-1}$$

At the second surface, part of the wave is again transmitted with amplitude $\sigma^2 E_0$ and part reflected with amplitude $\rho\sigma E_0$. The amplitudes of the successive transmitted waves are then

$$\left.\begin{array}{l} E_1 = \sigma^2 E_0 \\ E_2 = \sigma^2 \rho^2 E_0 \\ E_3 = \sigma^2 \rho^4 E_0 \\ \quad \cdot \\ \quad \cdot \\ \quad \cdot \\ E_N = \sigma^2 \rho^{2N-2} E_0 \end{array}\right\} \tag{7-2}$$

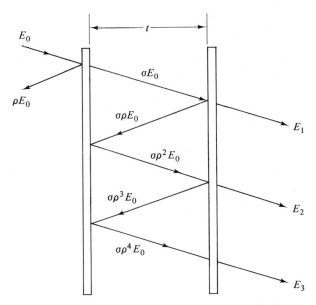

Figure 7-2. An incoming wave E_0 undergoes multiple reflections.

The phase differences between the various waves must be taken into account. When all the waves are in phase, there is a maximum in the total transmitted amplitude, but if they alternate in phase by 180° then the net amplitude is zero. As shown in Fig. 7-3, the path difference Δ between two adjacent transmitted components is

$$\Delta = 2l - p \qquad (7\text{-}3)$$

where

$$l = \frac{t}{\cos \theta}$$

and

$$p = 2d \sin \theta = 2(t \tan \theta) \sin \theta$$

Substituting into Eq. (7-3) yields

$$\Delta = 2t \cos \theta \qquad (7\text{-}4)$$

There will be a maximum in the transmitted value of E when $\Delta = m\lambda$. The phase difference, which is given by $\delta = 2\pi\Delta/\lambda$, is

$$\delta = \left(\frac{2\pi}{\lambda}\right)2t \cos \theta \qquad (7\text{-}5)$$

When the various transmitted beams overlap, either because they are wide or because they are brought to a common focus by a lens, the total

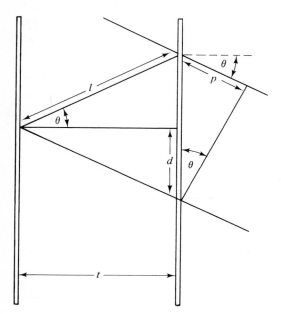

Figure 7-3. Geometrical construction for calculating path difference of two components.

amplitude E_T, taking the phase of E_1 as reference, is

$$E_T = E_1 + E_2 + E_3 + \ldots + E_N$$
$$= \sigma^2 E_0 + \sigma^2 \rho^2 E_0 e^{-i\delta} + \sigma^2 \rho^4 E_0 e^{-2i\delta} + \ldots + \sigma^2 \rho^{2N-2} E_0 e^{-i(N-1)\delta}$$

or

$$E_T = \sigma^2 E_0 \{ 1 + \rho^2 e^{-i\delta} + \rho^4 e^{-i2\delta} + \ldots + \rho^{2N-2} e^{-i(N-1)\delta} \} \qquad (7\text{-}6)$$

With the aid of the expansion

$$\frac{1}{1-x} = 1 + x + x^2 + \ldots \qquad (7\text{-}7)$$

Eq. (7-6) becomes

$$E_T = \frac{\sigma^2 E_0}{1 - \rho^2 e^{-i\delta}} \qquad (7\text{-}8)$$

The physically observed quantity is the intensity I_T, which is given by

$$I_T = \tfrac{1}{2} E_T E_T^* \qquad (7\text{-}9)$$

from Eq. (5-15). Using Eq. (7-8) to evaluate Eq. (7-9) yields

$$I_T = \frac{\tfrac{1}{2} \sigma^4 E_0^2}{1 + \rho^4 - 2\rho^2 \cos \delta} \qquad (7\text{-}10)$$

This can be rewritten using $I_0 = \tfrac{1}{2} E_0^2$ and the identity

$$\sin^2 \frac{\delta}{2} = \frac{1 - \cos \delta}{2} \qquad (7\text{-}11)$$

as

$$I_T = \frac{\sigma^4 I_0}{(1 - \rho^2)^2 \left\{ 1 + \dfrac{4\rho^2 \sin^2 \delta/2}{(1 - \rho^2)^2} \right\}} \tag{7-12}$$

In Eq. (7-1) let $T = \sigma^2$, $R = \rho^2$, and $T = 1 - R$, where T and R are the coefficients for energy transmission and reflection, respectively. With these substitutions, Eq. (7-12) becomes

$$I_T = \frac{I_0}{1 + \dfrac{4R \sin^2 \delta/2}{(1 - R)^2}} \tag{7-13}$$

This equation together with Eq. (7-5) shows that the intensity transmitted through the interferometer is a function of four variables: the *reflectivity* R of the coated surfaces, their separation t, the angle θ with respect to the normal, and the wavelength λ.

Under certain conditions the phase factor δ will be an integral multiple of 2π and complete transmission will take place, even though each of the two surfaces has a very high reflectivity! Waves reflected from each surface are 180° out of phase and cancel. Either surface alone would reflect most of the light and transmit only a small part, but both surfaces together transmit everything! Under other conditions, the phase factor δ will be an odd multiple of π, and very little light will be transmitted.

The sharpness of the fringes is determined by the quantity $4R/(1 - R)^2$. For example, when $R = 0.5$, this quantity has a value 8.0 and Eq. (7-13) becomes

$$\frac{I_T}{I_0} = \frac{1}{1 + 8 \sin^2 \delta/2}$$

indicating that the fractional transmitted intensity varies between 1 and $\frac{1}{9}$. The angle of observation varies from 0° at the center of the pattern to some maximum, and the function $\sin^2 \delta/2$ varies between the limits of 0 and 1. This sets up a periodic pattern or *fringe system*. Increasing the reflection coefficient to $R = 0.9$, causes the transmitted intensity to vary between 1 and $\frac{1}{361}$. Thus the higher the reflectivity, the greater the fringe contrast. This is illustrated in Fig. 7-4 where the ratio I_T/I_0 is plotted as a function of δ for several values of R. We see that higher reflectivity gives narrower fringes.

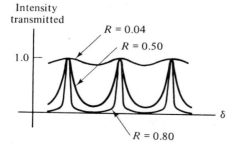

Intensity transmitted

$R = 0.04$

$R = 0.50$

1.0

δ

$R = 0.80$

Figure 7-4. The shape of fringes set up in a Fabry-Perot interferometer for different values of the reflectivity of the plates.

Note that for low reflectivities the fringes are broad, as in the Michelson interferometer. This is because the interference essentially involves only two beams. For a low value of R, the higher-order beams are so weak that they can be neglected. At high reflectivities, more beams interfere, and the maxima drop off more quickly.

Problem 7-1

Normally the useful range of a Fabry-Perot interferometer is restricted to about the first ten fringes. Beyond that, the fringes are spaced more and more closely together and are harder to read. The spacing between plates is usually greater than 0.1 mm. From this information, calculate the minimum wedge angle the outer faces must have so that the rays reflected from them do not overlap the interference pattern and give unwanted side images.   ■■

7-2 Resolution

If a source emits at two wavelengths, designated as λ and $\lambda + \Delta\lambda$, a separate fringe pattern will be set up for each of these values. When the wavelengths are close, the fringe systems will essentially overlap and the two will be inseparable, but if they are far enough apart, they can be distinguished. We would like to know how close the two wavelengths can be and still be resolved by the interferometer. One widely accepted criterion is shown in Fig. 7-5. Adjacent curves of intensity vs phase angle intersect at the half-power points. Their sum has a noticeable dip which can just be seen by the eye. By Eq. (7-5), the maxima of the two curves have values

$$\delta_1 = \left(\frac{2\pi}{\lambda}\right) 2t \cos \theta_1 \tag{7-14a}$$

$$\delta_2 = \left(\frac{2\pi}{\lambda + \Delta\lambda}\right) 2t \cos \theta_2 \tag{7-14b}$$

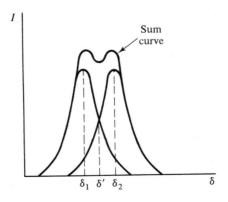

$\delta_1 \;\; \delta' \;\; \delta_2$ $\qquad\qquad\qquad$ δ

Figure 7-5. The criterion for resolution of two spectral lines.

At δ' both curves are at half maximum, and for each the denominator in Eq. (7-13) equals 2.0. This condition is satisfied when the second term in the denominators is unity, or when

$$\sin^2 \frac{\delta'}{2} = \frac{(1 - R)^2}{4R} \tag{7-15}$$

For sharp fringes, the value of $\delta'/2$ is near zero or a multiple of π, and the sine function can be approximated by its argument, giving

$$\delta' = \frac{1 - R}{\sqrt{R}} \tag{7-16}$$

Now $\delta_1 - \delta_2$ is twice δ' so that

$$\delta_1 - \delta_2 = \frac{2(1 - R)}{\sqrt{R}} \tag{7-17}$$

For values of θ_1 and θ_2 near zero, the cosine terms in Eqs. (7-14) can be replaced by unity, and the expression for each δ can then be substituted into Eq. (7-17), giving

$$\frac{4\pi t}{\lambda} \frac{\Delta\lambda}{\lambda} = \frac{2(1 - R)}{\sqrt{R}}$$

or

$$\frac{\Delta\lambda}{\lambda} = \frac{\lambda}{2\pi t} \frac{1 - R}{\sqrt{R}} \tag{7-18}$$

The separation of two wavelengths $\Delta\lambda$ that can just be resolved according to our criterion depends on the wavelength, plate separation, and the plate reflectivities. The quantity $\lambda/\Delta\lambda$ is termed the *chromatic resolving power*.

Problem 7-2

What chromatic resolving power must a device have to resolve two modes of a helium-neon laser 500 megahertz apart? The average wavelength is 632.8 nm. Assume that the maximum possible reflectivity is 0.99. What plate separation is required to attain this resolution? ■

Eq. (7-18) seems to indicate that any desired resolution can be attained if either the reflectivities of the plates or their separation is made large enough. Actually, this is not true because there are limiting factors on the magnitudes of both of these quantities. The principal limitation, however, is associated with the fact that Eq. (7-18) was derived with the implicit assumption that the plates are perfectly flat and we ought to investigate what we mean by perfectly flat. Since this would take us too far from our main topic*, let us obtain an approximate answer from Eqs. (7-14), which indicate that a variation in t produces a variation in δ. If t varies by a fractional wavelength λ/M, then δ

*See, for example, reference 7-1.

varies by $4\pi/M$. Therefore the resolution described by Eq. (7-18) implies that $\delta_1 - \delta_2$ (the separation between peaks) in Eq. (7-17) is greater than $4\pi/M$ (the width of one line). Hence

$$\frac{4\pi}{M} < \frac{2(1 - R)}{\sqrt{R}}$$

or

$$\frac{1}{M} < \frac{(1 - R)}{2\pi\sqrt{R}} \tag{7-19}$$

This shows that the flatness of the plates must be commensurate with the reflectivity in high-resolution interferometers.

As an example, let us calculate the reflectivity and flatness of a Fabry-Perot interferometer required to resolve two components of the Hg green line at 546.1 nm. The components are separated by $\Delta\sigma = 0.10 \, \text{cm}^{-1}$ the plate separation is to be 1 mm. (See Eq. (6-8) for the definition of the wave number σ.) Then

$$\Delta\sigma = \Delta\left(\frac{1}{\lambda}\right) = \left|\frac{\Delta\lambda}{\lambda^2}\right| = 0.1$$

Substituting this into Eq. (7-18), we obtain

$$0.1 = \frac{1}{2\pi \times 10^{-1}} \frac{1 - R}{\sqrt{R}}$$

and solving for R yields

$$R = 0.93$$

By Eq. (7-19)

$$\frac{1}{M} = \frac{0.07}{6.3} = \frac{1}{90}$$

Thus, the reflectivity must be 0.93, and the inner surfaces of the plates must be flat to within a factor of $\lambda/90$ to resolve the two components of the Hg green line. Increasing the separation t of the plates reduces the requirements on both of these parameters.

A related factor that must be taken into consideration is the alignment of the plates. Not only must they be flat to λ/M across the working area, but they must also be aligned parallel to each other to this same extent. There are limitations on the plate separation as well as on the effective reflectivity. The separation times the number of effective reflections that a beam undergoes in the interferometer must not be greater than the temporal coherence length of the light being investigated; otherwise the fringes will wash out. A practical reason for keeping the plate separation small is that the diameter of the rings is inversely proportional to t. The larger the first diameter, the easier the pattern is to read.

Problem 7-3

How flat must the plates be in Problem 7-2? ▬

7-3 Free Spectral Range

When the difference in wavelengths of two components of a spectrum is large enough, the displacement of one pattern with respect to another can be larger than the fringe separation (see Fig. 7-6) and the orders, defined in connection with Eq. (6-5), are said to overlap. This makes the pattern more difficult to interpret. (When the components of a spectrum are all displayed adjacent to each other and with no intervening lines, the interpretation is straightforward.) The *free spectral range* is defined as the range in frequency between orders. For large plate separations, the fringes from adjacent orders are close together, and the free spectral range is small. Consequently the requirement for ease in interpretation (small separation) is the opposite of the requirement for high resolution (large separation). These two factors must be balanced in selecting an instrument for a particular application.

The free spectral range can be calculated if we first recall that the condition for a maximum in the transmitted light (a fringe), as obtained from

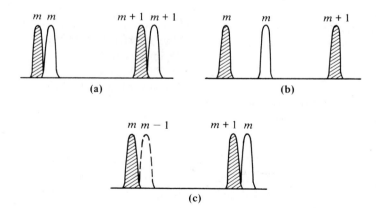

Figure 7-6. Two spectral components are shown for various plate separations. In (a), the separation is small and the components are just resolved. In (b), the separation increases. In (c), the separation is increased further. The mth order of one line overlaps the $m + 1$ order of the other line.

Eq. (7-4), is

$$m\lambda = 2t \cos \theta \qquad (7\text{-}20)$$

The condition that the $(m + 1)$st order of λ coincides with the mth order of $\lambda + \Delta\lambda$ is

$$(m + 1)\lambda = m(\lambda + \Delta\lambda) \qquad (7\text{-}21)$$

which can be rearranged to yield

$$\Delta\lambda = \frac{\lambda}{m} \qquad (7\text{-}22)$$

where $m \gg 1$. When the angle of incidence is almost $90°$, $\cos \theta = 1$. Substituting the expression for m from Eq. (7-22) into Eq. (7-20) yields

$$\Delta\lambda = \frac{\lambda^2}{2t} \qquad (7\text{-}23)$$

As an example, an interferometer at the wavelength 546.1 nm with a spacing of 1 mm has a free spectral range* of

$$\Delta\lambda = 0.15 \text{ nm}$$
$$= 5 \text{ cm}^{-1}$$

7-4 Procedure for Alignment of a Fabry-Perot Interferometer

An interferometer can be aligned using a three-step process. First a coarse adjustment is made by bringing multiple images into coincidence. A diffuse source such as a sodium lamp covered by a piece of ground glass is placed behind the interferometer (Fig. 7-7). Then an object (a pin or a spot on the glass) is viewed through the center of the interferometer. When the interferometer is out of adjustment, multiple images will be seen because each reflection will be displaced slightly (Fig. 7-8(a)), and the interferometer is adjusted so that these images coincide (Fig. 7-8(b), (c)). At the completion of this step, a ring pattern can be seen.

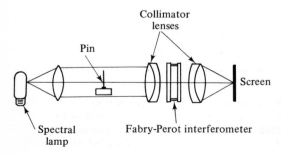

Collimator
lenses

Pin

Screen

Spectral
lamp

Fabry-Perot interferometer

Figure 7-7. Arrangement for a Fabry-Perot interferometer. The ring pattern appears on the observation screen.

*In wave numbers, $\Delta(1/\lambda) = 1/2t$ cm^{-1}. The reader should verify this.

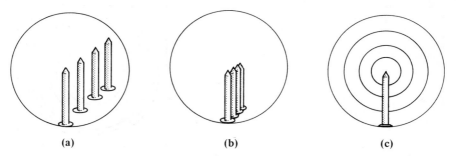

(a)   (b)   (c)

Figure 7-8. A pin placed behind the interferometer is used to make initial adjustments. In (a), multiple images are displaced. As the interferometer is brought into alignment (b), the images come closer together. When they are superimposed (c), the ring pattern can be seen.

Second, an intermediate adjustment is made by keeping the ring size constant as the eye is swept across the aperture. If, as indicated in Fig. 7-9, the plates are inclined at a slight angle, the separation at one edge is greater than in the center. As the eye moves toward that edge, the fringes grow larger. By adjusting the plates so the ring size remains constant as the eye moves from side to side and up and down, the separation can be set to about a twentieth of a wavelength.

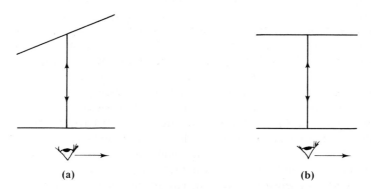

(a)   (b)

Figure 7-9. As the eye in (a) moves to the right the fringes get smaller indicating that the plates are tilted. In (b), the pattern is stationary indicating that the plates are parallel.

Third, the fine adjustment is made by observing the rings and making them as sharp as possible. A source such as a laser or single-isotope low-pressure lamp with a narrow spectrum is needed for this final stage of adjustment. The pattern can be magnified by means of a telescope. Small adjustments usually produce a considerable sharpening of the rings (Fig. 7-10).

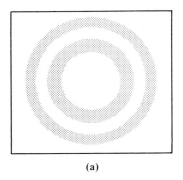

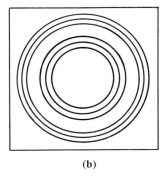

(a)   (b)

Figure 7-10. The output of a He-Ne laser operating in three modes is examined by a Fabry-Perot interferometer. In (a), the instrument is not aligned well. In (b), the instrument is aligned and the three operating wavelengths of the laser are observed.

7-5   Scanning Interferometers

The early method of taking data with the Fabry-Perot interferometer was to photograph the output pattern with a camera. If the focal length of the camera lens is f and the angular separation of two rings is $\Delta\theta$, then their spacing on the film is $f\,\Delta\theta$. Lenses of long focal length are employed so that the fringe spacing will be large, facilitating the interpretation. However, under these conditions the light falling on the film is weak, and consequently the exposure time must be long. During the exposure, the temperature of the spacer in the interferometer must be held constant, sometimes to 0.1°C, otherwise the spacing will change and the resolution will decrease.

One way around the difficulty is to use a scanning interferometer and a photomultiplier. The effective plate separation is varied and the wavelength that gives a maximum at the center of the pattern sweeps through the free spectral range, while a lens focuses the interferometer pattern onto a screen. A pinhole is placed at the focus of the lens and aligned so that only the center of the pattern is transmitted, the photomultiplier being placed behind the pinhole. Photomultipliers (see Chapter 16) often are two or more orders of magnitude more sensitive than film. The arrangement is shown in Fig. 7-11.

The optical path between the plates can be changed if n or t is varied. Early scanning interferometers worked by varying n. The interferometer was placed in a vacuum-tight enclosure which was then evacuated. As air or some other gas was let into the enclosure, the index of refraction of the region between plates changed. Although the index for air is close to unity (1.000 290), the difference between this value and $n = 1.000\ 000$ for a vacuum is usually large enough so that the interferometer is swept through several free spectral ranges.

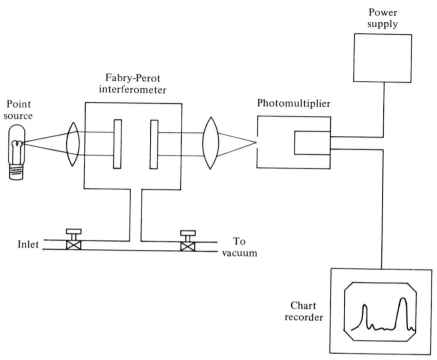

Figure 7-11. A scanning Fabry-Perot interferometer. Changes in pressure produce changes in the index of refraction and the optical path between the plates.

Modern scanning interferometers work by varying t. One plate is mounted on a ring of piezoelectric material.[7-2] When such a material is subjected to an electric field, it expands or contracts, changing changes the spacing t. The voltage required to sweep the interferometer through one free spectral range can be set equal to the sawtooth signal available from a commercial oscilloscope. The sweep frequency of the interferometer is then synchronized with the scan frequency of the oscilloscope. When the photomultiplier output is connected to the input of the oscilloscope, the spectrum is viewed directly (Fig. 7-12).

Problem 7-4

A scanning interferometer has an average plate separation of 1 mm. The separation varies half of a wavelength as 150 volts is applied to a piezoelectric crystal. What is the free spectral range? Can the instrument resolve two laser modes spaced 120 megahertz apart?

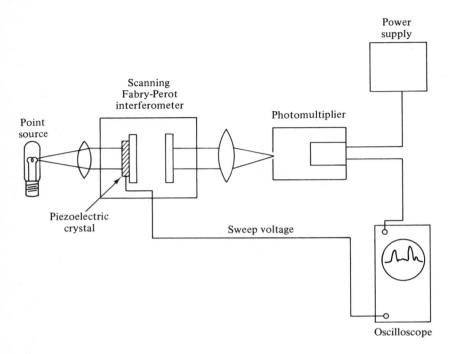

Figure 7-12. A piezoelectrically-driven scanning Fabry-Perot interferometer.

7-6 Confocal Interferometers

A variation of the Fabry-Perot interferometer uses curved instead of plane mirrors. It was invented by P. Connes, who realized that it would be easier to align such an instrument. The confocal configuration is usually used, i.e., the separation is equal to the radius of curvature of the mirrors. The disadvantage is that the incoming light must be collimated and parallel to the axis of the interferometer to within about 1°. Otherwise third-order aberrations (see Chapters 2–3) become noticeable and reduce the resolution. The instrument is particularly useful for examining the spectrum of a laser because the output is usually very well collimated (0.010 or less).

7-7 Interference Filters

An inexpensive kind of Fabry-Perot interferometer acts as a narrow band-pass filter, and it can be used to isolate a spectral line. To make this filter, a glass substrate is first coated with a thin, partially transmitting metallic film. Then a layer of material having a low index of refraction and a thickness equal to half the wavelength desired for peak transmission is added. Another metallic layer is added on top of this dielectric layer, and then a second piece of glass is used to cover the films. After this, the edges of the sandwich are sealed (Fig. 7-13). The width of the transmitted spectrum depends on the reflectivity of the metal layers and can be calculated from our previous theory. A typical value for the half-width at half maximum transmission is 10.0 nm. This can be reduced to about 5.0 nm by using the second-order interference maximum. In this case, transmission will take place at $2\lambda_0$ as well as at λ_0.

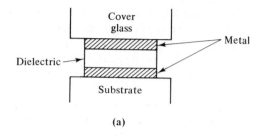

(a)

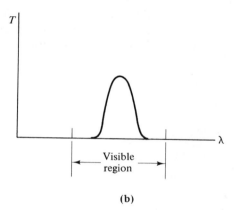

(b)

Figure 7-13. An inexpensive interference filter and its transmission curve for normal incidence.

7-8 Interferometric Measurement of the Thickness of the Mica Molecule

A general feature of multiple-beam interference is that the greater the number of beams, the sharper are the fringes. As we have seen, this is exploited in high-resolution spectroscopy by having an instrument in which the separation between two surfaces is constant and the reflectivities of the surfaces are high. Alternatively, if the spectrum is known, multiple-beam interference can be used to measure minute variations in the thickness of an object.

One of the most amazing applications of this concept was achieved by S. Tolansky, who used it to measure the thickness of the mica molecule.

Mica is a mineral that occurs in sheets. It is very easy to cleave mica parallel to its flat surface. One method is to insert a pin in the side of a piece to start the process, and then simply pull the sheets apart. A close examination of the surface of mica reveals that it has a number of "lines" on it, which are steps in the surface (see Fig. 7-14). The change in height at each step is some integer times the thickness of the unit cell of the mica crystal.

The surface of a sheet of mica having a step is coated with a metallic film of high reflectivity. Then it is placed against a similarly coated optical flat with the two metallic surfaces in contact. The assembly is illuminated by monochromatic light, and the fringe pattern is photographed using a microscope. There will be a discontinuity in the ring pattern at each step seen on the mica surface; the corresponding shift can be measured to a fraction of a fringe. Sometimes light of two different wavelengths is used so that the order number of the fringe shift can be determined.

Tolansky measured a large number of steps in this manner with an accuracy of about $\lambda/100$. The lowest common denominator of the measurements was 2.0 nm, which agrees well with determinations of the dimensions of the mica unit cell using X-ray diffraction. An interesting feature of this experiment is that the measuring stick is the wavelength of light, and yet measurements to one-hundredth of a wavelength are possible using interference techniques.

Figure 7-14. A step in the surface of a sheet of mica.

Problem 7-5

Design an interference filter which can isolate the mercury green line at 546.1 nm from the yellow line at 582.0 nm. What chromatic resolving power is necessary? The filter should have as high a transmission as possible. What reflectivity is needed if $t = \lambda_0$ (i.e., a first-order filter)? If $t = 2\lambda_0$ (a second-order filter)?

Problem 7-6

For the interference filter in Problem 7-5 to have a half-width at a half-maximum of 10.0 nm, what reflectivity is required? Can it be made using metallic films?

Problem 7-7

How flat must the surfaces of the glass substrate be for an interference filter in Problem 7-5? For Problem 7-6?

References

7-1 J. Sládková, *Interference of Light*, Iliffe Books, London (1968).

7-2 A. J. Dekker, *Solid State Physics*, Prentice-Hall, Englewood Cliffs, N.J. (1957).

7-3 S. Tolansky, *High Resolution Spectroscopy*, Methuen, London (1947).

General References

A. Melissinos, *Experiments in Modern Physics*, Academic Press New York (1966).

R. W. Wood, *Physical Optics*, Macmillan, New York (1934).

8

THIN FILMS

8-1 Reflection from a Dielectric Surface

When a light wave is incident on a surface of glass or other dielectric, it is a common observation that part is reflected and part is transmitted. It is found experimentally that the fraction of the total energy which is reflected depends on the index of refraction n of the material. In Fig. 8-1, an incident wave with amplitude E_0 falls upon a dielectric surface. We designate the reflected wave amplitude as ρE_0, and the transmitted wave amplitude by σE_0; the indices of refraction are as shown. Then the coefficient of reflection ρ depends on the *difference* in the indices of refraction on each side of the boundary, and it is given by the expression[5-1]

$$\rho = \frac{n_0 - n_1}{n_0 + n_1} \tag{8-1}$$

where the reflectivity R, the fraction of energy reflected, is equal to ρ^2, as previously indicated in Sec. 7-1.

According to Eq. (8-1), ρ is negative when the wave goes from a lower- to higher-index medium; this signifies a 180° phase change for the reflected wave. This was experimentally verified by Wiener in 1890, who inclined a thin photographic film at a small angle to a metal surface and illuminated the film from the back with a plane wave. Like the eye, photographic film res-

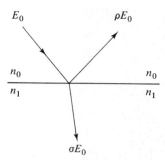

Figure 8-1. Reflection and transmission of the wave amplitude E_0 at a dielectric interface.

ponds to the square of the amplitude of the electric field. Wiener found, upon developing the film, that the exposed region consisted of a series of bands separated by unexposed areas, indicating that standing waves had been set up. At the point where the film was in contact with the surface, there was no darkening, that is, a node existed at the surface and required that the reflected wave be 180° out of phase with the incident wave. In the most general case p is complex, and the phase change can take any value between 0° and 360°; but we shall restrict our attention to real values. Corresponding values of the reflectivity R in air then come from Eq. (8-1). Glass, for example, has an index of refraction of approximately 1.5. This yields

$$R = p^2 = \frac{(-0.5)^2}{(2.5)^2} = 0.04$$

Problem 8-1

How does the reflectivity of water ($n = 1.33$) compare to that of glass ($n = 1.50$)? ■

Diamond has one of the highest known indices of refraction for a transparent material: $n = 2.4$, giving a reflectivity of 0.17. Cutting a diamond so that it has many facets increases the likelihood of a reflection in the direction of an observer. The cutting plus the high reflectivity makes the diamond sparkle. Rhinestones are made of glass, and even though they may be cut identically to a diamond, they do not sparkle as much because of the lower reflectivity. Another transparent material with a high index of refraction to use in place of diamonds in jewelry is rutile (TiO_2) with $n = 2.6$, which is even higher than diamond. However, this material is brittle and cracks easily, whereas diamond is very hard. Zircon ($ZrSiO_4$), for which $n = 1.9$, is durable and frequently used as an inexpensive replacement for diamonds.

According to Eq. (8-1), it is the difference in indices of refraction that accounts for reflection. If the index is the same on each side of a surface, there will be no reflection even though the materials are different. An illustra-

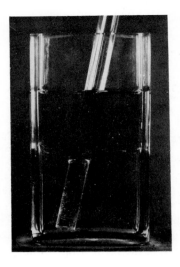

Figure 8-2. A pyrex rod is invisible in a solution whose index of refraction equals that of the rod.

tion of this is shown in Fig. 8-2. The beaker contains a solution of benzene and alcohol. The proportions are adjusted so that the index of refraction of the solution equals that of pyrex ($n = 1.4$). When the pyrex rod is immersed it becomes invisible.

In an old Hollywood movie, the heroine hid the jewels from some robbers by dropping them into a bottle of gin. When the robbers looked in the bottle they saw nothing. After they left and at the dramatic climax of the film, she produced them from the bottle and received everyone's congratulations. The story is implausible because the index of refraction of liquids varies between 1.3–1.5. If the jewels were diamonds, they would have reflected a noticeable amount of light even when immersed in the liquid. Since the audience saw them disappear as they were dropped into the bottle, the conclusion must be drawn that the producers used rhinestones in the movie. Along the same lines, H.G. Wells exploited the essence of Eq. (8-1) in the science fiction classic, *The Invisible Man.* When the central character of the story drank a liquid that changed his index of refraction to unity, he became invisible.

8-2 Reflection from Two Surfaces

We will now consider reflection from two parallel surfaces separated by a distance t (Fig. 8-3). The amplitude of the incident wave is taken as E_0. At the first interface, the coefficients of amplitude transmission and reflectance are σ_1 and ρ_1, respectively, and at the second interface, they are σ_2 and ρ_2, respectively. By Eq. (8-1)

$$-\rho_1 = \frac{n_0 - n_1}{n_0 + n_1}, \qquad \rho_2 = \frac{n_1 - n_2}{n_1 + n_2} \tag{8-2}$$

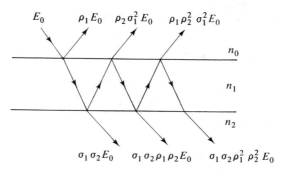

$E_0 \quad \rho_1 E_0 \quad \rho_2 \sigma_1^2 E_0 \quad \rho_1 \rho_2^2 \sigma_1^2 E_0$

n_0

n_1

n_2

$\sigma_1 \sigma_2 E_0 \qquad \sigma_1 \sigma_2 \rho_1 \rho_2 E_0 \qquad \sigma_1 \sigma_2 \rho_1^2 \rho_2^2 E_0$

Figure 8-3. Reflection from two parallel surfaces.

where the negative sign of $-\rho_1$ means that it is measured **inside** the layer of index n_1. (As explained below, we want to calculate the **transmitted** wave.) The phase difference introduced by two consecutive reflections inside the middle layer is, by Eq. (7-5),

$$\delta_1 = \left(\frac{2\pi n_1}{\lambda}\right) 2t \cos \theta \tag{8-3}$$

Problem 8-2

To verify Eq. (8-3), we must alter Eq. (7-3) to take into account the reduction in the velocity of light when it travels through glass. The effective path length becomes $2n_1 l$ and is called the *optical path length*, whereas $2l$ is simply the geometric path. Combine this fact with Snell's law to compute the phase shift. ∎

In this situation, it will be easier for us to calculate the expression for the transmission coefficient and then apply conservation of energy to obtain the expression for the reflection coefficient. The amplitudes of the transmitted components are

$$\left. \begin{array}{l} E_1 = \sigma_1 \sigma_2 E_0 \\ E_2 = \sigma_1 \sigma_2 \rho_1 \rho_2 E_0 \\ E_3 = \sigma_1 \sigma_2 \rho_1^2 \rho_2^2 E_0 \\ \quad \vdots \\ \quad \vdots \\ \quad \vdots \end{array} \right\} \tag{8-4}$$

and the total transmitted amplitude E_T is

$$E_T = \sigma_1 \sigma_2 E_0 (1 + \rho_1 \rho_2 e^{-i\delta} + \rho_1^2 \rho_2^2 e^{-i2\delta} + \ldots) \tag{8-5}$$

which can be written, using Eq. (7-7), as

$$E_T = \frac{\sigma_1 \sigma_2 E_0}{1 - \rho_1 \rho_2 e^{-i\delta}} \tag{8-6}$$

The time average of the transmitted intensity is

$$I_T = \tfrac{1}{2} E_T E_T^*$$

or using $I_0 = \tfrac{1}{2} E_0^2$ gives

$$I_T = \frac{I_0 \sigma_1^2 \sigma_2^2}{1 + \rho_1^2 \rho_2^2 - 2\rho_1 \rho_2 \cos \delta} \tag{8-7}$$

In the absence of absorption, the reflection coefficient R and the transmission coefficient T add to unity, or

$$R = 1 - T \tag{8-8}$$

where

$$T = \frac{I_T}{I_0} \tag{8-9}$$

Then Eqs. (8-7), (8-8), and (8-9) give

$$R = \frac{\rho_1^2 + \rho_2^2 - 2\rho_1 \rho_2 \cos \delta}{1 + \rho_1^2 \rho_2^2 - 2\rho_1 \rho_2 \cos \delta} \tag{8-10}$$

8-3 Antireflecting Layers

Equation (8-10) will now be used to calculate the reflecting properties of several two-surface combinations. An important application is the *antireflecting* film. Zero reflectivity can be obtained by depositing a suitable film on a substrate of glass or other dielectric. The thickness of the film is adjusted so that

$$n_1 t = \frac{\lambda}{4} \tag{8-11}$$

The film is referred to as a *quarter-wave film*. From Eq. (8-3) at normal incidence, we have

$$\delta_1 = \pi, \qquad \cos \delta_1 = -1 \tag{8-12}$$

Under this condition, the reflectivity of the film calculated from Eq. (8-10) is

$$R = \frac{(\rho_1 + \rho_2)^2}{(1 + \rho_1 \rho_2)^2} \tag{8-13}$$

Substituting the values for ρ_1 and ρ_2 from Eq. (8-2) yields

$$R = \frac{n_1^2 - n_0 n_2}{n_0 n_2 + n_1^2} \tag{8-14}$$

From this expression it is apparent that the reflectivity of the combination is zero if n_1, the index of the middle layer, is the geometric mean of the other two indices, that is, if

$$n_1 = \sqrt{n_0 n_2} \tag{8-15}$$

To make a glass surface ($n_2 = 1.5$) nonreflecting in air ($n_0 = 1.0$), it must be coated with a material of index of refraction $n_1 = 1.22$. Unfortunately,

n for most solids varies between 1.5 and 2.2, and there is no known solid having this low an index of refraction, although cryolite (Na_3AlF_6) with $n = 1.33$ comes close to satisfying this requirement. According to Eq. (8-14), the reflectivity of a glass surface coated with a quarter-wave film of cryolite would be 0.008. This is a considerable reduction from 0.04, the reflectivity of the uncoated surface. Cryolite has the disadvantage of being soft and not very durable. Usually magnesium fluoride (MgF_2) with $n = 1.384$ is used for such antireflecting films. A value $R = 0.012$ is obtained for a glass surface coated with a quarter-wave film of MgF_2, and λ_0 is given by

$$nt = \frac{\lambda_0}{4}$$

For wavelengths other than λ_0, the value of δ differs from π, and the reflectivity increases. A graph of R vs λ for normal incidence is shown in Fig. 8-4 with λ_0 equal to 550 nm, the center of the visible region of the spectrum.

Such films reflect violet and red much more than yellow and green. Thus, one can quickly check a camera lens to see if it is coated by determining the color of light reflected from it.

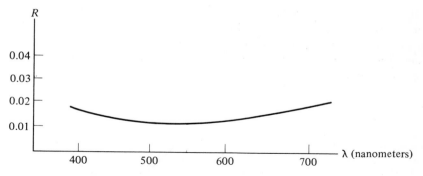

Figure 8-4. The reflectivity of a surface coated with quarter wave film of MgF_2.

8-4 High Reflectivity Layers

In the previous section we found that when the index of refraction of a film is less than that of the substrate, the reflectivity of the combination is less than that of the substrate alone. Next, let us consider how to obtain an increase in reflectivity. Figure 8-5 indicates that a wave reflected at the first surface undergoes a 180° change in phase if it encounters a medium of higher index, but the wave reflected from the lower surface does not undergo a phase shift because it passes to a medium of lower index. The two reflected waves should be in phase for the reflectivity to be a maximum, and the film thickness should be a quarter wavelength so that $\delta = \pi$.

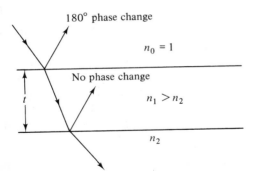

Figure 8-5. The phase changes upon reflection.

In one popular application, quarter-wave films are used to increase the reflectivity of costume jewelry. Rhinestones or other glass objects ($n = 1.5$) are coated with silicon monoxide SiO ($n = 2.0$). The reflectivity of such a combination can be calculated from Eq. (8-13), giving $R = 0.20$ (where p_2 is negative in this case). This represents a considerable increase over $R = 0.04$ for a single glass surface. The thickness of the coating is varied so that $4nt$ is equal to various values of λ in the visible region; different regions of the object then reflect different colors. Alternatively, when the coated objects are viewed at angles other than the normal incidence, the colors will change.

Figure 8-6 shows the reflectivity of such a surface as a function of δ, and Table 8-1 gives the reflectivity for glass surfaces coated with various quarter-wave films.

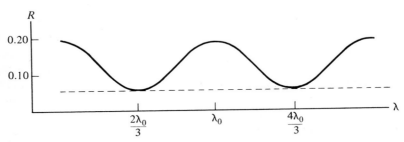

Figure 8-6. The reflectivity of a glass surface ($n = 1.5$) coated with SiO ($n = 2.0$).

Table 8-1
Thin Films on Glass

Film	n	Maximum Value of R	Remarks
SiO	2.0	0.20	Must be of high purity.
ZnS	2.3	0.37	—
TiO$_2$	2.6	0.40	Requires high temperature for deposition.
Sb$_2$S$_3$	3.2	0.57	Colored brownish yellow.

8-5   Reflections from a Thin Slab of Dielectric

Equation (8-10) can be used to calculate the reflectivity of a thin piece of dielectric such as glass (Fig. 8-7). In this case,

$$n_0 = n_2 = 1.0 \text{ and } n_1 = 1.5$$

According to Eq. (8-2)

$$\rho_1 = -\rho_2 \tag{8-16}$$

and from Eq. (8-10)

$$R = \frac{2\rho_1^2[1 + \cos \delta]}{1 + \rho_1^4 + 2\rho_1^2 \cos \delta} \tag{8-17}$$

This function is zero when $\delta = \pi$ and odd multiples thereof. It has a maximum value when $\delta = 0$ or a multiple of 2π. The maximum is equal to

$$R = \frac{4\rho_1^2}{(1 + \rho_1^2)^2} \tag{8-18}$$

For glass, $\rho_1^2 = 0.04$ giving $R = 0.15$. The reflectivity of the two-surface combination depends on the value of δ, which in turn is determined by the thickness, angle of inclination, and wavelength.

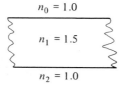

$n_0 = 1.0$

$n_1 = 1.5$

$n_2 = 1.0$

Figure 8-7. A thin sheet of dielectric.

For surfaces of low individual reflectivity, R of the combination varies between zero and approximately four times the value for a single surface. This is another interesting interference effect. When the reflections from each surface are 180° out of phase, they cancel exactly. When they are in phase they add, producing an amplitude approximately twice as large as would be produced by a single surface and an intensity approximately four times as large. The average value of R is what we would expect without interference: about twice that for a single surface. The function R of Eq. (8-18) is shown in Fig. 8-8.

It should be stressed that the temporal coherence length of the incident light must be longer than the distance between surfaces for interference to take place between waves reflected from the different surfaces. This requirement is satisfied when the light source is a single emission line from an ordinary spectral lamp and when the thickness of the middle layer is less than a millimeter. It is also satisfied when the source is a laser and the surfaces are centimeters apart.

On the other hand, the condition is not satisfied by a broad-band or a white-light source, and with a piece of window glass, no interference takes

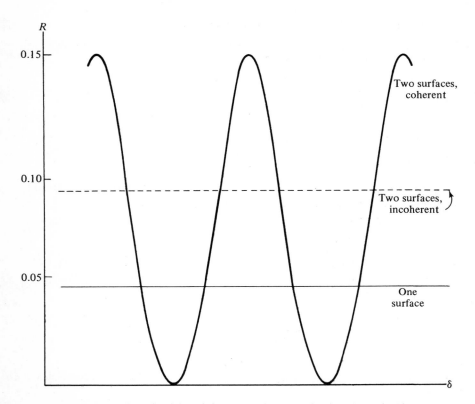

Figure 8-8. The reflectivity of the two-surface combination shown in Fig. 8-7.

place between light reflected from the front and back surfaces. The total reflected intensity there is simply the incoherent superposition of two waves; this is approximately twice the intensity of one alone.

8-6 Multiple-Layer Thin Films

Up to this point we have treated reflections from one and from two surfaces. We were able to solve the two-surface problem, even though it was considerably more complex than the single surface, because we were able to obtain an analytical expression for the infinite series that represents the sum of partial waves. We might anticipate that the general N-surface problem would be still more difficult because it involves summing up N infinite series, each term of which is itself a series. A few of the paths of reflected beams from a multiple-layer combination are shown in Fig. 8-9 to indicate the complexity of the problem.

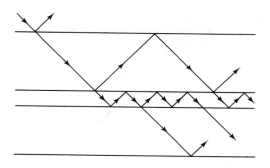

Figure 8-9. Multiple-layer films have many reflections.

For many years it was thought that the N-layer problem was intractable, much like the N-body problem in mechanics. A few simple cases had been solved by inspection, but no analytical expression existed for the general case. In 1937, Rouard discovered that a multiple-layer film could be analyzed by representing each layer by a 2×2 matrix M_j which has the form

$$M_j = \begin{pmatrix} \cos \delta_j & \dfrac{i \sin \delta_j}{n_j} \\ in_j \sin \delta_j & \cos \delta_j \end{pmatrix} \tag{8-19}$$

where we define the phase shift as one-half the value specified by Eq. (8-3), or

$$\delta_j = \frac{2\pi}{\lambda} n_j t_j \cos \theta_j \tag{8-20}$$

Then the effect of a combination of layers is obtained if we simply take the product of the matrices representing each layer. Special matrices

$$\begin{pmatrix} n_0 & -1 \\ n_0 & +1 \end{pmatrix} \quad \text{and} \quad \begin{pmatrix} 1 \\ n_f \end{pmatrix}$$

represent the top and bottom media. The reflectivity of a thin-film combination is given by

$$R = \left| \frac{a}{b} \right|^2 \tag{8-21}$$

where a and b are obtained from

$$\begin{pmatrix} a \\ b \end{pmatrix} = \begin{pmatrix} n_0 & -1 \\ n_0 & +1 \end{pmatrix} \begin{pmatrix} \cos \delta_j & i\dfrac{\sin \delta_j}{n_j} \\ in_j \sin \delta_j & \cos \delta_j \end{pmatrix} \begin{pmatrix} 1 \\ n_f \end{pmatrix} \tag{8-22}$$

It is understood that the middle term represents the product of one such matrix M_j for each layer.

8-6a A Two-Layer Antireflecting System

The application of the matrix method to a given film combination is straightforward. Usually we want to know the reflectivity of a system over a range of wavelengths, and a separate calculation must be made at each value of λ or δ. Calculations can be performed manually for simple systems, but computers are now used for the more complicated systems.

As an illustration of the procedure, the reflectance as a function of λ will be calculated for a two-layer antireflection system originally developed by Turner. We saw in Section 8-3 that a single-layer quarter-wave film on a glass substrate reduces the reflectivity from 4% to about 1% at the center of the visible region and to about 2% at the ends. The two-layer system of Fig. 8-10

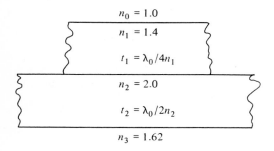

$$n_0 = 1.0$$
$$n_1 = 1.4$$
$$t_1 = \lambda_0/4n_1$$
$$n_2 = 2.0$$
$$t_2 = \lambda_0/2n_2$$
$$n_3 = 1.62$$

Figure 8-10. A two-layer antireflecting combination designed by Turner.

reduces the reflectivity considerably more. This system can be used on any glass substrate; we shall perform the calculation for flint glass ($n = 1.62$). The lower layer is a half-wave film of high-index material ($n = 2.0$) and the upper layer is a quarter-wave film of low-index material ($n = 1.4$). The phase factors needed for the matrices are

$$\delta_1 = \frac{2\pi}{\lambda} t_1 n_1 = \frac{\pi \lambda_0}{2\lambda}$$

and

$$\delta_2 = \frac{2\pi}{\lambda} t_2 n_2 = \frac{\pi \lambda_0}{\lambda}$$

For the central wavelength $\lambda = \lambda_0$

$$\delta_1 = \frac{\pi}{2}, \qquad \sin \delta_1 = 1, \qquad \cos \delta_1 = 0$$

$$\delta_2 = \pi, \qquad \sin \delta_2 = 0, \qquad \cos \delta_2 = -1$$

Substituting these values into Eq. (8-22) yields

$$\begin{pmatrix} a \\ b \end{pmatrix} = \begin{pmatrix} 1 & -1 \\ 1 & 1 \end{pmatrix} \begin{pmatrix} 0 & \dfrac{i}{1.41} \\ i(1.41) & 0 \end{pmatrix} \begin{pmatrix} -1 & 0 \\ 0 & -1 \end{pmatrix} \begin{pmatrix} 1 \\ 1.62 \end{pmatrix} = \frac{i}{1.41} \begin{pmatrix} 0.38 \\ 3.62 \end{pmatrix}$$

The reflectivity R given by Eq. (8-21) is then $R = 0.01$.

At two other wavelengths, $\lambda = 1.24\lambda_0$ and $\lambda = 0.89\lambda_0$, the calculation yields $R = 0$. A portion of the curve is shown in Fig. 8-11.

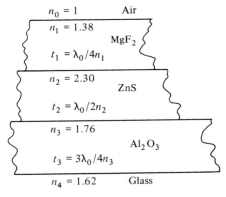

Figure 8-11. The reflectivity of the combination shown in Fig. 8-10.

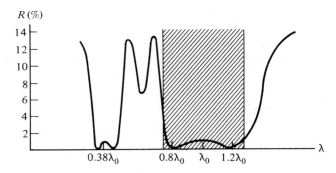

Figure 8-12. A three-layer antireflection coating.

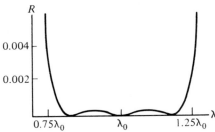

Figure 8-13. The reflectivity of the coating shown in Fig. 8-12.

8-6b   A Three-Layer Antireflecting System

One can surmise by inspection of Figs. 8-4 and 8-11 that the number of minima in the reflection curve (for a given order of interference) is equal to the number of layers. A three-layer system is therefore expected to have a third minimum; this could be centered at λ_0 by a suitable choice of the film parameters. Such a system was also designed by Turner (Fig. 8-12).

The results of performing the calculation for this system are shown in Fig. 8-13. The reflectivity is reduced to less than 0.05 % throughout almost the entire visible region.

8-6c   Multiple-Layer, High-Reflectivity Systems

The reflectivity of a surface can be made equal to 0.99 or higher if we increase the number of dielectric layers deposited on it. Films of such high reflectivities are useful in laser cavities or in high-resolution Fabry-Perot interferometers. Metallic films are not suitable for many applications; their reflectivities are not high enough, and for $R > 0.8$, they absorb most of the nonreflected portion of the incoming wave.

A high-reflectivity system is made by stacking pairs of quarter-wave films consisting of alternating high- and low-index material. A pair of films having $\delta_L = \delta_H = \pi/2$ is represented by the matrix

$$\begin{pmatrix} 0 & -\dfrac{i}{n_H} \\ -in_H & 0 \end{pmatrix}\begin{pmatrix} 0 & -\dfrac{i}{n_L} \\ -in_L & 0 \end{pmatrix} = \begin{pmatrix} -\dfrac{n_L}{n_H} & 0 \\ 0 & -\dfrac{n_H}{n_L} \end{pmatrix} \qquad (8\text{-}23)$$

and a stack of N such pairs by

$$\begin{pmatrix} \left(-\dfrac{n_L}{n_H}\right)^N & 0 \\ 0 & \left(-\dfrac{n_H}{n_L}\right)^N \end{pmatrix} \qquad (8\text{-}24)$$

The reflectivity in air of such a stack deposited on a substrate having M_f can be calculated if we evaluate Eq. (8-21) for this system:

$$\begin{pmatrix} a \\ b \end{pmatrix} = \begin{pmatrix} 1 & -1 \\ 1 & 1 \end{pmatrix}\begin{pmatrix} \left(-\dfrac{n_L}{n_H}\right)^N & 0 \\ 0 & \left(-\dfrac{n_H}{n_L}\right)^N \end{pmatrix}\begin{pmatrix} 1 \\ n_f \end{pmatrix} \qquad (8\text{-}25)$$

which yields

$$R = \left[\frac{n_f - (n_L/n_H)^{2N}}{n_f + (n_L/n_H)^{2N}}\right]^2 \qquad (8\text{-}26)$$

As N becomes large the reflectivity approaches unity. Theoretically, the reflectivity can be made as close to unity as desired if a sufficiently high number of pairs are used. In practice there is always some absorption and scattering from a film combination. In the present state of the art, the practical limit on R is about 0.999.

A typical reflectance curve for a 31-layer system used as a mirror in a He-Ne laser is reproduced in Fig. 8-14.

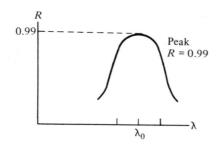

Figure 8-14. The reflectivity of a typical mirror used on a He-Ne laser.

Problem 8-3

A thin-film stack consists of alternating layers of MgF_2 and TiO_2, each $\lambda_0/4$ thick. How many pairs must be deposited on a glass substrate ($n = 1.50$) to attain $R = 0.99$ at λ_0? How many must be deposited to attain $R = 0.999$? ∎

8-7 Metallic Films

In metallic films of high reflectivity, as much light is absorbed as transmitted. For example, at 700 nm a beam divider made with a thin silver film on glass reflects, transmits, and absorbs approximately equal parts of the incident energy. At higher reflectivities, there is proportionally more absorption relative to transmission.

A thick layer of silver has one of the highest reflectivities of any metal ($R = 0.96$) and is frequently used for mirrors. Silver films have the disadvantage of tarnishing when exposed to air, and aluminum ($R = 0.92$) is used as a substitute. Although its reflectivity is lower, it forms a transparent layer of Al_2O_3 on its surface when exposed to air that protects the film from further oxidization.

The reflectivity and transmission of a thin metallic film depends on its thickness, as shown in Fig. 8-15.

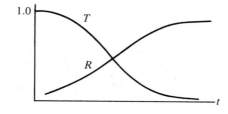

Figure 8-15. The dependence on thickness of the reflectivity R and transmission of a thin aluminum film.

8-8 The One-Way Mirror

The one-way mirror consists of a glass coated with a partially transmitting metallic film. Actually, the transmission and reflection are the same in both directions. When the one-way mirror is examined in a well-illuminated room, a great deal of light is reflected from it. If the region behind the mirror is dark, the small amount of light coming through it will be masked by the reflected light.

Viewed from the dark side, however, the situation is reversed. Very little light is reflected, and most of the light seen by an observer is transmitted from the other side.

8-9 Methods for Depositing Thin Films

The two most widely used methods of depositing thin films are vacuum evaporation and chemical deposition.

An evaporator, illustrated in Fig. 8-16, typically has a mechanical pump capable of producing a vacuum of 10^{-2} Torr (mm Hg) and a diffusion pump capable of 10^{-6} Torr or more. The material to be deposited is placed inside the chamber in "boats" or crucibles which are surrounded by heating coils. The substrate is suspended above the source and the chamber evacuated. The vacuum must not only be low enough so that the mean free path of an

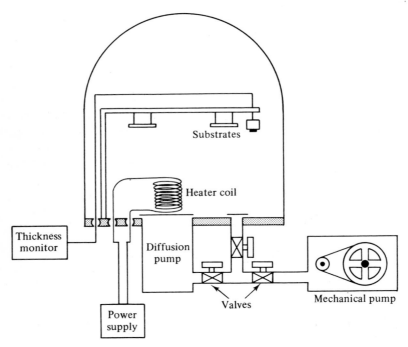

Figure 8-16. An evaporator for vacuum deposition of thin films.

atom is longer than the distance from source to substrate, but it must also be sufficiently low that gas atoms do not collide with the surface and adhere to it during deposition. This latter condition typically requires a vacuum on the order of 5×10^{-6} Torr. Heating coils above the substrate are used to bake contaminants off the surface before deposition. Another method for cleaning the surface is by ion bombardment using a glow discharge. After a suitable pressure has been reached and the substrate cleaned, the heater is turned on and the material is evaporated. A film-thickness monitor measures the thickness of the layer as it is being deposited. Some commonly used materials are listed in Table 8-2.

Table 8-2
Some Commonly Used Materials for Thin Films*

Material	Melting Temperature at 760 mm of Hg	Vaporization Temperature at 10^{-4} mm of Hg	Index of Refraction
Al	660°C	1010°C	
Ag	961°C	1105°C	
Au	1062°C	1132°C	
MgF$_2$	1266°C	925°C	1.38
Na$_3$AlF$_6$	1000°C	1480°C	1.33
LiF	870°C	1180°C	1.39
SiO	1702°C	850°C	2.0
SiO$_2$	1710°C	1025°C	1.44
Al$_2$O$_3$	2045°C	1550°C	1.66

* Source Master Equipment Catalogue 1973–4, Sloan Technology Corp. Santa Barbara, Calif.

For materials with high melting temperatures, electron beams are used to obtain concentrated heating of the material surface.

Some films can be deposited by chemical means. These films usually are a product of chemical reactions, and they adhere to the substrate as they are precipitated. For many years, glassblowers have silvered the insides of dewars, christmas tree ornaments, and other objects by Brashears process. (The formula is given in early editions of the *Handbook of Chemistry and Physics*.) A solution with complex silver ions is cooled and an activating agent added. When the mixture is placed in contact with a warm surface, the reaction proceeds rapidly and silver is deposited on the surface. In this manner, high-quality coatings can be deposited at inaccessible places.

A number of metal oxides such as SiO_2 and TiO_2 form complex solutions with isopropyl and other alcohols. When a substrate is dipped into the solution and removed, a thin film of the solution adheres to it. In the presence of moisture, the alcohol evaporates, and a residue is left behind on the surface. Upon baking, hydroxyl groups form H_2O, which evaporates, and a durable film is formed.

Perhaps the most important requirement in depositing thin films is that the substrate be very clean; otherwise the film will not adhere to it. Beautiful films have been known to curl up during the night and spoil a perfectly good experiment because the substrate had not been properly cleaned.

General References

O.S. Heavens, *Optical Properties of Thin Films*, Dover, N.Y. (1965).

M. Francon, *Modern Applications of Physical Optics*, Wiley, N.Y. (1963).

9

INTERFERENCE BY
DIVISION OF WAVE FRONT

9-1 Introduction

In division of amplitude, as described in earlier Chapters 6, 7, and 8, a wave at one point is divided into two or more parts by a beam splitter, a time delay is introduced into each part, and the parts are recombined, resulting in an interference pattern. The connection between fringe visibility of a pattern produced by division of amplitude and the coherence length of the source light was discussed in Chapter 6.

Division of wave front, in which two or more portions of a wave front are combined, will be described in this chapter. The relative phases of the portions determine whether they interfere constructively or destructively, and the visibility of the fringes produced by division of wave front will be related to the *spatial coherence* distance produced by the source light.

Division of wave front leads naturally into diffraction, for the two phenomena are closely related. When parts of a wave front come from different openings and combine, we have *interference;* but when parts of a wave front come from a single opening and combine, we have *diffraction.*

9-2 The Two-Slit Experiment

The two-slit experiment illustrates division of wave front. It was first performed in 1802 by Thomas Young, who used pinholes. In repeating the experiment, slits are usually used because they transmit more light and the resulting pattern is easier to observe. The arrangement used by Young (Fig. 9-1) involves sunlight passing through a pinhole, P_1, which acts as a point source. This assures that the wave on the right-hand side of the pinhole is spatially coherent. A screen with two adjacent pinholes, P_2 and P_3, divides the wave front into two portions, and the openings act as mutually coherent point sources of spherical waves. At a spot on the screen where the path difference for the two waves is zero, they are in phase, and a maximum in the intensity exists. If the path difference is $\lambda/2$, the two waves are 180° out of phase, and a minimum exists. The alternating maxima and minima constitute a fringe system, each of the maxima corresponding to a different order of interference.

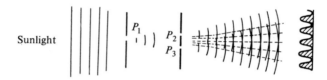

Figure 9-1. Young's two-pinhole experiment. The separation between P_2 and P_3 is greatly exaggerated.

We shall calculate the interference pattern produced by a single point source emitting monochromatic waves; later this will be generalized to polychromatic and extended sources. In Fig. 9-2, a monochromatic point source S is located on the line perpendicular to the midpoint between the two slits, S_1 and S_2. The separation of the slits is a. We wish to calculate the intensity distribution of the pattern which would be observed on a screen at a distance d from the slits.

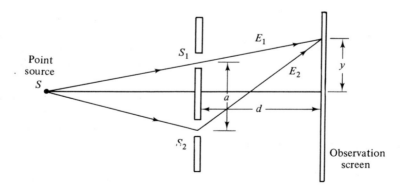

Figure 9-2. The geometry for calculating the path difference for the two waves coming from S_1 and S_2 respectively.

Upon arriving at a point located a distance y from the center of the screen, the two waves have traveled over paths whose difference Δ is given by

$$\Delta = \sqrt{d^2 + (y + a/2)^2} - \sqrt{d^2 + (y - a/2)^2} \tag{9-1}$$

Using $d \gg y + a/2$ and the binomial expansion

$$\sqrt{1 + x^2} \sim 1 + \frac{x^2}{2} \tag{9-2}$$

Eq. (9-1) becomes

$$\Delta = d\left\{1 + \frac{(y + a/2)^2}{2d^2} - 1 - \frac{(y - a/2)^2}{2d^2}\right\} \tag{9-3}$$

$$= \frac{ay}{d}$$

The phase difference between the two waves is given by

$$\delta = \frac{2\pi}{\lambda}\Delta$$

$$= \frac{2\pi}{\lambda}\frac{ay}{d} \tag{9-4}$$

Taking the phase of the wave designated by E_1 on the screen as reference, we write

$$E_1 = E_0$$

and

$$E_2 = E_0 e^{i\delta} \tag{9-5}$$

The total amplitude E_T is then

$$E_T = E_1 + E_2 \tag{9-6}$$

As we have already mentioned in connection with Eq. (5-15), the quantity seen by the eye or photographed by a film is proportional to the time average of the intensity I_T, which is given by

$$I_T = \frac{1}{2}E_T E_T^* \tag{9-7}$$

Combining Eqs. (9-5), (9-6), and (9-7) and substituting $I_0 = \frac{1}{2}E_0 E_0^*$ yields

$$I_T = 2I_0(1 + \cos \delta) \tag{9-8}$$

or

$$I_T = 4I_0 \cos^2 \frac{\delta}{2} \tag{9-9}$$

The pattern on the screen therefore consists of fringes whose amplitude varies as the square of the cosine of the position (Fig. 9-3). (As will be shown in the next chapter, the finite width of the slits leads to a diffraction pattern given by the envelope of the curve.)

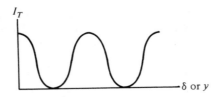

Figure 9-3. The interference fringes in a two-slit experiment.

Problem 9-1

If a pair of slits 0.1 mm apart are illuminated by light of 435.8 nm or 4358 Å, what will be the separation of the maxima of the interference pattern viewed on a screen 1 meter away? What slit separation produces maxima 1 mm apart? ■■

9-2a Polychromatic Sources

When the incident light contains more than one wavelength, each will set up a separate fringe system with its individual maximum at the center of the pattern. When white light is used, the central maximum will be white and the adjacent minima are black, whereas nearby fringes will be colored. Fringes farther away will be washed out. Figure 9-4 illustrates the fringes produced by red, green, and blue light.

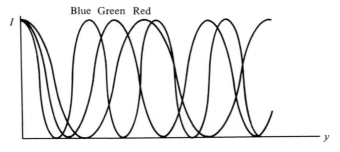

Figure 9-4. The fringes produced by three wavelengths of light in a two-slit experiment.

9-2b Extended Sources

We will now consider the pattern that is produced by an extended incoherent source; such a source can be represented by a sum of points. When it emits incoherently, each point radiates independently of its neighbors, and there are no fixed phase relations between radiating centers. The observed intensity from the source is the sum of the intensities from each point; this is proportional to N, the number of points. When the source is coherent, on the other hand, the amplitudes add; the net amplitude is proportional to

N, and the intensity is proportional to N^2, as demonstrated in section 5-6.

In Fig. 9-5, the extended source is located at a distance D from the two slits. A point S_j is located a distance u from the center of the source at S_0. Each such point on the source will produce an interference pattern on the observation screen that will be slightly displaced with respect to the pattern produced by S_0. If the source is large enough, the peaks in the interference pattern produced by one point will fill in the valleys produced by another point, and the combined pattern will wash out. An alternative way of stating this is that the visibility of the fringe system will decrease.

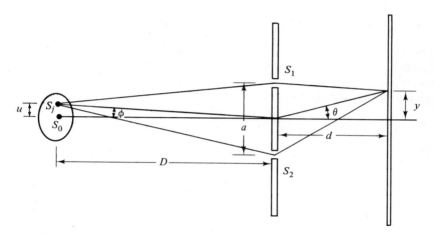

Figure 9-5. An extended source can be represented by a series of points S_j. The source is located a distance D from the two slits.

Using Eq. (9-3), we can express the path difference for the two beams passing through S_1 and S_2 as

$$\Delta = \frac{ay}{d} + \frac{au}{D} \tag{9-10}$$

and the phase difference is $\delta = 2\pi\Delta/\lambda$ as above. Then the contribution to the intensity on the screen from the jth point, from Eq. (9-8), is given by

$$dI_j(y) = 2I(u)(1 + \cos \delta)\, du \tag{9-11}$$

For incoherent sources, the total intensity at point y is the sum of the intensities from each point j on the sources; therefore

$$I_T(y) = 2 \int_{-\infty}^{\infty} I(u)(1 + \cos \delta)\, du \tag{9-12}$$

Using the trigonometric identity

$$\cos (A + B) = \cos A \cos B - \sin A \sin B \tag{9-13}$$

we can rewrite Eq. (9-12) as

$$I_T(y) = 2 \int_{-\infty}^{\infty} I(u) \, du + 2 \int_{-\infty}^{\infty} I(u) \cos\left(\frac{kay}{d}\right) \cos\left(\frac{kau}{D}\right) du$$

$$- 2 \int_{-\infty}^{\infty} I(u) \sin\left(\frac{kay}{d}\right) \sin\left(\frac{kau}{D}\right) du \tag{9-14}$$

There are a number of different experimental conditions that can be employed in two-slit experiments, which must be specified in order to proceed further in simplifying Eq. (9-14).

9-2c Monochromatic Source with Fixed Slit Separation

The first case to be considered is an extended quasimonochromatic incoherent source and fixed slit separation a. Under these conditions the terms in (kay/d) are constants and can be brought outside the integrals. Then, with the aid of the substitutions

$$\theta = \frac{kay}{d} \tag{9-15}$$

$$\alpha = \frac{ka}{D} \tag{9-16}$$

$$P = 2 \int_{-\infty}^{\infty} I(u) \, du \tag{9-17}$$

$$S = 2 \int_{-\infty}^{\infty} I(u) \sin \alpha u \, du \tag{9-18}$$

$$C = 2 \int_{-\infty}^{\infty} I(u) \cos \alpha u \, du \tag{9-19}$$

Eq. (9-14) can be written as

$$I(y) = P + C \cos \theta - S \sin \theta \tag{9-20}$$

This form is easy to evaluate for a given source. (Note the similarity between the above treatment and that of Chapter 6.) The first source we will consider is a step function of the form $I(u) = I_0$ for $u \le |u_0|$ and $I(u) = 0$ elsewhere. (See Fig. 9-6). Since this is an even function, Eq. (9-18) gives $S = 0$; the other two terms become

$$P = 4I_0 u_0 \tag{9-21}$$

and

$$C = 4I_0 u_0 \frac{\sin \alpha u_0}{\alpha u_0} \tag{9-22}$$

Substituting these in turn into Eq. (9-20) yields

$$I(y) = 4I_0 u_0 \left[1 + \frac{\sin \alpha u_0}{\alpha u_0} \cos\left(\frac{kay}{d}\right) \right] \tag{9-23}$$

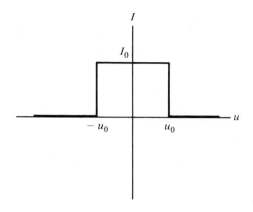

Figure 9-6. A step-function source.

Figure 9-7 shows a graph of this function for a particular value of αu_0. The visibility V of the fringes is defined as (see Chapter 7)

$$V = \frac{I_{max} - I_{min}}{I_{max} + I_{min}} \tag{9-24}$$

For the fringes described by Eq. (9-23), Problem 9-2 will show that

$$V = \left| \frac{\sin \alpha u_0}{\alpha u_0} \right| \tag{9-25}$$

Note that this is a constant. We see that a source described by the step function produces fringes with the same spacing as does the point source (Eq. 9-8); however, their visibility is decreased by the factor $(\sin \alpha u_0)/\alpha u_0$, a function which depends on the angular size of the source. Measuring the visibility of the fringes enables us to calculate a value for the angular size of the source.

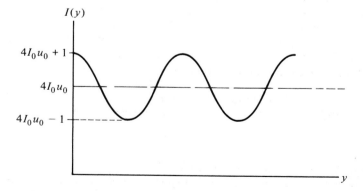

Figure 9-7. The fringes in a two-slit experiment produced by an extended source. Note that their visibility is less than that for a point source.

For a given slit separation a, we can define the coherence angle φ_{coh} subtended by the source as that angle for which $V = \frac{1}{2}$. From Eq. (9-25), we see that this condition holds when

$$\alpha u_0 = 1.9$$

where φ_{coh} is, by definition, $2u_0/D$. Then by Eq. (9-16)

$$\varphi_{\text{coh}} = \frac{2u_0}{D} \sim 0.6\frac{\lambda}{a} \tag{9-26}$$

Alternatively, given an angle subtended by the source, a coherence width a_{coh} can be defined for the wave front as that distance at which $V = \frac{1}{2}$; therefore

$$a_{\text{coh}} = 1.9\frac{D}{ku_0} = 0.3\frac{\lambda D}{u_0} \tag{9-27}$$

We have arbitrarily taken the value of V to be 0.5 in our definitions of coherence. Other values could just as easily have been selected, such as 0.9, $\pi/4$, or e^{-1}. Different values of V selected for the definition of coherence determine the numerical value of the constant in Eqs. (9-26) and (9-27), but the essential results of the calculations remain the same regardless of the value of V selected.

9-2d Temporal Coherence and Spatial Coherence

In Chapter 6 we described the Michelson interferometer and found that the instrument can be used to measure the coherence length. This was an indication of the lengths of the wave trains being emitted by a source. The coherence length is inversely related to the width of the spectrum; that is, the more monochromatic the source, the longer the temporal coherence length.

The double-slit experiment can be used to measure the spatial coherence interval of a wave front. This is an indication of the lateral extent of correlations in the phase of the wave front, and it is inversely related to the angular diameter of an incoherent source.

9-2e Measurement of Stellar Diameters

As will be shown in Chapter 10, the angular resolution $\Delta\theta$ attainable by a lens is given by

$$\Delta\theta = 1.22\frac{\lambda}{D} \tag{9-28}$$

The resolution of a lens is limited by the size of its aperture. Michelson realized that a modified two-slit arrangement could be used to measure the angular diameter of a star more accurately than with a telescope lens. In 1920 he built the stellar interferometer (Fig. 9-8) based on this idea. In the stellar

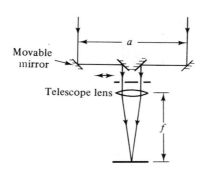

Figure 9-8. The Michelson stellar interferometer. The separation of the mirrors, a, can be varied while the fringe pattern is observed.

interferometer, the effective aperture of a telescope is increased by mounting two mirrors on a beam in front of the lens. In one version the mirrors were 6 meters apart, and the aperture of the lens of the accompanying telescope was $2\frac{1}{2}$ meters. The effective aperture of the telescopic lens was thus increased by a factor of about 3.0.

In practice, the theoretical resolution of a large telescope lens is never reached because of atmospheric turbulence. Large telescopes have greater light-gathering power but not greater resolution. The interference pattern of the two-slit system is also affected by turbulence. The pattern moves laterally, but the fringes remain visible. The actual enhancement in resolution of the $2\frac{1}{2}$-meter lens using the stellar interferometer was approximately a factor of ten.

Taking advantage of the effective increase in aperture obtained by mounting the mirrors on a beam requires considerable experimental skill; vibrations and temperature changes produce variations in the interference pattern. When the instrument is used, the distance a between the slits is varied while the center of the interferometer pattern (at $y = 0$) is observed. Since stars are weak sources, measurements are taken in white light whenever possible. Recall that the visibility of fringes produced by a white point source is high at the center of the pattern. To numerically evaluate the visibility of the fringe system produced at the observation screen, we use Eq. (9-20) with $y = 0$ and treat a as the variable. Since $y = 0$, then $\theta = 0$ and Eq. (9-20) becomes

$$I(a)_{y=0} = P + C \qquad (9\text{-}29)$$

To evaluate the quantities P and C, we represent the source by the function of Fig. 9-6. This gives

$$P = 4I_0u_0 \qquad (9\text{-}30)$$

and

$$C = 4I_0u_0 \frac{\sin (ku_0a/D)}{(ku_0a/D)} \qquad (9\text{-}31)$$

Hence

$$I(a) = 4I_0u_0\left[1 + \frac{\sin (ku_0a/D)}{(ku_0a/D)}\right] \qquad (9\text{-}32)$$

The visibility of the fringe system, given by

$$V = \left| \frac{\sin{(ku_0 a/D)}}{(ku_0 a/D)} \right| \tag{9-33}$$

is illustrated in Fig. (9-9).

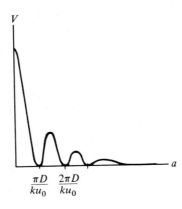

Figure 9-9. The visibility of fringes produced by a step function source as a function of a; the effective separation of the two slits.

As a is increased from zero, the central fringe becomes less distinct; at $a = \pi D/ku_0$ it has zero visibility. Measuring the value of a at which this occurs permits the calculation of the angular diameter as

$$\varphi = \frac{2u_0}{D} = \frac{2\pi}{a_{min}k} = \frac{\lambda}{a_{min}} \tag{9-34}$$

The first star measured in this manner was Betelgeuse, for which $\varphi = 0.047$ seconds of arc. The distance of the star was then determined by triangulation. (Measurements of its apparent position are taken six months apart.) The diameter was found to be 300 times the diameter of the sun! This is larger than the size of the earth's orbit. Only a few of the closest giant stars have been measured by this method. For better resolution, the distance a would have to be increased, but mechanical difficulties in holding the mirrors stationary have caused problems. However, the diameters of several moons of the planets have been measured. It should also be noted that our analysis here is based on hypothetical, one-dimensional objects; a more refined treatment will be given in Chapter 10.

Problem 9-2

(a) Verify Eq. (9-25) by considering the expressions for I_{max} and I_{min} when $\sin \alpha u_0$ is positive and when it is negative.

(b) Justify Eq. (9-33) in a similar fashion.

Problem 9-3

A radio telescope operative at $\lambda = 20$ cm has an antenna 100 m in diameter (roughly the size of a football field). How does its angular resolution compare to that of a 100-in. optical telescope? ▬

9-2f Double Stars

The separation of double stars can also be measured using the Michelson stellar interferometer. When the distance between stars is much greater than their diameters, their intensity distribution can be represented by two delta functions (Fig. 9-10) as

$$I(u) = I_0 \delta(u_0 + b) + A I_0 \delta(u_0 - b) \tag{9-35}$$

Evaluating Eq. (9-20) for this intensity distribution yields

$$I(a) = I_0\{(1 + A) + (1 + A) \cos \theta \cos \alpha b_0 - (1 - A) \sin \theta \sin \alpha b_0\} \tag{9-36}$$

The visibility function is

$$V = \left| \frac{1}{1 + A} \sqrt{(1 + A^2) + 2A \cos 2\alpha b_0} \right| \tag{9-37}$$

which is displayed in Fig. 9-11. The maximum value $V = 1$ occurs when $\cos 2\alpha b_0 = 1$. The minimum value $V = (1 - A)/(1 + A)$ occurs when $\cos 2\alpha b_0 = -1$. When the slits are close together (small a), $V = 1$, indicating that

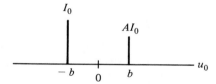

Figure 9-10. The idealized intensity distribution of a double star.

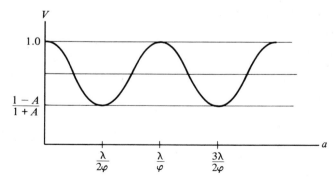

Figure 9-11. The visibility function produced by the doublet shown in Fig. 9-10.

the fringes are sharp. As the slit separation *a* increases, the visibility decreases to a minimum at

$$a = \frac{\lambda}{2\varphi} \qquad (9\text{-}38)$$

9-3   Other Examples of Division of Wave Front

Lloyd's mirror is shown in Fig. 9-12. A wave front is emitted from a point source, with one portion traveling directly from the source to the observation screen and the other portion reflected from the mirror onto the screen. In the region of the overlap, an interference pattern is set up.

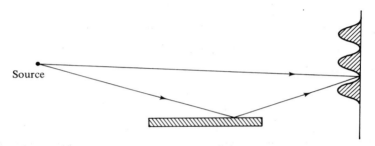

Figure 9-12. Lloyd's mirror. Interference takes place between the direct and the reflected waves.

Problem 9-4

A point source is located one centimeter above a large sheet of glass. One meter from the source, a white card is placed perpendicular to the glass, and Lloyd's fringes are seen on the card. What is the spacing of the fringes on the card if $\lambda = 546.1$ nm? Can they be easily observed with the unaided eye? What happens to the spacing if the source is lowered to a position 1 mm above the glass? ∎

The *Fresnel biprism* is illustrated in Fig. 9-13. A plane wave is incident

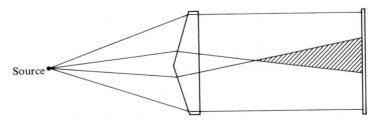

Figure 9-13. The Fresnel biprism.

on the prism, the upper and lower halves being refracted in different directions. A fringe system is set up where they overlap.

The *Nelson experiment* demonstrates that the light emitted from a laser is spatially coherent. One end of a ruby laser rod is coated with a high reflectivity silver film (Fig. 9-14). Two openings are left in the coating so that a small portion of the light can be emitted, and a screen is placed a short dis-

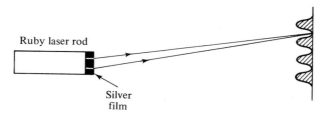

Figure 9-14. The Nelson experiment.

tance away. This indicates that light emitted from different regions of the extended source is spatially coherent. If the same experiment is repeated using a fluorescent source or an ordinary spectral lamp such as Hg or Na, no fringes will be seen because, during the observation time, there are no fixed correlations in phase between the light coming through the two slits.

General References

Lipson and Lipson, *Optical Physics*, Oxford University Press, London (1970).

R. W. Wood, *Physical Optics*, Macmillan, New York (1934).

C. L. Andrews, *Optics of the Electromagnetic Spectrum*, Prentice-Hall, Englewood Cliffs, N.J. (1960).

PART

III

THE FOURIER ANALYSIS

APPROACH

TO PHYSICAL OPTICS

This part presents some of the most important facets of modern optics: diffraction phenomena, the Fourier transform properties of lenses, spatial filtering, and image processing. To some extent it is the culmination of image formation which was developed in the earlier chapters. In Part I, we regarded light as a ray using simple geometry as the basis for our treatment of lenses and lens systems. Now we must go one step further and realize that we ignored the finite wavelength of light. To take this into account requires that we study diffraction and that we consider the equation which governs the propagation of an electromagnetic wave and its solutions. The application of these solutions (which we shall keep as simple as possible) to diffraction effects in turn leads us to the use of the Fourier transform, which has already been encountered in connection with the Michelson interferometer. During the process of showing how a lens acts as an instrument for generating Fourier transforms, we shall see how simple (in principle) it is to alter the result by removing high- or low-frequency components, which at times is the purpose of spatial filtering of the image. We shall also see that the nature of the transform is unavoidably affected by lens aberrations, thus bringing us back to ideas first considered in Chapters 2 and 3. Hence Chapter 10, although long, covers a set of topics which naturally group together. Chapter 11 deals with holography, which may be considered as an application of diffraction and interference.

10

THE FOURIER ANALYSIS
APPROACH TO PHYSICAL OPTICS

10-1 Introduction

We pointed out in an earlier chapter that light is one type of electromagnetic wave. If a wave is radiated in free space, it travels at constant velocity and without alteration until it strikes some sort of detector or absorbing surface. On the other hand, if the wave is partially blocked by an opaque barrier or an aperture in a metal plate, the apparent direction of travel will be altered. As we shall see in the development which follows, this bending of the path is called *diffraction*.

When a plane containing an opening is placed in front of a screen and illuminated, a careful examination of the shadow reveals the following:

1. If the screen is far from the opening, an outline of the light source is seen. The edges of this image are fuzzy, consisting of alternating light and dark bands.

2. If the screen is close to the opening, the outline of this aperture is now seen, again with alternating light and dark bands.

Diffraction is actually a kind of interference. Waves from one part of the opening interfere with waves from another part to produce the diffraction pattern; the maxima and minima which appear in the pattern indicate that interference is taking place. We shall demonstrate that *it is the finite size of an aperture which is responsible for diffraction effects;* thus, an unrestricted wave does not produce a diffraction pattern.

Huyghens was an early proponent of the wave theory of light. In 1678 he explained the occurrence of double refraction in calcite by assuming that light was a wave and proposing a mechanism for wave propagation in crystals. *Huyghens' principle*[10-1],[10-2] states that each point on a wave front acts as a source of new waves. In a given time, the new waves advance a certain distance, and the envelope of all these secondary wavelets (obtained by drawing a tangent surface) yields the new wave front.

In 1818, Fresnel realized that Huyghens' principle could be used not only to describe double refraction in crystals but also to calculate diffraction patterns of apertures. A knowledge of the wave pattern across a slit (or other aperture) and the employment of this principle permit the pattern on a screen behind the slit to be determined.

Our approach to this topic will be to first give an introductory description of diffraction effects, involving Fresnel's ideas expressed in terms of an integral. Then we will note that the diffraction pattern far from an object is the Fourier transform of the object.* Looking at diffraction from this new point of view, we shall see that an image can be filtered so as to enhance certain of its features. It is remarkable that people have been applying the basic diffraction integral for over one hundred years but that only recently has it been realized that it could be put to active use to process optical images.

There are two classes of diffraction that are normally of interest, corresponding to different approximations used in the evaluation of the diffraction integral. The field distribution far from the object (at infinity) is called the *Fraunhofer diffraction pattern* and is mathematically simpler. The field distribution near the object is called the *Fresnel diffraction pattern.* Its calculation, which is quite complicated, is performed using Fresnel integrals.

The Fraunhofer pattern can be observed by placing a screen at a long distance from the object (as an approximation to infinity). Another way the Fraunhofer pattern can be seen is by placing a double convex lens behind the object; the pattern then appears at the focal plane of the lens. All the diffracted rays at a given angle are focused to a point by the lens, which is equivalent to looking at them from infinity. It is important to realize however that the diffraction pattern arises from the aperture, and not the lens; the lens is used only for convenience in observing the pattern.

10-2 Fraunhofer Diffraction

Let us use Huyghens' principle to find the amplitude $A(x', y')$ of a wave at a point P with coordinates x' and y' on a screen (Fig. 10-1), given its

---

*An introductory treatment of Fourier series, integrals, and transforms is given in the Appendix.

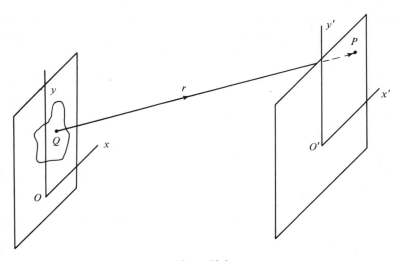

Figure 10-1

amplitude $A(x, y)$ at the point Q with coordinates x and y in an aperture. The screen is located at a long distance from the aperture, so that $r = \overline{QP}$ is large compared to the dimensions of the opening. As shown in connection with Eq. (7-5), the phase change δ at P with respect to Q is

$$\delta = \frac{2\pi}{\lambda} r = kr$$

where the relation between k and λ is given by Eq. (5-2). Because of the large value of r, we can assume that every point (x, y) in the aperture is approximately equidistant from the point (x', y') on the screen, where the complex amplitude will be given by an exponential term, in accordance with Eq. (5-6). Hence, we simply multiply $A(x, y)$ by the area $dS = dx\,dy$ of an elementary section of the aperture and by a phase factor $\exp(i\delta) = \exp(ikr)$ to obtain the expression

$$dA(x', y') = CA(x, y)e^{ikr}\,dx\,dy$$

where the constant C incorporates any necessary dimensional or scale factors. In accordance with Huyghens' principle, the total amplitude on the screen is the sum of all the individual contributions from the aperture, so that

$$A(x', y') = C \int A(x, y)e^{ikr}\,dS \tag{10-1}$$

In accordance with our assumption about the relation between the aperture size and r, we can absorb $A(x, y)$ into C and obtain the simpler expression

$$A = C \int e^{ikr}\,dS \tag{10-2}$$

where A refers to the amplitude at the screen. Equation (10-2) is the *Fraunhofer diffraction equation*. This expression gives the amplitude of the wave; to find the *intensity I* (as in Section 5-2), we take the square of A, for the intensity is a measure of the energy carried by the wave. This is analogous to the situation in electric circuits: we find the power from the square of the voltage or current, either of which specifies the amplitude.

10-3 Fraunhofer Diffraction by a Single Slit

Suppose we have a slit of width w and of length l (Fig. 10-2). Let us stipulate that the slit is narrow, so that $w \ll l$. We choose axes in the customary manner. Consider two parallel rays in the XOZ plane making an angle θ to the Z axis, as shown. At some point on the upper ray at a distance r from the slit, we write

$$r = r_0 + x \sin \theta \qquad (10\text{-}3)$$

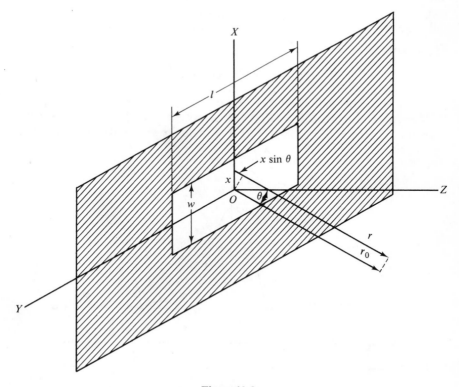

Figure 10-2

where r_0 is the corresponding value of r on the lower ray (the one for which $x = 0$). The element of area in the slit is taken as

$$dS = l\, dx \tag{10-4}$$

Then Eq. (10-2) becomes

$$A = Ce^{ikr_0}l \int_{-w/2}^{w/2} e^{ikx \sin \theta}\, dx$$

Performing the integration gives

$$A = Ce^{ikr_0}l \frac{e^{ikx \sin \theta}}{ik \sin \theta}\bigg|_{-w/2}^{w/2}$$

$$= Ce^{ikr_0}l \frac{\sin \{(kw \sin \theta)/2\}}{k \sin \theta}$$

$$= C'\left(\frac{\sin \alpha}{\alpha}\right) \tag{10-5}$$

where

$$\alpha = \tfrac{1}{2}kw \sin \theta$$

and

$$C' = Ce^{ikr_0}\frac{wl}{2}$$

The intensity I by Eq. (5-15) is

$$I = \frac{1}{2}AA^* = I_0\left(\frac{\sin^2 \alpha}{\alpha^2}\right) \tag{10-6}$$

where $I_0 = \tfrac{1}{2}C'C'^*$ is the value of I for $\alpha = 0$; $S = wl$ is the area of the slit.

The curve of $(\sin \alpha/\alpha)^2 = (I/I_0)$ vs α is plotted in Fig. 10-3. The principal maximum occurs at $\theta = 0$, or $\alpha = 0$, and there are secondary maxima which

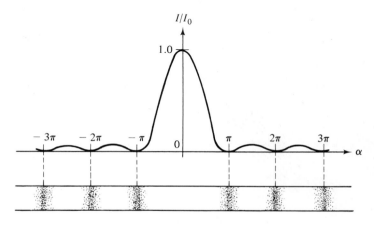

Figure 10-3

can be located by differentiating Eq. (10-6). That is,

$$\frac{dI}{d\alpha} = \frac{2}{I_0}\left(\frac{\sin\alpha}{\alpha}\right)\left(\frac{\alpha\cos\alpha - \sin\alpha}{\alpha^2}\right) = 0$$

or

$$\alpha = \tan\alpha$$

This transcendental equation may be solved numerically or graphically, and we find that the roots are

$$\alpha = 0, 1.43\pi, 2.46\pi, \ldots$$

The minima correspond to $\sin\alpha = 0$ (but $\alpha \neq 0$) which gives

$$\alpha = m\pi, \qquad m = \pm 1, \pm 2, \ldots$$

or

$$\sin\theta = \frac{m\lambda}{w} \tag{10-7}$$

For small angles, this becomes

$$\theta = \frac{m\lambda}{w}$$

and the bright central band has a total angular width of $2\lambda/w$. The diffraction pattern therefore consists of this intense band, with weaker bands on either side (Fig. 10-3). A narrow slit will then produce a wide, but weak, central maximum which shrinks and becomes brighter if w is increased.

For a rectangular or square aperture, we integrate over both x and y and obtain

$$\frac{I}{I_0} = \left(\frac{\sin^2\alpha}{\alpha^2}\right)\left(\frac{\sin^2\beta}{\beta^2}\right) \tag{10-8}$$

where the definition of β is analogous to that of α. The two-dimensional diffraction pattern of a rectangle is therefore a series of orthogonal light and dark bands.

10-4 Fraunhofer Diffraction by a Circular Aperture

Let us change the slit in Section 10-3 to a circular opening of radius R (Fig. 10-4). We take the element of area as a strip parallel to the y axis of width dx and length $2\sqrt{R^2 - x^2}$ (Fig. 10-5). If we use Eq. (10-3) again, Eq. (10-2) becomes

$$A = 2Ce^{ikr_0}\int_{-R/2}^{R/2} e^{ikx\sin\theta}\sqrt{R^2 - x^2}\,dx \tag{10-9}$$

Letting

$$u = \frac{x}{R}$$

and

$$p = kR\sin\theta$$

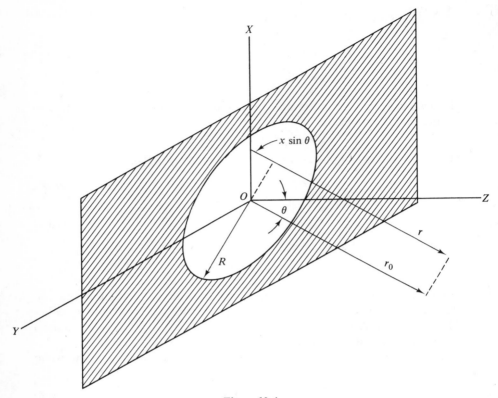

Figure 10-4

Eq. (10-9) is converted to

$$A = 2Ce^{ikr_0} R^2 \int_{-1}^{1} e^{i\rho u}\sqrt{1 - u^2}\, du \qquad (10\text{-}10)$$

The integral in this expression may be written as

$$\int_{-1}^{1} e^{i\rho u}\sqrt{1 - u^2}\, du = \int_{-1}^{1} \cos(\rho u)\sqrt{1 - u^2}\, du + i \int_{-1}^{1} \sin(\rho u)\sqrt{1 - u^2}\, du \qquad (10\text{-}11)$$

The second term on the right vanishes because $\sin(\rho u)\sqrt{1 - u^2}$ is an odd function, and the first term becomes

$$\int_{-1}^{1} \cos(\rho u)\sqrt{1 - u^2}\, du = 2 \int_{0}^{1} \cos(\rho u)\sqrt{1 - u^2}\, du$$

since the integrand is an even function. This new integral can be evaluated numerically, or it may be expressed as a *Bessel function J_1*, defined[10-3] by

$$J_1(z) = \frac{2z}{\pi} \int_{0}^{1} (1 - t^2)^{1/2} \cos(zt)\, dt \qquad (10\text{-}12)$$

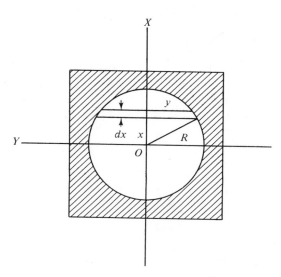

Figure 10-5

Combining this with Eq. (10-10) gives

$$A = \frac{2Ce^{ikr_0}\pi R^2 J_1(\rho)}{\rho} \tag{10-13}$$

Further, when $\theta = 0$, then $\rho = 0$, and Eq. (10-12) shows that

$$\left.\frac{J_1(\rho)}{\rho}\right|_{\rho=0} = \frac{2}{\pi}\int_0^1 (1-u^2)^{1/2}\cos(\rho u)\,du\Big|_{\rho=0} = \frac{2}{\pi}\int_0^1 (1-u^2)^{1/2}\,du$$

$$= \frac{2}{\pi}\left[\sqrt{\frac{1-u^2}{2}} + \frac{1}{2}\arcsin u\right]\Big|_0^1 = \frac{1}{2} \tag{10-14}$$

Since the intensity is $I = \frac{1}{2}A^2$, then the intensity I_0 for $\theta = 0$ is

$$I_0 = \frac{1}{2}(Ce^{ikr_0}S)^2$$

where S is the area of the aperture. By Eq. (10-13)

$$\frac{I}{I_0} = \left[\frac{2J_1(\rho)}{\rho}\right]^2 \tag{10-15}$$

Problem 10-1

(a) Find $J_1(\rho)$ for $\rho = 0, 0.5, 1.0, \ldots, 10.0$ by numerical integration, verifying Fig. 10-6.

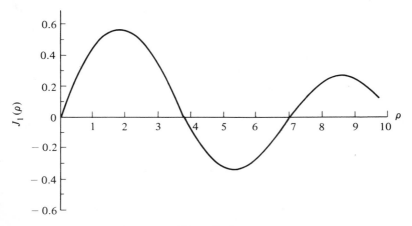

Figure 10-6

(b) Use this result to find $[2J_1(\rho)/\rho]^2$ as part of the same computer program, verifying Fig. 10-6.

(c) Compare Fig. 10-7 with Fig. 10-3.   ■■■

Using the results of Problem 10-1, this expression is plotted in Fig. 10-7, and the alternating light and dark circles corresponding to this intensity pattern are also shown. The bright central area is known as the *Airy disc.* The first dark ring corresponds to the first zero of $J_1(\rho)$, for which $\rho = 3.832$ (Fig. 10-6). Since

$$\sin \theta = \frac{\rho}{kR}$$

we have approximately

$$\theta = \frac{3.83\lambda}{2\pi R} = \frac{1.22\lambda}{D} \tag{10-16}$$

where D is the diameter of the aperture. This formula is of great practical importance because the image formed by a lens in a camera, for example, is actually the Fraunhofer diffraction pattern of each point in the object.

If D is the diameter of the lens (or the entrance pupil), then Eq. (10-16) also gives the minimum angular separation of two points whose Airy discs might hopefully be distinguished in the image plane. This follows from the fact that, if the angular spacing is $1.22\lambda/D$, then the central maximum of one disc corresponds to the first zero of the other one (Fig. 10-8). We realize that this condition is similar to that of Section 7-2. It is called the *Rayleigh criterion*, and depends on the ability of the eye to separate two discs placed so that the center of one is on the poorly defined periphery of another. It is to some extent arbitrary, but represents a reasonable approximation of the eye's image-resolving ability.

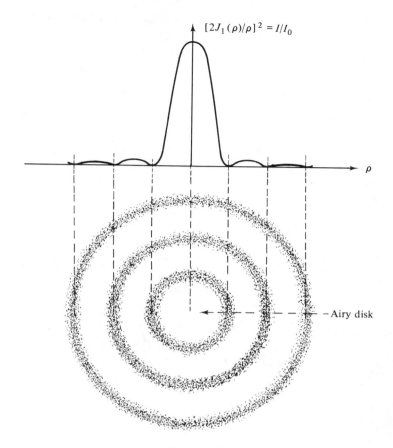

Figure 10-7

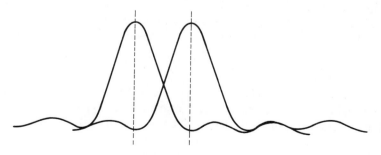

Figure 10-8

Problem 10-2

How big should the diameter of a lens be in order to just resolve the limbs (two points on the ends of a diameter) of the moon? ▰

We should realize at this point that diffraction effects place limitations on image quality, in addition to those due to aberrations. For example, a large double convex lens of radius R and focal length f used as the objective of a telescope will have a significant amount of chromatic aberration. In fact, we have seen in Chapter 4 that the transverse chromatic aberration x_C is given by

$$x_C = -\frac{RV}{2} \tag{4-82}$$

where V is the dispersive power. Let us assume that the image spread due to chromatic aberration and the image spread due to diffraction are additive, and find the relation between R and f which gives the sharpest image.

By (10-16), we obtain the spread x_D in the image due to diffraction from

$$\theta = \frac{x_D}{f} = \frac{1.22\lambda}{D} \quad \text{or} \quad x_D = \frac{1.22\lambda f}{2R}$$

Adding the magnitudes of x_D and x_C and then minimizing this sum by differentiating and equating the result to zero gives

$$\frac{dx}{dR} = -\frac{1.22\lambda f}{2R^2} + \frac{V}{2} = 0$$

or

$$R = \sqrt{\frac{1.22\lambda f}{V}} \tag{10-17}$$

Problem 10-3

The light-gathering power of a lens is called its *speed*. This property should depend on the area of a simple lens or the area of the entrance pupil for a system. It should also depend on the solid angle subtended by the image at the lens. This angle increases as the image comes closer to the lens. The speed is inversely proportional to the magnitude f' of the focal length. Hence, the speed S is defined as the ratio of the diameter to the focal length, or

$$S = \frac{d}{f'} \tag{10-18}$$

It is customary to express speed in terms of *f-numbers*, which are the reciprocals of S. For example, a miniature camera typically has a lens for which $f = 5.0$ cm and $d = 1.8$ cm. Then

$$S = \frac{1.8}{5.0} = 0.36$$

and

$$f\text{-number} = \frac{1}{S} = f : 2.8$$

If the diameter is increased to 2.5 cm, the f-number decreases to $f : 2$, which is a rather fast lens. Note that since the area is proportional to the diameter squared, doubling the f-number raises the speed by a factor of 4.0.

Show that the diameter of the Rayleigh disc for a lens, when expressed in units of 10^{-6} m (one micrometer is 10^{-6} m), approximately equals the f-number. ■

Problem 10-4

Lipson and Lipson[10-4] point out that the involved analysis which produced Eq. (10-16) can be replaced with a simple approximation. Consider a rectangular slit whose length is equal to the diameter of the circular opening in Fig. 10-5. Let the two openings have the same area. Show that the first minimum in the diffraction pattern for the rectangular slit is specified by the relation

$$\theta = \frac{1.27\lambda}{D} \tag{10-19}$$

This represents a deviation of about 5% from the value given by Eq. (10-16). ■

Problem 10-5

Let the rectangular opening of Fig. 10-3 be replaced with an infinite sheet of transparent material of nonuniform density. We shall specify the light-transmission properties along the x direction by the Gaussian function

$$f(x) = Be^{-x^2/b^2} \tag{10-20}$$

where B and b are constants. Show that the diffraction pattern is also Gaussian and, therefore, does not show the secondary maxima of the rectangular and circular apertures. This is a consequence of the fact that a Gaussian opening does not terminate abruptly; its transmission properties decay exponentially on either side of the center, and there is no sharply defined edge. ■

Problem 10-6

A metal plate has two rectangular slits arranged as shown in Fig. 10-9. Verify that

$$\frac{I}{I_0} = \left(\frac{\sin \alpha}{\alpha}\right)^2 \cos^2 \gamma \tag{10-21}$$

where

$$\alpha = \tfrac{1}{2}kw \sin \theta \tag{10-22}$$

$$\gamma = \tfrac{1}{2}kd \sin \theta \tag{10-23}$$

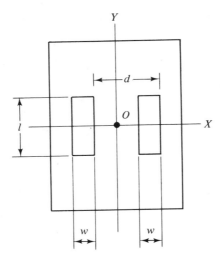

Figure 10-9

Equation (10-13) is the solid curve of Fig. 10-10. This curve may be interpreted as the function $[(\sin \alpha)/\alpha]^2$ of Eq. (10-6) combined with $\cos^2 \gamma$. That is, the behavior for a single slit is the envelope in Fig. 10-10, and the structure in the curve is contributed by the interaction of the two slits. ∎

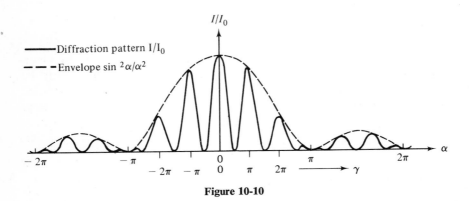

Figure 10-10

10-5 Apodization

The diffraction patterns of Figs. 10-3 and 10-7 show that most of the energy passing through an opening appears in the central portions of the image rather than in the secondary peaks. In situations for which the total usable energy is critical, such as observations involving very faint stars, it is desirable to increase the energy in the central section at the expense of the secondary peaks. This can be accomplished if the transmission properties of the opening are changed, somewhat along the lines indicated by Problem 10-5. The process of redistributing the energy is called *apodization*, from the Greek word meaning "without feet," and it refers to the reduction of the secondary maxima. As an example, the rectangular slit of Fig. 10-2 can be covered with a glass plate containing a coating of variable density and arranged so that the transmission of light is the greatest at the center (i.e., along the y axis) and falls to zero at the edges in accordance with the relation

$$A(x) = \cos\left(\frac{\pi x}{w}\right) \tag{10-24}$$

where $A(x)$ is the *aperture or pupil function*. It specifies the ratio of the transmitted amplitude to the incident amplitude. As Eq. (10-24) indicates

$$A = \begin{cases} 1 & \text{at} \quad x = 0 \\ 0 & \text{at} \quad x = \pm\dfrac{w}{2} \end{cases}$$

Equation (10-2) for this situation then becomes

$$A = Ce^{ikr_0}l \int_{-w/2}^{w/2} \cos\left(\frac{\pi x}{w}\right)e^{ikx \sin \theta}\, dx \tag{10-25}$$

Problem 10-7

(a) Evaluate the integral in Eq. (10-25) to obtain

$$\frac{I}{I_{0A}} = \left(\frac{\pi^2 \cos \alpha}{\pi^2 - 4\alpha^2}\right)^2 \tag{10-26}$$

where I_{0A} is the value of the intensity for the apodized slit when $\alpha = 0$.
(b) Show that

$$I_{0A} = 1.625 I_0 \tag{10-27}$$

where I_0 refers to the original rectangular slit.
(c) Verify the diffraction pattern of Fig. 10-11 with a numerical calculation. ∎

As Fig. 10-11 shows, the use of the absorption plate has put more energy into the central maximum and reduced the height of the secondary maxima. This agrees with what Problem 10-5 indicates: reducing the sharpness of the transition from the transparent to the opaque region has caused a redistribution. Note, however, that only an exponential transition, which requires an infinite distance to go from $A(x) = 1$ to $A(x) = 0$, will put all the energy into the central maximum.

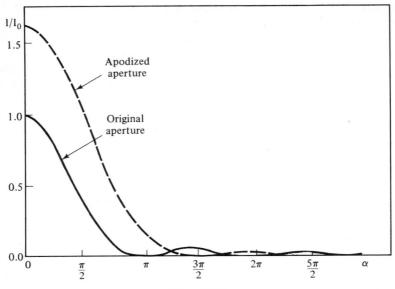

Figure 10-11

10-6   Fresnel Diffraction

Waves whose behavior may be expressed by the relation

$$\mathbf{A} = \mathbf{A}_{\max} e^{\pm i(\mathbf{k} \cdot \mathbf{r} - \omega t)} \tag{5-7}$$

were described as plane waves in Chapter 5. When a wave is generated by a point source and is observed close to the source, it is a *spherical wave* rather than a plane wave and the spatial dependence of its amplitude varies as $\{\exp(ikr)\}/r$. This makes the intensity fall off as $1/r^2$, where r is the distance from the source, agreeing with the results of our study of illumination in Chapter 4. Such a wave results in *Fresnel diffraction*, which occurs whenever the source or the observation point is close to the diffracting body. As an example, let us consider a plane wave passing through a slit. With the aid of Fig. 10-2 we find that the diffraction pattern is a consequence of the phase shift $(x \sin \theta)$ between the two rays shown. When we move the observation point close to the source S as shown in Fig. 10-12, the phase shift must be measured by the path difference between the central ray SOS' and the ray STS' at a distance R from the axis. The length of this latter ray is

$$r + r' = \sqrt{h^2 + R^2} + \sqrt{h'^2 + R^2} \tag{10-28}$$

If R is small compared to h and h' (S or S' is close to O, but not so close that

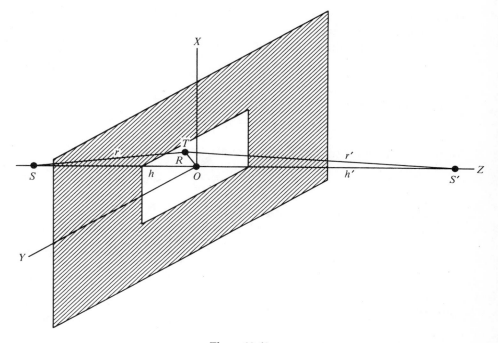

Figure 10-12

R seems large), we may write the first term on the right as

$$\sqrt{h^2 + R^2} = h\sqrt{1 + \left(\frac{R}{h}\right)^2} \sim h\left(1 + \frac{R^2}{2h^2}\right)$$

If we use a similar expression for the second radical, Eq. (10-28) becomes

$$r + r' = (h + h') + \frac{R^2}{2}\left(\frac{1}{h} + \frac{1}{h'}\right)$$

Letting

$$x^2 + y^2 = R^2$$

and defining a length L by

$$\frac{1}{L} = \frac{1}{h} + \frac{1}{h'}$$

we obtain

$$r + r' = (h + h') + \frac{1}{2L}(x^2 + y^2)$$

The integral Eq. (10-2), with a path length of $(r + r')$, becomes

$$A = Ce^{ik(h+h')} \iint e^{ik(x^2+y^2)/2L}\, dS \qquad (10\text{-}29)$$

The change in the factor $1/(r + r')$ is small compared to the exponential term and can be neglected.

If we introduce the dimensionless variables

$$u = x\sqrt{\frac{k}{\pi L}}, \qquad v = y\sqrt{\frac{k}{\pi L}} \tag{10-30}$$

Eq. (10-29) becomes

$$A = D \int e^{i\pi u^2/2}\, du \int e^{i\pi v^2/2}\, dv \tag{10-31}$$

where D is a new constant. These integrals are converted by the identity

$$\int_0^w e^{i\pi w^2/2}\, dw = \int_0^w \cos\left(\frac{\pi w^2}{2}\right) dw + i \int_0^w \sin\left(\frac{\pi w^2}{2}\right) dw \tag{10-32}$$

The two terms on the right, designated as

$$\left.\begin{aligned}
C(w) &= \int_0^w \cos\left(\frac{\pi w^2}{2}\right) dw \\
S(w) &= \int_0^w \sin\left(\frac{\pi w^2}{2}\right) dw
\end{aligned}\right\} \tag{10-33}$$

are known as the *Fresnel integrals*, and may be evaluated numerically (See Problem 10-8a). If we plot $S(w)$ vs $iC(w)$, we obtain the *Cornu spiral* shown in Fig. 10-13.

To apply these results to the case of the rectangular slit, we must evaluate Eq. (10-29) over the area enclosed by the aperture. Consider a rectangular opening whose edges are determined by the coordinates x_1, x_2, y_1, and y_2. Then the limits of integration are u_1, u_2 and v_1, v_2 as obtained from Eq. (10-30). In the limiting case of no diffraction at all, we have $u_1 = -\infty$, $u_2 = \infty$ and $v_1 = -\infty$, $v_2 = \infty$, i.e., the opening has been removed. From Fig. 10-13:

$$C(\infty) = S(\infty) = 0.50$$
$$C(-\infty) = S(-\infty) = -0.50$$

Hence

$$\begin{aligned}
A &= D \int_{-\infty}^{\infty} e^{i\pi u^2/2}\, du \int_{-\infty}^{\infty} e^{i\pi v^2/2}\, dv \\
&= D[C(u) + iS(u)]\,|_{-\infty}^{\infty}\, [C(v) + iS(v)]\,|_{-\infty}^{\infty} \\
&= D[(1 + i)(1 + i)] \\
&= D(1 + i)^2
\end{aligned}$$

and

$$I_0 = AA^* = 4D^2 \tag{10-34}$$

where I_0 is the intensity for no diffraction and A^* is the complex conjugate of A.

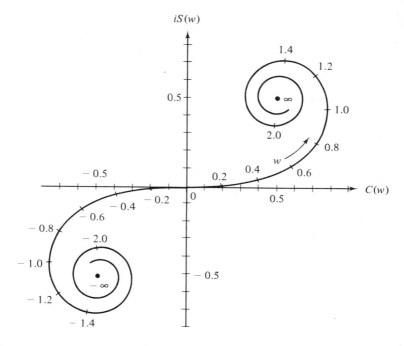

Figure 10-13

For the case of a long slit parallel to the x axis, we keep $u_1 = -\infty$ and $u_2 = \infty$, but let v_1 and v_2 specify the edges of the slit. Finally, we consider a straightedge located at $v_2 = v$, and let $v_1 = -\infty$. Then

$$A = D \int_{-\infty}^{\infty} e^{i\pi u^2/2} \, du \int_{-\infty}^{v} e^{i\pi v^2/2} \, dv$$

$$= D(1 + i)[C(v) + iS(v) + 0.5 + 0.5i]$$

For an observation point S' on the z axis, the value $v = 0$ means that the source S, the diffracting edge, and S' all lie on a straight line, that is, S' lies right on the boundary of the shadow. Using

$$C(0) = S(0) = 0.0$$

gives

$$A = D(1 + i)[0.5(1 + i)]$$

$$= 0.5A_0$$

where A_0 is the unobstructed amplitude, and

$$I = 0.25I_0 \tag{10-35}$$

Moving the edge so that v becomes positive is equivalent to looking at an area on the screen which is fully illuminated. To find the exact values of

$C(v)$ and $S(v)$ corresponding to this position, we can use the computer output, but if we are satisfied with an estimate, we can compute the definite integrals graphically. For example, if we wish to find the integral

$$\int_{-\infty}^{+1.0} e^{i\pi v^2/2}\, dv = \int_{-\infty}^{1.0} \left[\cos\left(\frac{\pi v^2}{2}\right) + i\sin\left(\frac{\pi v^2}{2}\right) \right] dv$$

we draw a vector on the Cornu spiral (Fig. 10-13) from $w = -\infty$ to $w = 1.0$. The projection $\mathfrak{R}$ on the real axis equals the real part of the integral above, since integrating from $-\infty$ to 0 and from 0 to 1.0 is equivalent to the limits of $-\infty$ to 1.0 and the imaginary part $\mathfrak{I}$ is obtained similarly. This process is indicated in Fig. 10-14. The intensity obtained in this way for values of v

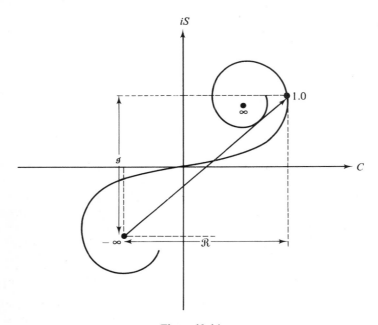

Figure 10-14

corresponding to both the illuminated region ($v > 0$) and the shadow region ($v < 0$) is shown in Fig. 10-15. Note that the diffraction causes a small but finite amount of light to appear in the shadow region.

Problem 10-8

(a) Write a program for evaluating the integrals in Eq. (10-33), verifying Fig. 10-13.

(b) A distant FM transmitter operating at 100 megahertz, a water tower of 10-m diameter, and a radio receiver lie in a straight line. The tower and

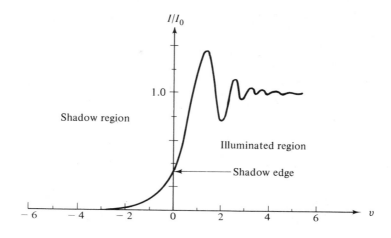

Figure 10-15

receiver are 1000 m apart. Determine the effect of the obstacle on the received signal in an antenna normal to the line joining the transmitter and receiver.

Let us make clear again the distinction between Fraunhofer diffraction and Fresnel diffraction. In Fig. 10-2, we considered two parallel rays passing through the slit and we calculated the diffraction pattern *far* from the slit, using the approximation expressed by equation (10-3) and also assuming that the aperture subtended a very small angle at the point of observation. The path difference depended on $x \sin \theta$, that is, it is a linear function of x. *Fraunhofer diffraction corresponds to a plane wave front approaching the object and to a linear exponent in the diffraction integral.* In Fig. 10-12, on the other hand, we dealt with a point source S and a point of observation S', either or both of which were *close* to the aperture. We can think of this as corresponding to a spherical wave. At these distances, $r + r'$ differs from $h + h'$, by an amount $(x^2 + y^2)/2L$. The exponent in the diffraction integral in Eq. (10-29) is now quadratic in x (and y) rather than linear, and the integrals themselves must be evaluated numerically. Hence, Fraunhofer diffrac-

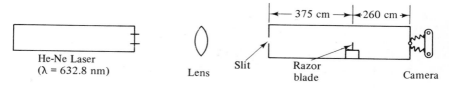

Figure 10-16

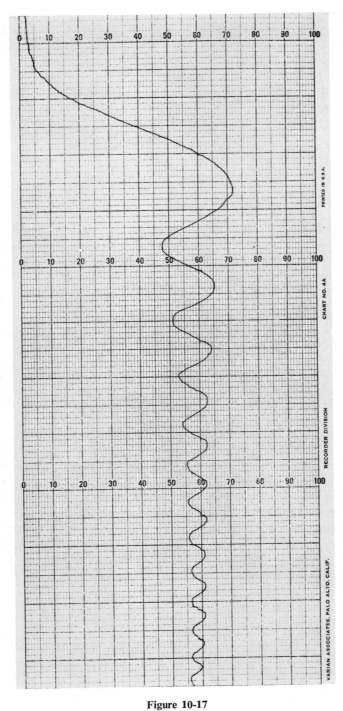

Figure 10-17

234

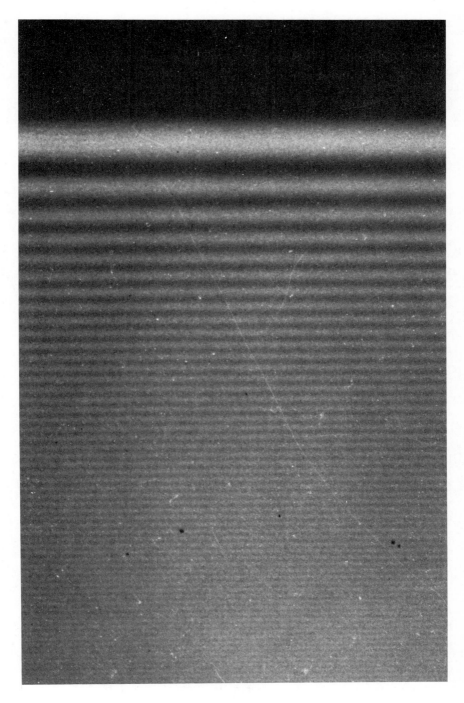

Figure 10-18

tion is observed when both the source and observation point are at long distances from an aperture. (A spherical wave of large radius is approximately equivalent to a plane wave) and Fresnel diffraction is observed when either is a short distance from an aperture.

As an example of the latter, consider the experiment of George[10-5] shown in Fig. 10-16. The lens converts the laser output into a spherical wave, and we are able to see the Fresnel pattern of the edge of the razor blade at a distance which is enormous in comparison to the wavelength ($\lambda = 632.8 \times 10^{-9}$ m) simply because the laser output retains its wave shape almost indefinitely. The intensity of the pattern produced on the film as a function of distance is measured with an instrument called a densitometer. The results shown in Fig. 10-17 agree with the theoretical curve (Fig. 10-15). A photograph of the illuminated region (Fig. 10-18) shows the light and dark bands which correspond to the densitometer curve. A slightly different type of Fresnel experiment was done by Boyer and Fortin.[10-7] They focused a laser with a microscope objective and observed the diffraction pattern created by a wire 0.1 mm in diameter at a distance of 3 m. Their measurements agreed with the theoretical calculation using the Fresnel integrals (rather than the Cornu spiral) to within 4%. An equally simple way of photographing Fraunhofer patterns, using a sodium light source and a camera with a 40-cm telephoto lens, is described by Graham;[10-6] his results are found to agree with Figs. 10-3 and 10-10.

10-7 More Exact Diffraction Theories

The treatment of diffraction given so far has been based on Huyghens' principle and the results we obtained agree with experimental observation. It has long been realized, however, that this theory is lacking in two respects. First, the phase of the reradiated wavelets had to be shifted by $\pi/2$ to give results which were correct under all circumstances, and second, it was predicted that the secondary wavelets would generate the same diffraction pattern in the backward as in the forward direction, an effect which does not occur. If we arbitrarily include an *obliquity factor* $(1 + \cos \theta)$ in Eq. (10-2), where θ is the angle between the direction of propagation of the original wave and the direction of the diffracted wave, then the backward pattern is completely eliminated, for the obliquity factor causes the diffraction integral to vanish when $\theta = \pi$ or $\cos \theta = -1$. Further, it indicates that there is a decrease of amplitude with angle in the forward direction, since $(1 + \cos \theta)$ has its maximum value for $\theta = 0$. Thus, the form of this obliquity factor appears to be very logical.

Kirchhoff in 1882 derived a diffraction integral somewhat like Eq. (10-2), but containing the correct phase and angular terms, by starting with the wave equation, Eq. (5-8). We need not consider his derivation here but simply mention that it involves the use of well-known vector identities (Green's theorems in particular). An alternate approach, involving somewhat more complicated situations, was given by Sommerfeld[10-8]. Neither of these theories, however, takes into account the fact that light is a form of electromagnetic radiation which interacts with material media via its electric and magnetic vectors or fields and that polarization of the light must also be considered. Experiments with microwaves indicated that there are some results which scalar diffraction theory cannot explain and which require the polarization to be taken into account. For details on both the scalar and the vector theory of diffraction, we refer the reader to the standard treatise of Born and Wolf.[5-1]

10-8 Fourier Analysis and Fourier Optics

10-8a Basic Principles

It is generally realized that light waves and radio waves are both forms of electromagnetic radiation; they both can be described by solutions to the wave equation, Eq. (5-8), and possess many properties in common. It is not generally realized, however, that there is a close analogy between the behavior of alternating currents in circuits containing inductance or capacitance and the behavior of light as it passes through an optical system. This parallel situation can be developed on the basis of Fourier analysis, as discussed in the Appendix. We shall briefly review here the aspects of this subject which help clarify the analogy to be developed.

Let us start with the periodic pulse of unit amplitude and wavelength 2π illustrated in Fig. A-3. Equation (A-8), written as

$$y = 0.5 + \frac{2}{\pi} \sum_{n=0}^{\infty} \frac{(-1)^n}{(2n+1)} \cos(2n+1)x \qquad (10\text{-}36)$$

indicates that this repeating square wave is the sum of a series of cosine terms, each of successively increasing propagation constant n and decreasing amplitude. The decomposition of a train of square waves into its pure sinusoidal components is called a *Fourier analysis* or *spectral decomposition* of the original wave. We can completely specify the nature of the square wave by simply listing the frequencies and amplitudes of the sinusoidal components. This information can be determined electrically, in principle, by passing the wave through a tunable band-pass filter so that each component can be detected and measured separately. By using a rejection filter, we can remove specific components and alter the shape of the wave train.

If we have only a single pulse (Fig. A-1), then the Fourier series in Eq. (10-36) is not valid, and we must express the Fourier decomposition in terms of an integral. That is, instead of summing over a discretely spaced set of components of the form $\cos x$, $-\frac{1}{3}\cos 3x$, $\frac{1}{5}\cos 5x, \ldots$, we use a set of functions whose frequencies are infinitesimally close, and the single pulse is then written as a Fourier integral. To be specific, let the symmetric pulse have unit height and width d, rather than π, so that the definition of this function is

$$f(x) = \begin{cases} 1 & \text{for} \quad -\frac{d}{2} < x < \frac{d}{2} \\ 0 & \text{for} \quad |x| > \frac{d}{2} \end{cases} \tag{10-37}$$

and its Fourier transform is

$$F(k) = \int_{-d/2}^{d/2} e^{-ikx}\, dx = \frac{1}{-ik} e^{-ikx} \Big|_{-d/2}^{d/2}$$

$$= \frac{2}{k} \sin\left(\frac{kd}{2}\right) = d\frac{\sin(kd/2)}{(kd/2)}$$

$$= d\frac{\sin(\pi d/\lambda)}{(\pi d/\lambda)} \tag{10-38}$$

The resulting function of (d/λ) is known as the *sinc function* and is defined by

$$\text{sinc}\,(x) = \frac{\sin \pi x}{\pi x} \tag{10-39}$$

as already plotted in Fig. A-13(c). We have previously encountered it in connection with Fraunhofer diffraction by a slit, the amplitude being

$$A = C'\,\text{sinc}\,\alpha \tag{10-5}$$

where

$$\alpha = \frac{\pi w \sin \theta}{\lambda}$$

Hence, we expect the amplitude of the diffraction pattern to be the Fourier transform of the aperture function $A(x, y)$, and this we shall see is the case. A slit of finite width causes a plane wave to interfere with itself in such a way that it generates the Fourier spectrum of the aperture function. If the slit is simply an opening, we are finding the transform of $f(x)$ as given by Eq. (10-37), but if there is apodization or phase shift, then the Fourier transform corresponds to a more complicated aperture function.

The diffraction integral that we use Eq. (10-1) can also be written as

$$A(x', y') = C \iint A(x, y)e^{-ikr}\, dx\, dy$$

without changing the resulting intensity. For a rectangular opening, for example, we have

$$A(x', y') = C \int_{-\infty}^{\infty} \int_{-\infty}^{\infty} A(x, y)e^{-ik(x+y)}\, dx\, dy \tag{10-40}$$

As shown in the Appendix, Eq. (A-24), the Fourier transform of the function $f(x, y)$ is

$$\mathcal{F}[f(x, y)] = F(k_x, k_y) = \int\limits_{-\infty}^{\infty} \int\limits_{-\infty}^{\infty} f(x, y)e^{-i(k_x x + k_y y)} \, dx \, dy \qquad (10\text{-}41)$$

Hence, Eq. (10-40) will have this form if we let $k_x = k_y = k$ and if $C = 1$. Since the constant does not usually have the value unity, we should say that the intensity of the Fraunhofer diffraction pattern is proportional (rather than identical) to the Fourier transform of the aperture function. Thus, *diffraction phenomena provide a method for optically computing Fourier transforms.*

Problem 10-9

(a) Justify the limits specified in Eq. (10-40).

(b) Eq. (10-40) expresses the amplitude $A(x', y')$ of the diffraction pattern at a point (x', y') on the screen in terms of an integral involving the amplitude $A(x, y)$ over the aperture. On the other hand, the Fourier transform definition, Eq. (10-41), implies that the amplitude of the diffracted wave at the receiving screen should be a function of k. Show that the calculation of Section 10-3 permits us to write $A(x', y')$ as $A(k)$, thus matching the definition. We shall show immediately below that when a lens is part of a diffraction system, both k and the coordinates x', y' are explicitly involved.   ■

When we consider the effect of a lens on a plane wave, the situation becomes more complicated. Suppose a ray of light leaves a lens at a point $B(x, y)$ and strikes the focal plane at a point $B'(x', y')$. To simplify the geometrical calculation which follows, assume for the moment that we are dealing with a meridional problem. Let the point B be in the unit plane of a simple lens (Fig. 10-19). The path difference Δr between a ray with a slope

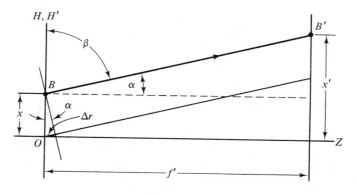

Figure 10-19

α leaving B and another ray from the origin is related to α through the expression

$$\sin \alpha = \frac{\Delta r}{x}$$

or

$$\Delta r = x \cos \beta$$

If x is small, then

$$\tan \alpha = \frac{x' - x}{f'} \sim \frac{x'}{f'}$$

But

$$\tan \alpha \sim \sin \alpha = \cos \beta$$

hence,

$$\Delta r = \frac{xx'}{f'} \tag{10-42}$$

Generalizing to skew rays, Eq. (10-42) will have the form

$$\Delta r = \frac{xx' + yy'}{f'} \tag{10-43}$$

The diffraction integral in Eq. (10-40) then becomes

$$A(x', y') = C \iint P(x, y)e^{-ik(xx'+yy')/f'} \, dx \, dy \tag{10-44}$$

where we have introduced a *pupil function* $P(x, y)$ to explicitly indicate that the lens affects the amplitude of the light passing through it.

Let us define the *spatial frequencies* v_x and v_y associated with x' and y', respectively, by

$$\left. \begin{array}{l} v_x = \dfrac{kx'}{f'} = \dfrac{2\pi x'}{\lambda f'} \\[2mm] v_y = \dfrac{ky'}{f'} = \dfrac{2\pi y'}{\lambda f'} \end{array} \right\} \tag{10-45}$$

so that

$$A(v_x, v_y) = C \iint P(x, y)e^{-i(v_z x + v_y y)} \, dx \, dy \tag{10-46}$$

By explicitly showing the functional dependence of A on v_x and v_y, we see that *the amplitude of the diffracted wave is the Fourier transform of the pupil function*. Hence, if a lens possessed only the paraxial focusing properties expressed in Eq. (10-43), it would serve as a device which takes a uniform plane wave and uses it to form an image in the focal plane whose amplitude is the two-dimensional Fourier transform of $P(x, y)$. The variables v_x and v_y are called spatial frequencies because they play a role analogous to that of the angular frequency ω in the Fourier transform definition

$$F(\omega) = \int_{-\infty}^{\infty} f(t)e^{-i\omega t} \, dt \tag{A-28}$$

Although ω has units of radians/second and $\nu = \omega/2\pi$ has units of cycles/second or hertz, the quantities ν_x and ν_y are expressed in terms of cycles/meter, as Eqs. (10-45) indicate.

10-8b   An Example of a Phase Object

A simple example which illustrates the Fourier transform interpretation of Eq. (10-40) or Eq. (10-46) is the effect on the diffraction pattern of a phase change in the incident wave. We saw in Chapter 8 that two adjacent waves emerging from a thin film of thickness t and index n will have a phase difference δ given by

$$\delta = \left(\frac{2\pi n}{\lambda}\right) 2t \cos\theta \tag{8-3}$$

where θ is the angle of incidence or reflection. If the beam is normal ($\theta = 0$), then $\delta = 4\pi n t/\lambda$ and for a 180° phase shift (or an integral multiple of this value), the thickness is specified by the relation

$$\frac{(2m + 1)\lambda}{4} = nt$$

where $m = 0, 1, 2, \ldots$. Let us use a square aperture of edge d at the entrance pupil of a lens and cover the central part of it with a film of thickness $t = \lambda/4n$, which produces a 180° phase shift in the incoming wave (Fig. 10-20(a)). Then the pupil function is specified as (Fig. 10-20(b))

$$P(x, y) = \begin{cases} -1 & \text{when} \quad -\dfrac{d}{4} \leq x, y \leq \dfrac{d}{4} \\[2ex] +1 & \text{when} \quad \begin{cases} -\dfrac{d}{2} \leq x, y \leq -\dfrac{d}{4} \\[2ex] \dfrac{d}{4} \leq x, y \leq \dfrac{d}{2} \end{cases} \end{cases} \tag{10-47}$$

Evaluating Eq. (10-46) for a single variable x or y with $P(x, y)$ as given by Eq. (10-47), we obtain an intensity which varies with distance along either the x' or y' directions in the image plane in accordance with the relation

$$I = C\left[\frac{\sin u}{u} - \frac{\sin(u/2)}{u/2}\right]^2 \tag{10-48}$$

where $u = \nu_x d/2$ or $\nu_y d/2$. Since I vanishes for $u = 0$, the constant cannot be set equal to I_0; instead, we arbitrarily let $C = 1.0$ at the first maximum I_1 in I, as shown in Fig. 10-20(c). The effect of placing a phase-shifting film over the central portion of the aperture is thus opposite to apodization; energy in the diffraction pattern is moved away from the central maximum.

Problem 10-10

Verify Eq. (10-48). ■

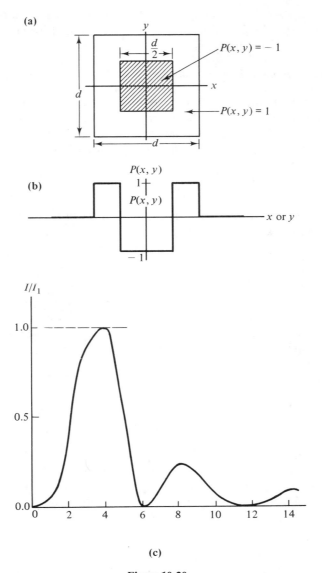

Figure 10-20

It is easy to see that if we had evaluated the diffraction integral Eq. (10-40) for the Fraunhofer pattern of an opening whose aperture function is of the form of Eq. (10-47), we would have obtained a result like Eq. (10-48). That is, the presence or absence of the lens does not change the fact that the 180° phase shift in the central thin film causes the energy to be shifted towards the edges of the diffraction pattern.

10-8c The Imaging Properties of a Lens—The Field Approach

The aim of this section is to derive the mathematical expression which describes how a lens forms an image given the amplitude distribution, $A(x, y)$, of an object. In other words, we will calculate the field behind a lens given the field of an object. This will involve some tedious mathematical manipulations. Unfortunately, they are required for a rigorous development of the theory and a sound understanding of how to apply the theory to real-world situations.

The use of a thin film is not the only way in which an optical system can produce a phase difference; the varying thickness of the lens from axis to margin also causes such a shift. To compute this quantity, consider the asymmetric lens of Fig. 10-21, which shows a ray of light traveling a slant distance T_1' from surface 1 to surface 2, with an inclination angle α_1'. If α_1' is small, then

$$T_1' = t_1' - \overline{V_1 A_1} - \overline{A_2 V_2}$$

or

$$T_1' = t_1' - \{r_1 - \sqrt{r_1^2 - (x_1^2 + y_1^2)}\} - \{-r_2 - \sqrt{r_2^2 - (x_2^2 + y_2^2)}\}$$

$$= t_1' - r_1\left\{1 - \sqrt{1 - \frac{x_1^2 + y_1^2}{r_1^2}}\right\} + r_2\left\{1 - \sqrt{1 - \frac{x_2^2 + y_2^2}{r_2^2}}\right\} \quad (10\text{-}49)$$

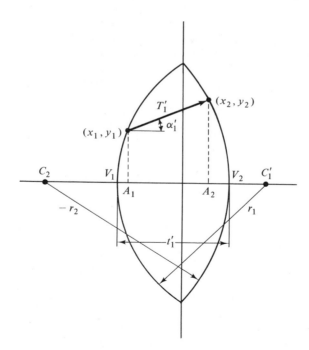

Figure 10-21

Note that the above approximation is not as severe as the paraxial one, where we assumed that $T_1' \sim t_1'$. We may also use the relations

$$x_1 \sim x_2 = x$$
$$y_1 \sim y_2 = y$$
(10-50)

and a binomial approximation to obtain

$$T_1' = t_1' - r_1 \left\{ \frac{x_1^2 + y_1^2}{2r_1^2} \right\} + r_2 \left\{ \frac{x_2^2 + y_2^2}{2r_2^2} \right\}$$

$$= t_1' - \left(\frac{1}{r_1} - \frac{1}{r_2} \right) \left(\frac{x^2 + y^2}{2} \right)$$
(10-51)

Recall from Chapter 1 that the focal length of a thin lens is given by the relation

$$\frac{1}{f'} = (n_1' - 1) \left[\frac{1}{r_1} - \frac{1}{r_2} \right]$$
(1-63)

Hence, Eq. (10-51) becomes

$$T_1' = t_1' - \frac{(x^2 + y^2)}{2f'(n_1' - 1)}$$
(10-52)

To see how this determines the effect of the lens on the wave passing through it, consider a plane wave striking vertex V_1 (Fig. 10-22). There is a variable phase delay as the wave travels through the lens; the greater the thickness of glass it transverses, the larger the delay. The phase delay inside the lens therefore depends on the index of refraction and on T_1', whereas outside it depends only on the distance $(t_1' - T_1')$ between the wave front entering at V_1 or emerging at V_2 and the glass surfaces. Hence, the plane wave

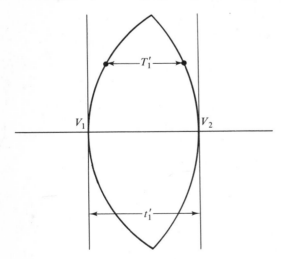

Figure 10-22

has the form $\exp\{ik[n'_1 T'_1 + (t'_1 - T'_1)]\}$, and by Eq. (10-52)

$$\exp[ik\{n'_1 T'_1 + (t'_1 - T'_1)\}] = \exp[ik\{(n'_1 - 1)T'_1 + t'_1\}]$$

$$= e^{in_1'kt_1'}e^{-ik(x^2+y^2)/2f'} \tag{10-53}$$

This result consists of two factors: a constant term which we may drop, since we are concerned with relative rather than absolute intensities, and another term involving $x^2 + y^2$. The amplitude of the wave as it travels through an optical system will be proportional to $\exp(ikr)$ and the variable term in (10-53) will modify this to $\exp[i(kr + \delta)]$, where δ is the phase shift introduced by the lens. Hence, the amplitude of the wave emerging at V_2 is related to that of the incident wave at V_1 by the expression

$$A(x_2, y_2) = A(x_1, y_1)e^{-ik(x_1^2 + y_1^2)/2f'} \tag{10-54}$$

If an apodizing or phase shifting device is present in the entrance pupil of the lens, then a pupil function $P(x_1, y_1)$ as introduced in Eq. (10-44) should be included above, but for simplicity, we shall consider an unmodified lens.

We are now ready to describe the behavior of a wave as it passes from object to lens and through the lens to the image plane, using a method due to Smith[10-9]. The reader is warned at this point that the algebra which follows is formidable and there appears to be no simpler way of obtaining the extremely important conclusions we shall reach. It is convenient to work in the reverse direction; that is, we shall first regard the wave front which passes through the exit pupil of the lens as a Huygens source whose amplitude is $A(x_2, y_2)$ and from it, we compute the amplitude $A(x', y')$ at the image plane. The diffraction integral (10-1) then becomes

$$A(x', y') = C \int A(x_2, y_2)e^{ikr} dS_2 \tag{10-55}$$

The slant distance T'_2 between the point (x_2, y_2) and the point (x', y'), for which the spacing in the z direction is t'_2, is

$$T'_2 = \sqrt{(x' - x_2)^2 + (y' - y_2^2) + t_2'^2}$$

$$= t'_2\sqrt{1 + \left(\frac{x' - x_2}{t'_2}\right)^2 + \left(\frac{y' - y_2}{t'_2}\right)^2}$$

and a binomial expansion gives

$$T'_2 = t'_2\left[1 + \frac{1}{2}\left(\frac{x' - x_2}{t'_2}\right)^2 + \frac{1}{2}\left(\frac{y' - y_2}{t'_2}\right)^2\right] \tag{10-56}$$

Since $T'_2 = r$ for this situation, Eq. (10-55) becomes

$$A(x', y') = C \int A(x_2, y_2)e^{ik[(x'-x_2)^2 + (y'-y_2)^2]/2t_2'} dS_2 \tag{10-57}$$

where the term $e^{ikt_2'}$ has been incorporated into the constant C and the notation dS_2 means that we are integrating over the exit pupil.

Next, we incorporate the phase delay in passing through the lens, as expressed by Eq. (10-54). Equation (10-57) is then

$$A(x', y') = C \int e^{-ik(x_1{}^2 + y_1{}^2)/2f'} A(x_1, y_1) e^{ik[(x'-x_2)^2 + (y'-y_2)^2]/2t_2'} \, dS_2$$

(10-58)

Finally, we use a relation like Eq. (10-57) to connect the amplitude $A(x, y)$ at the object with the amplitude $A(x_1, y_1)$ at the entrance pupil, obtaining the expression

$$A(x', y') = C \iint A(x, y) e^{-ik(x_1{}^2 + y_1{}^2)/2f'}$$
$$\times \, e^{ik[(x_1-x)^2 + (y_1-y)^2]/2t_1} e^{ik[(x'-x_2)^2 + (y'-y_2)^2]/2t_2'} \, dS_2 \, dS \quad (10\text{-}59)$$

where dS refers to an integration in the plane of the object.

For a reasonably thin lens, we can assume that x_1 is approximately equal to x_2 and similarly for y_1 and y_2; thus we express the coordinates at the lens as

$$x_1 = x_2 = x_L$$
$$y_1 = y_2 = y_L$$

Considering just the x terms in Eq. (10-59), we then have

$$e^{-ikx_L{}^2/2f'} e^{ik(x_L-x)^2/2t_1} e^{ik(x'-x_L)^2/2t_2'}$$
$$= e^{i(kx_L{}^2/2)[(1/t_1) + (1/t_2') - (1/f')]} e^{-ikx_L[(x/t_1) + (x'/t_2')]} e^{ikx'^2/2t_2'} e^{ikx^2/2t_1} \quad (10\text{-}60)$$

This is integrated with respect to the variable dx_L, where

$$dS_2 = dS_L = dx_L \, dy_L$$

and the limits can be taken as $-\infty, \infty$, since there is no contribution to the integral beyond the edges of the lens. Using the standard definite integrals[10-10]

$$\int_0^\infty \begin{Bmatrix} \sin \\ \cos \end{Bmatrix} (ax^2 + 2bx) \, dx = \frac{1}{2} \left(\cos \frac{b^2}{a} \mp \sin \frac{b^2}{a} \right) \sqrt{\frac{\pi}{2a}}$$

we obtain

$$\int_{-\infty}^\infty e^{i(ax^2 + 2bx)} \, dx = (1 + i) \sqrt{\frac{\pi}{2a}} e^{-ib^2/a} \quad (10\text{-}61)$$

and this in turn gives

$$e^{ikx'^2/2t_2'} \int_{-\infty}^\infty \left[\int_{-\infty}^\infty e^{i(kx_L{}^2/2)[(1/t_1) + (1/t_2') - (1/f')] - ikx_L[(x/t_1) + (x'/t_2')]} \, dx_L \right] e^{ikx^2/2t_1} \, dx$$

$$= e^{ikx'^2/2t_2'} \int_{-\infty}^\infty \left[(1 + i) \sqrt{\frac{\pi}{ik[(1/t_1) + (1/t_2') - (1/f')]}} \right.$$
$$\left. \times \, e^{-(ik/2)[(x/t_1) + (x'/t_2')]^2/[(1/t_1) + (1/t_2') - (1/f')]} e^{ikx^2/2t_1} \right] dx$$

$$= C e^{(ikx'^2/2t_2')[1 - [t_1 f'/(t_2' f' + t_1 f' - t_1 t_2')]]}$$

$$\times \int_{-\infty}^\infty e^{(ikx^2/2t_1)[1 - [t_2' f'/(t_2' f' + t_1 f' - t_1 t_2')]] - ik[xx'f'/(t_2' f_2' + t_1 f' - t_1 t_2')]} \, dx \quad (10\text{-}62)$$

where C contains all the constants.

We see that the quadratic exponential terms in front of the integral will cancel if

$$t_1 f' = t_2' f' + t_1 f' - t_1 t_2'$$

or

$$f' = t_1 \qquad (10\text{-}63)$$

Similarly, the quadratic terms inside the integral cancel if

$$t_2' f' = t_2' f' + t_1 f' - t_1 t_2'$$

or

$$f' = t_2' \qquad (10\text{-}64)$$

Under these conditions, the integral represented by Eq. (10-62), with the corresponding integral over the variables in the y direction, will cause Eq. (10-59) to reduce to

$$A(x', y') = C \int_{-\infty}^{\infty} \int_{-\infty}^{\infty} A(x, y) e^{-ik(xx'+yy')/f'} \, dx \, dy \qquad (10\text{-}65)$$

where C again incorporates all the constant factors outside the integral. By Eq. (10-45), this may be written as

$$A(v_x, v_y) = C \int_{-\infty}^{\infty} \int_{-\infty}^{\infty} A(x, y) e^{-ik(v_x x + v_y y)/f'} \, dx \, dy \qquad (10\text{-}66)$$

Thus, if C is taken as unity and the lens does not have an apodization or absorption screen, the field at the focal plane F is converted into its Fourier transform in the focal plane F'. The relation between the fields at other locations is more complicated. As Eq. (10-62) indicates, there are in general quadratic terms present. Physically these lead to Fresnel-type diffraction and much more complicated transformations.

10-8d Summary of Fourier Optics

To summarize, we have demonstrated the following optical methods of producing two-dimensional Fourier transforms:

(a) The Fraunhofer diffraction pattern at a screen at infinity is the Fourier transform of the aperture function of an opening, as shown by Eq. (10-40).

(b) The Fraunhofer diffraction pattern at the focal distance f' from a lens is the Fourier transform of its pupil function as shown by Eq. (10-46).

(c) An object at f is converted into its Fourier transform at f' by a lens as shown by Eqs. (10-63), (10-64), and (10-66).

10-9 Spatial Filtering

10-9a Basic Principles, Periodic Objects

The previous section showed how and why an aperture or a simple lens can compute a Fourier transform, and the similarity with the corresponding electronic problem was discussed. It was also suggested that it should be possible to remove components of the Fourier spectrum and alter the resulting pattern. This process is called *spatial filtering*, and the basic idea goes back to the experiments of Abbe[10-11] and Porter.[10-12]

To understand its significance, let us use the simple approach of Brown.[10-13] It is shown in Problem 10-6 that the diffraction pattern due to a pair of slits of width w and spacing d is

$$\frac{I}{I_0} = \text{sinc}^2\,\alpha \cos^2 \gamma \tag{10-21}$$

where

$$\alpha = \tfrac{1}{2}kw\sin\theta, \qquad \gamma = \tfrac{1}{2}kd\sin\theta$$

The function $\text{sinc}^2\,\alpha$, which defines the envelope of the curve in Fig. 10-10, is known as the *shape factor* and is due to the individual slits, while the fine structure depends on the spacing of the slits. If we now generalize to a one-dimensional array of N slits of width w, spaced at a distance d, the corresponding diffraction integral is (Fig. 10-23)

$$\int_{-\infty}^{\infty} e^{ikx\sin\theta}\,dx = \int_{0}^{w} e^{ikx\sin\theta}\,dx + \int_{d}^{w+d} e^{ikx\sin\theta}\,dx + \cdots$$

$$+ \int_{(N-1)d}^{(N-1)d+w} e^{ikx\sin\theta}\,dx$$

$$= \left(\frac{e^{ikw\sin\theta}-1}{ik\sin\theta}\right)[1 + e^{ikd\sin\theta} + \cdots + e^{ik(N-1)d\sin\theta}]$$

Using the formula for the sum of a geometric series with a ratio r and having n terms gives

$$S = \frac{a(1-r^n)}{1-r} = \frac{1 - e^{iNkd\sin\theta}}{1 - e^{ikd\sin\theta}}$$

for the second term on the right. Using α and γ as in Eq. (10-21), the relative amplitude of the diffraction pattern becomes

$$\frac{A}{A_0} = \left(\frac{e^{2i\alpha}-1}{2i\alpha/w}\right)\left(\frac{1 - e^{2iN\gamma}}{1 - e^{2i\gamma}}\right)$$

But

$$e^{2i\alpha} - 1 = (e^{i\alpha} - e^{-i\alpha})e^{i\alpha}$$

$$= 2e^{i\alpha}i\sin\alpha$$

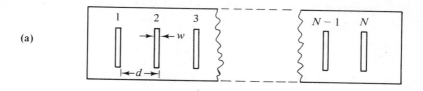

(a)

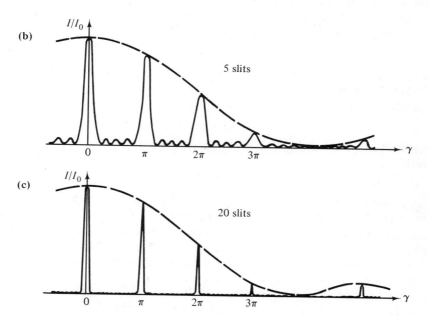

(b)

(c)

Figure 10-23

with similar expressions for the exponentials involving γ; therefore we obtain

$$\frac{A}{A_0} = \frac{e^{i\alpha}\sin\alpha}{\alpha/w}\frac{e^{iN\gamma}\sin N\gamma}{e^{i\gamma}\sin\gamma}$$

$$= 2we^{i\alpha}e^{i(N-1)\gamma}\frac{\sin\alpha}{\alpha}\frac{\sin N\gamma}{\sin\gamma}$$

To find the intensity, we multiply A by its complex conjugate and absorb all the constants into I_0. In addition, we explicitly incorporate a factor of N^2 into the expression for I/I_0 so that $I \rightarrow I_0$ when $\theta \rightarrow 0$. The final result is then

$$\frac{I}{I_0} = \text{sinc}^2\,\alpha\left(\frac{\sin N\gamma}{N\sin\gamma}\right)^2 \tag{10-67}$$

which has the shape factor of Eq. (10-21), as we might expect.

The other term, the *grating factor* $[(\sin N\gamma)/(N \sin \gamma)]^2$, is a generalization of $\cos^2 \gamma$ in Eq. (10-21), for when $N = 2$, we have

$$\frac{\sin 2\gamma}{2 \sin \gamma} = \cos \gamma$$

which is a well-known identity. Plots of Eq. (10-67) for $N = 5$ and $N = 20$ are shown in Figs. 10-23(b) and (c). We see that the central lobe of sinc α, as shown in Fig. 10-3, is converted into a number of sharp maxima, so that the diffraction pattern consists of a large number of very narrow, closely spaced lines. These principal maxima occur for values of γ which make the denominator $\sin \gamma$ of the grating factor vanish. That is

$$\gamma = m\pi, \qquad m = 0, \pm 1, \pm 2, \ldots \qquad (10\text{-}68)$$

The maxima are called spectra of order m as used in connection with Eq. (6-5); for example, the central peak is the zero-order spectrum. Equation (10-68) is equivalent to

$$\tfrac{1}{2}kd \sin \theta_m = m\pi$$

where the θ_m are the angles with respect to the normal, and this relation, in turn, becomes

$$d \sin \theta_m = m\lambda \qquad (10\text{-}69)$$

which we recognize as being similar to the Bragg law as used by crystallographers. This should not be surprising, since a single crystal is simply a diffraction grating for X rays and for electrons. The intensity, as given by Eq. (10-67), and the reinforcement condition as in Eq. (10-69) are approximations, valid only for small angles, since we have neglected the obliquity factor which should appear in the diffraction integral.

We have previously shown that an object at the focal point F of a thin lens results in a Fraunhofer pattern (i.e., a Fourier transform) at a distance f' in image space. In the case of a grating used as the object (Fig. 10-24), the diffraction integral Eq. (10-46) enables us to find how f' determines the size and spacing of the bright bands in the diffraction pattern.

Problem 10-11

(a) Show that the half-width x' of the central bright band in the plane through F' is

$$x' = \frac{f'\lambda}{Nd} \qquad (10\text{-}70)$$

(b) Show that the maximum corresponding to $\gamma = \pi$ is separated from the central maximum by a distance h_1, where

$$h_1 = \frac{f'\lambda}{d} \qquad (10\text{-}71)$$

Let a plane wave strike a grating normally, as shown in Fig. 10-24(a). By Eq. (10-69), the zero-order maximum corresponds to rays for which $\theta_0 = 0$, i.e., they are parallel to the optical axis. Such rays should intersect at the focal point F' for either a thin lens or one that is free of aberrations,

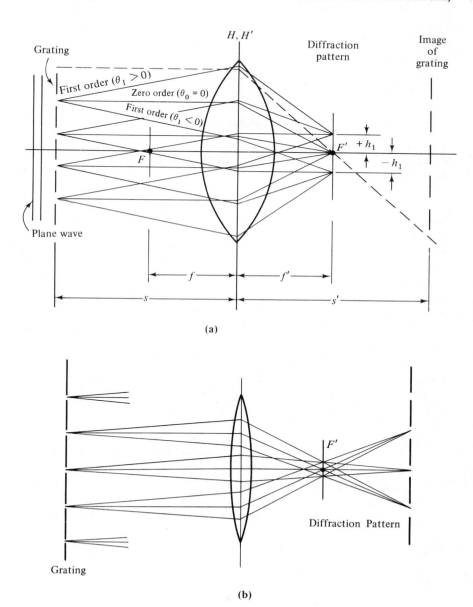

(a)

(b)

Figure 10-24

as shown. Next, we turn our attention to the rays corresponding to $m = 1$ and traveling with a positive slope $+\theta_1$. Although we demonstrated in the preceding section that the object and the Fraunhofer pattern are located at conjugate foci, the problem just below shows that the position of the grating in object space is arbitrary. That is, *for this special choice of an object*, the fact that the diffracted orders produce sets of parallel rays means that the Fourier transform appears in the focal plane specified by f' regardless of where the object is located. The path of the rays beyond F' was omitted for clarity in Fig. 10-24(a); in Fig. 10-24(b), we see how the three orders from a given object point recombine at the corresponding image point.

Problem 10-12

Use the matrix method to show that a set of parallel rays with slope $+\theta_1$ which strike a thin lens meet in the focal plane F' at a distance $\theta_1 f' = +h_1$ from the axis. ∎

To return to spatial filtering, suppose we place an opaque screen with a narrow slit in it at F', choosing the width of the opening to be slightly larger than $2h_1$. Then the two orders $m = 0$ and $m = \pm1$ shown will pass through the opening and all higher orders—which strike the plate at a distance greater than $\pm h_1$ from the axis—will be blocked. This has a profound effect on the image at s'. Figure 10-25(a) shows the image intensity as we normally expect to observe it; the bright regions have a width $w' = \beta w$, where β is the magnification, and a spacing $d' = \beta d$. The effect of the aperture is shown in Fig. 10-25(b). The rounded corners of the pattern indicate that the higher diffraction orders have been removed; this is a well-known phenomenon in Fourier analysis, as indicated in Fig. A-2 (Appendix) for example. These higher orders correspond to the sharp corners in a step function. To obtain a perfectly square corner, an infinite number of terms in the Fourier series must be included. For the higher values of γ, Eq. (10-45) shows that k is proportional to v_x, and γ is proportional to k by definition. We see therefore that the slit is removing the high spatial frequency components. Exactly the same thing can happen in an electric circuit; the effect of a low-pass filter on a square wave is to round the corners. On the other hand, if we replace the slit with an opaque strip of the same size and permit only the high spatial frequency components or high diffraction orders to reach the image plane, then we have an optical high-pass filter and obtain the image shown in Fig. 10-25(c). Now the edges are sharp and the central sections show the effect of removing the low-frequency components.

10-9b A Numerical Example

We can appreciate in a simple, quantitative way the effect of spatial filtering on an image by using an example of Gerrard.[10-14] Returning to the single

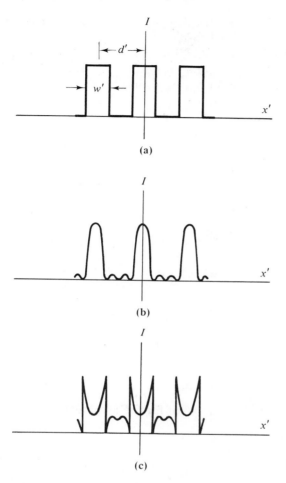

Figure 10-25

slit of width w, we recall that the relative amplitude of the diffraction pattern was given as

$$A = C' \operatorname{sinc} \alpha \qquad (10\text{-}5)$$

We shall combine this expression for the amplitude of a single slit, the modifications introduced by repeating the slit in a periodic fashion, and the arrangement of the orders indicated in Fig. 10-24 to compute the intensity of the pattern corresponding to the image of a diffraction grating. Since each value of m corresponds to a set of parallel rays striking the lens, we can say that a plane wave striking the grating normally is decomposed into an infinite set of plane waves leaving it, each with a particular amplitude A_m, phase φ_m, and angle of inclination θ_m, where $m = 0, \pm 1, \dots$. Each A_m will be determined by Eq. (10-5), θ_m by Eq. (10-69), and φ_m must be calculated from the geometry. Before doing so, let us find the total amplitude due to all the individual contributions through the use of what electrical engineers call a *phasor dia-gram*. Although these are commonly used to represent complex numbers, for our purposes here we need simply take A_m for $m = 0$ as the phase refer-ence axis, as shown in Fig. 10-26. The terms for $m = \pm 1, \pm 2, \ldots$ add vector-ially to this, with the horizontal components giving a contribution $A_m \cos \varphi_m$

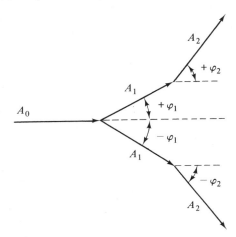

Figure 10-26

and the vertical components cancelling in pairs. The total amplitude at each image point of Fig. 10-24b is then

$$A = \sum_{m=-\infty}^{\infty} A_m \cos \varphi_m \tag{10-72}$$

where $\cos \varphi_0 = 1$ since $\varphi_0 = 0$. Using Eq. (10-5) for the amplitudes gives

$$\frac{A}{C'} = \sum_{m=-\infty}^{\infty} \operatorname{sinc} \alpha_m \cos \varphi_m$$

where

$$\alpha_m = \tfrac{1}{2} kw \sin \theta_m$$

or

$$\frac{A}{C'} = 1 + 2 \sum_{m=1}^{\infty} \operatorname{sinc} \alpha_m \cos \varphi_m \tag{10-73}$$

since $\theta_0 = 0$, sinc $0 = 1$, $\varphi_0 = 0$, and $\cos 0 = 1$.

It is convenient to use Eq. (10-69) in paraxial form, so that

$$\theta_m = \frac{m\lambda}{d} \tag{10-74}$$

Then

$$\alpha_m = \frac{\pi}{\lambda} w \theta_m = \frac{mw\pi}{d} \tag{10-75}$$

Turning to φ_m, this introduces a factor $\exp(ik\delta_m)$ in the analytical expression for a plane wave. If x is the distance of one of the slits in the grating from the axis, then $x \sin \theta_m \sim x\theta_m$ is the difference in path length or phase between a ray of order m leaving this slit and one at the center. This is analogous to the situation illustrated in Fig. 10-2. Hence

$$\varphi_m = k\delta_m = kx\theta_m$$
$$= \frac{mk\lambda x}{d} = \frac{2\pi mx}{d}$$

Since we wish to find the intensity in the image plane, we use the Gaussian lens equation in the form

$$\frac{1}{s} + \frac{1}{s'} = \frac{1}{f'}$$

This corresponds to a magnification

$$\frac{s'}{s} = \beta = \frac{x'}{x} = \frac{(s' - f')}{f'}$$

and

$$\varphi_m = \frac{2\pi mx'f'}{(s' - f')d} \tag{10-76}$$

Inserting Eq. (10-75) and Eq. (10-76) into Eq. (10-73) and squaring gives the image intensity as

$$\frac{I}{I_0} = \left[1 + 2\sum_{m=1}^{M} \text{sinc}\,\frac{mw\pi}{d} \cos \frac{2\pi mx'f'}{(s' - f')d}\right]^2 \tag{10-77}$$

where the upper limit M designates the number of terms which we use in the summation.

As a specific example, let the grating have 5000 lines per cm with opaque regions 50% wider than the slits. Then $d = 0.0002$ cm and $w = 0.8$ μm. If the grating is 1.2 cm from a lens for which $f' = 1.0$ cm, then $s' = 6.0$ cm, $\beta = s'/s = 5.0$, and the slit images have a width of 4.0 μm. By Eq. (10-76)

$$\varphi_m = \frac{2\pi mx'}{(5.0)(0.0002)} = 6283.1mx'$$

and from Eq. (10-75)

$$\alpha_m = 1.256m$$

The series to be evaluated is thus

$$\frac{I}{I_0} = \left[1 + 2\sum_{m=1}^{M} \frac{\sin(1.256m)}{1.256m} \cos(6283.1mx')\right]^2 \tag{10-78}$$

Problem 10-13

(a) Calculate I/I_0 in Eq. (10-78) for $M = 1, 2, 4, 6$, and 11, verifying Figs. 10-27(a)–(e).

(b) Apply Eq. (10-78) to the case of a high-pass spatial filter, obtaining a few curves such as in Fig. 10-25(c).

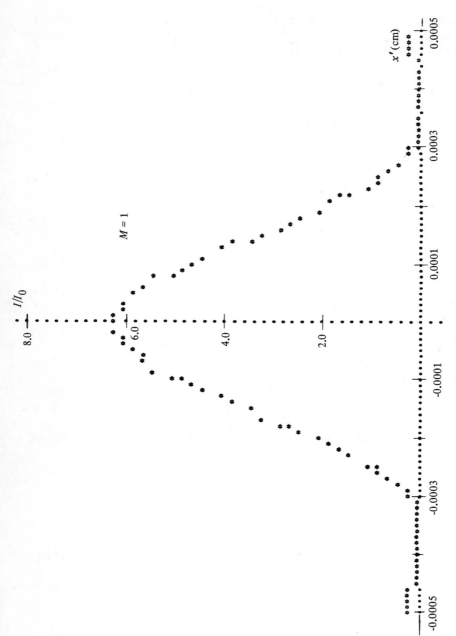

Figure 10-27a

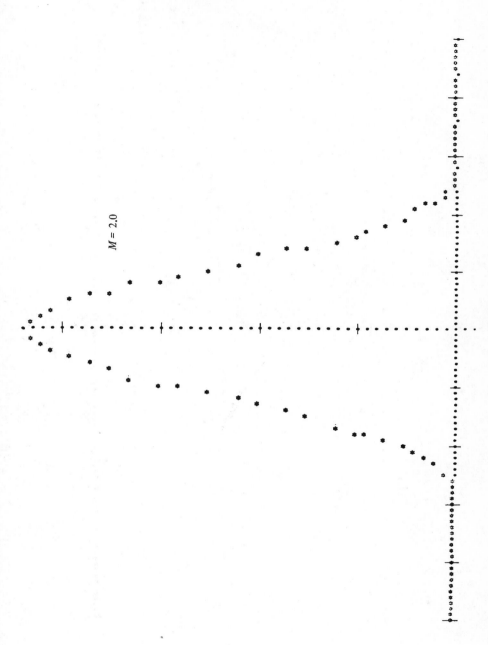

M = 2.0

Figure 10-27b

257

$M = 4$

Figure 10-27c

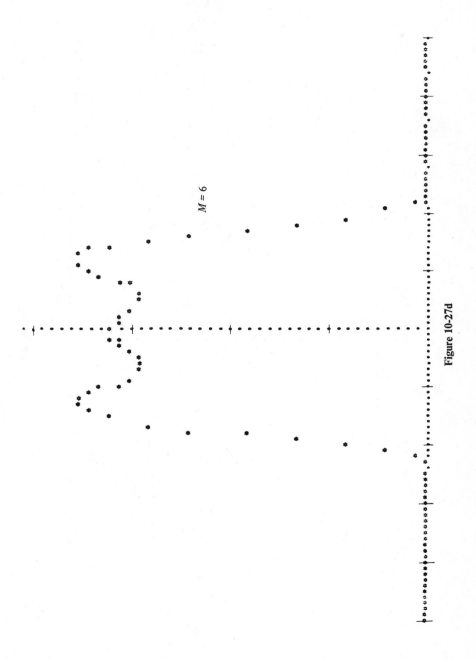

$M = 6$

Figure 10-27d

259

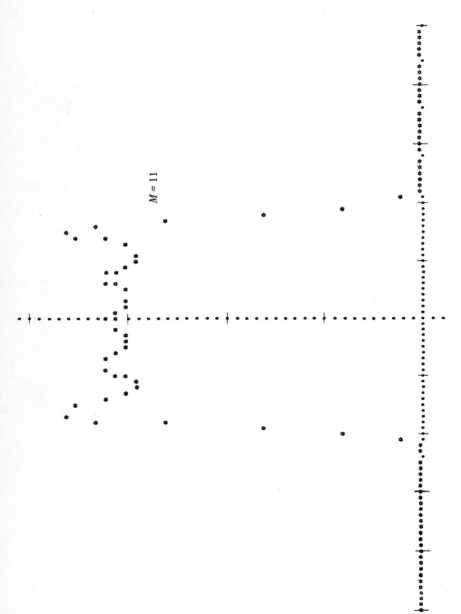

$M = 11$

Figure 10-27e

10-9c Image Processing

The results of Problem 10-13 show that spatial filtering can have a drastic effect on the nature of the image. A number of practical applications of spatial filtering are used in the field of image processing, as described by Shulman[10-15] and Francon.[10-24] One of the simplest is the smoothing of halftone photographs. This process, as used in newspapers for example, produces an image which consists of a regular array of dots of varying size. The dots are heavy in the dark areas of the picture, and vice versa. We can think of the picture as having two components: a carrier (the dots) and the picture (the slow variation in the dots' size). A filter in the form of a small aperture at the Fourier transform plane will then remove the two-dimensional diffraction pattern corresponding to the dots, since their sharp edges represent high spatial frequency components; the resulting image closely resembles the continuous-tone original photograph. A typical arrangement for accomplishing this image processing, due to Peckham, Hagler, and Kristiansen,[10-16] is shown in Fig. 10-28. A microscope objective (which has a small aperture

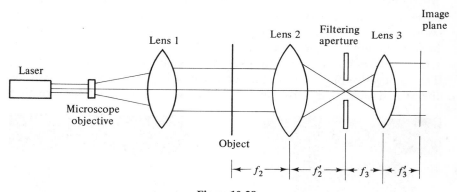

Figure 10-28

but is very highly corrected for aberrations) combined with lens 1 serves as a beam spreader. The object is at the focal point of lens 2, and lens 3 is at a distance equal to the sum of the focal lengths. Hence the diffraction pattern at F'_2, F_3 is transformed a second time, producing an image at F'_3. Note that when the object is not simply a grating (the halftone dots alone, however, would be equivalent to a two-dimensional grating), then the geometry of Fig. 10-28 is necessary. That is, the object should be at f_2, as required by Eqs. (10-63) and (10-64), and the two lenses should be spaced a distance $(f'_2 + f_3)$ from each other. The original photograph and the result of the spatial filtering are shown in Fig. 10-29.

A very interesting example has been described by Phillips.[10-17] Figure

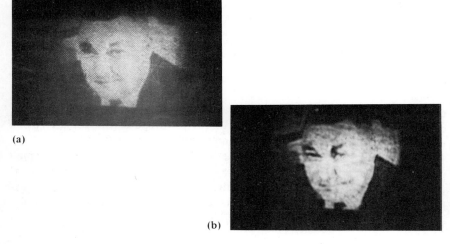

(a)

(b)

Figure 10-29

10-30(a) shows a self-portrait by Minnesota artist Karl E. Bethke, who painted regularly spaced acrylic squares onto canvas. The appearance of shading was obtained by varying the size of the squares, in a manner similar to the halftone process. It is possible to make a filter for the squares by using a pattern of squares having a uniform size and the same spacing as in the portrait. Then a negative photograph could be taken of its transform; this would be the desired filter. When properly positioned in the transform plane, the filter

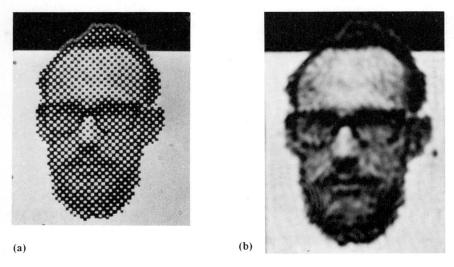

(a)   (b)

Figure 10-30

will block out only those components which make up the pattern of squares, passing both the higher and lower spatial frequency components of the original picture. Unfortunately, such a filter is difficult to construct and align. Therefore a pinhole, as before, is used as the filter, and the result is shown in Fig. 10-30(b). The image does suffer some degradation because high frequencies which make up small details are lost.

A practical application of these ideas is due to Watkins.[10-18] Integrated circuits consisting of a number of diodes, transistors, resistors, and other components are fabricated by photographic masking processes. A master mask is drawn by a draftsman, reduced in several steps to a size of about $1 \text{ mm} \times 1 \text{ mm}$, and then used to prepare a multiple mask by a step-and-repeat process. The final array consists of several hundred identical microcircuit masks, but a few of them may not be perfect copies due to inhomogeneities in the masking materials or failure to achieve exact reproduction, as seen in Fig. 10-31(a). Locating the defective masks by visual observation is an extremely difficult job, since each section has to be compared with the master under a microscope. Watkins used the fact that the diffraction pattern for an array of these masks is similar to that of an $M \times N$ array of rectangular openings so that it consists of two sets of sharp lines intersecting at right angles; that is, it will be an $M \times N$ array of tiny points [Fig. 10-31 (b)].

A filter consisting of an array of spots is placed at the focal plane F' of the lens performing the transform. For a perfect series of masks, the entire pattern is filtered out, and the inverse transform at the image plane is a photograph of very low intensity. However, a nonperiodic error such as a scratch or a contamination gives extra lines in the diffraction pattern which are not removed at the filter plane; this shows up in Fig. 10-31(c), which clearly reveals the defects in Fig. 10-31(a). Hence, a defective mask can be identified very quickly. To locate a particular bad element in an array, the original $M \times N$ matrix can be examined half at a time in the same way, and so on.

Interesting and unexpected effects are observed with multiple apertures. For simplicity, consider an $N \times N$ rectangular array of identical openings. The diffraction integral for any given opening is

$$A(x', y') = \int A(x, y)e^{ikr} \, dx \, dy$$

If we transform the coordinates in the plane of the array by the relations

$$x_1 = x + x_0$$
$$y_1 = y + y_0$$

and also

$$r_1 = r + r_0$$

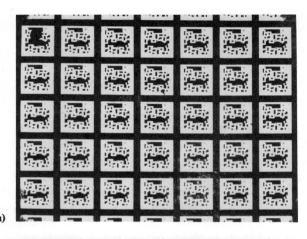

(a)

(b)

(c)

Figure 10-31

264

the integral becomes

$$A(x', y') = \int A(x_1 - x_0, y_1 - y_0)e^{ik(r_1 - r_0)} \, dx_1 dy_1$$

The phase factor $\exp(-ikr_0) = \exp(i\delta)$ can be removed from the integral, since it does not depend on x_1 and y_1. The function $A(x_1 - x_0, y_1 - y_0)$ is simply the original amplitude, but centered about a new origin. Hence, the diffraction pattern for each individual aperture is unchanged except for a displacement. When we sum over a row of N apertures, we then obtain

$$A(x', y') = \left(\sum_{j=1}^{N} e^{i\delta_j} \right) \int A(x_1 - x_0, y_1 - y_0)e^{ikr} \, dx_1 dy_1$$

If the apertures are randomly spaced, then the phase factors sum to a value of $\sqrt{N}$, as shown in Section 5-6. On the other hand, a regularly spaced array will give values ranging from N to zero, depending on the observation point (x', y').

We thus see that a regular array of N apertures has an intensity which is N^2 times that of a single aperture. Intuitively, we can visualize each opening adding a contribution of the proper phase to all the other openings. Equation (10-67) shows this to be true for the diffraction grating. In deriving this equation, we arbitrarily divided the intensity by a factor of N^2 in order to compare the pattern of the single slit with that of the grating.

On the other hand, the random array increases the intensity only by a factor of N, since there will be out-of-phase components as well. The same results apply to $N \times N$ arrays, the phase factors combining to give an N^2 enhancement for a perfectly periodic structure or a factor of N in the random case. For a periodic arrangement, but with a few shape defects, the result is an intensity increase by approximately N^2 and some extra peaks in the diffraction pattern. These effects are shown in Fig. 10-32, taken from Hoover.[10-19] Figure 10-32a shows the Fraunhofer pattern of 3000 randomly oriented rectangular openings and Fig. 10-32b is a 54 × 54 regular array of the same kind of opening. Note the clear, sharp spikes in the latter because of the enhanced intensity.

10-9d   The Abbe Theory of the Microscope

As another application, we return to the work of Abbe,[10-11] whose motivation was actually the theory of the resolving power of the microscope. Following Stone,[10-20] the relation between transform plane amplitude $A_{F'}(x_{F'})$ and object plane amplitude $A(x)$, from Eq. (10-65), is

$$A_{F'}(x_{F'}) = \int_{-\infty}^{\infty} A(x)e^{-ikxx_{F'}/f'} \, dx \qquad (10\text{-}79)$$

and this Fourier transform has an inverse of the form

$$A(x) = \int_{-\infty}^{\infty} A_{F'}(x_{F'})e^{ikxx_{F'}/f'} \, dx_{F'} \qquad (10\text{-}80)$$

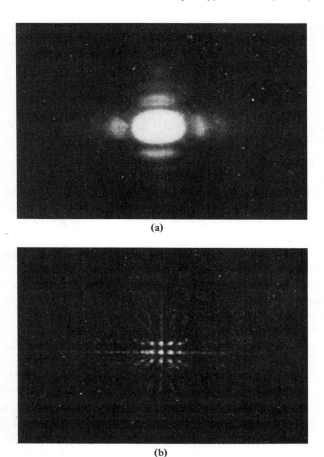

(a)

(b)

Figure 10-32

However, if the image is a faithful reproduction of the object with magnification β, then

$$A'(x') = A\left(\frac{x'}{\beta}\right) \tag{10-81}$$

where A' is the amplitude at the image plane. Eq. (10-80) combined with Eq. (10-81) gives

$$A'(x') = \int_{-\infty}^{\infty} A_{F'}(x_{F'})e^{ikx'x_{F'}/f'\beta}\, dx_{F'} \tag{10-82}$$

The amplitude in the transform plane for a single slit of width w will be

$$A_{F'}(x_{F'}) = C \operatorname{sinc} \alpha$$

$$= w\,\frac{\sin\,(kx_{F'}w/2f')}{(kx_{F'}w/2f')} \tag{10-83}$$

as obtained by integrating Eq. (10-79) and ignoring the constants. Using this in Eq. (10-82) gives

$$A'(x') = w \int_{-\infty}^{\infty} \frac{\sin (kw/2f')x_{F'}}{(kw/2f')x_{F'}} e^{(ikx'/f'\beta)x_{F'}} \, dx_{F'} \qquad (10\text{-}84)$$

Consider the identity

$$\int_{-\infty}^{\infty} \frac{\sin px}{x} e^{iqx} \, dx = \int_{-\infty}^{\infty} \frac{\sin px \cos qx}{x} \, dx + i \int_{-\infty}^{\infty} \frac{\sin px \sin qx}{x} \, dx$$

The second integral, involving an odd function, vanishes; the first integral—from tables[10-10]—becomes

$$2 \int_{0}^{\infty} \frac{\sin px \cos qx}{x} \, dx = \begin{cases} \pi & \text{for} \quad 0 < p < q \\ 0 & \text{for} \quad 0 < q < p \end{cases}$$

Hence

$$A'(x') = w \int_{-\infty}^{\infty} \frac{\sin (kw/2f')x_{F'}}{(kw/2f')x_{F'}} e^{(ikx'/f'\beta)x_{F'}} \, dx_{F'}$$

$$= \left.\begin{matrix} \dfrac{\pi f'}{k} \\ \\ 0 \end{matrix}\right\} \quad \text{if} \quad \begin{cases} \dfrac{\beta w}{2} < x' \\ \\ \dfrac{\beta w}{2} > x' \end{cases} \qquad (10\text{-}85)$$

Equation (10-85) simply specifies the magnified image of the positive half of the grating. Using a similar calculation for the other half, we obtain the sequence illustrated in Fig. 10-33, which shows the original slit, the Fourier transform, and the inverse transform of sinc α.

The calculation just given implies that the width w of the slit is fairly large. The function sinc α has its first minimum when

$$\frac{kx_{F'0}w}{2f'} = \pi$$

or

$$x_{F'0} = \frac{f'\lambda}{w}$$

where $x_{F'0}$ is the position of this minimum. A wide slit means that the entire function sinc α will essentially pass through the lens and reach the focal plane, but if w is very small, then only the central section can get through. That is, the lens edge acts as a low-pass spatial filter, and all detail in the image is lost. The transform at F' is simply a band of reasonably uniform intensity, and the inverse transform of this constant, just as in Eq. (10-83), is

$$A'(x') = R \text{ sinc } \alpha'$$

at the image plane, where

$$\alpha' = \frac{kRx'}{2\beta f'}$$

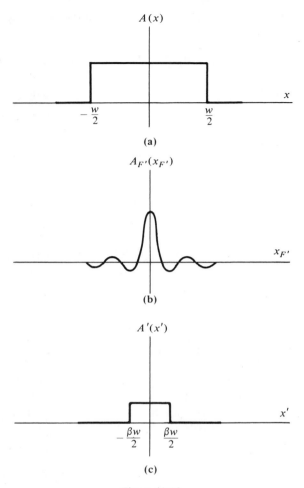

Figure 10-33

and R is the lens radius. The first minimum x'_0 of this pattern is then

$$x'_0 = \frac{2f'\beta\lambda}{R}$$

Hence, the final image is not a faithful reproduction, but a bright band whose size depends on the properties of the microscope objective being used.

Returning to the complete grating, the situation is very similar to what we have already discussed. Figure 10-26(a) represents the image of just one slit when the zero-order and the two first-order spectra are passed; the complete image is a repetition of this single one. We realize that the reproduction is poor, but it is distinguishable as a slit. If the spectra for $m = 1$ were also filtered out, Eq. (10-78) would reduce to $I = I_0$, and the image would again be a uniformly illuminated screen with no detail. Hence, as Abbe pointed out, for a microscope to resolve fine detail, the objective must be capable of passing at least the first two orders, and the more orders, the better the resolution, as Fig. 10-26 indicates. The spectra for $m = 1$ are spaced a distance $2f'\lambda/d$, and this represents the minimum spacing which the objective must be capable of resolving. If we examine the diffraction pattern with a microscope, then the objective serves as the spatial filter at the plane F' in Fig. 10-24, and its diameter D can be no smaller than $2f'\lambda/d$. Let θ be the half-angle subtended by the objective at the plane H, H'. Then

$$\sin \theta = \frac{R}{f'} = \frac{\lambda}{d} \qquad (10\text{-}86a)$$

High-power microscope objectives are usually of the oil-immersion type, and the theory of Chapter 1 shows that when the objective is immersed in a liquid of index n, Eq. (10-86) takes the form

$$n \sin \theta = \frac{\lambda}{d}$$

or

$$d = \frac{\lambda}{N.A.} \qquad (10\text{-}86b)$$

where $n \sin \theta$ is called the *numerical aperture (N.A.)* of the objective. Hence, *the minimum spacing d which can be resolved by a good microscope is on the order of the wavelength of light.* If an oil of high index and a large acceptance angle is used, the maximum practical value of $n \sin \theta$ is about 1.6.

It is interesting to note that Abbe's theory, which leads to Eq. (10-86), is based on the idea that at least one order must be passed by the lens. Earlier in Section 10-4, we calculated the diffraction pattern of a circular aperture and used the Rayleigh criteria to determine resolutions, Eq. (10-16). These two theories are based on quite different views of image formation, yet lead to the same result.

Problem 10-14

Use the facts that the smallest angular separation for which the human eye can resolve two points is about 1 minute of arc and the least distance of distinct vision is 25 cm to show that the theoretical maximum magnification for a microscope is on the order of 10^3. ∎

10-9e Contrast Improvement

We shall give a precise definition of the concept of contrast later in this chapter [see Section 10-11] but for the time being, let us simply regard it as a measure of the difference between the bright and dim portions of an illuminated object. For example, alternating black and white bands may be represented by the repeated square wave shown in Fig. 10-34(a), whereas two shades of gray would be represented by a pattern like Fig. 10-34(b). The maximum intensity is less than that in the top portion and the contrast is lower because the change in intensity is smaller.

We realize that Fig. 10-34(b) is like the square wave of Fig. A-3 of the Appendix, but shifted upward by an amount I_c. This is equivalent to changing the constant (or DC) term in Eq. (A-8) from 0.5 to a larger value. If we can eliminate this shift, then the square wave would move down so that its minimum rests on the x-axis, greatly enhancing the contrast. We can even reverse the contrast if the maximum is brought down to the axis and the minimum lies below it. These changes may be accomplished by altering the constant term in the series of Eq. 10-73; we either filter it out or pass it through a film producing a 180° phase reversal, which will change the sign. Fig. 10-24a shows that a filter which will remove the plane wave for which $m = 0$, this being the constant term in the summation, is a disc of radius less than h_1. This term would have been larger if we had started with an object of lower contrast; the grating which we used is assumed to give complete transmission through the slits and none at the opaque regions.

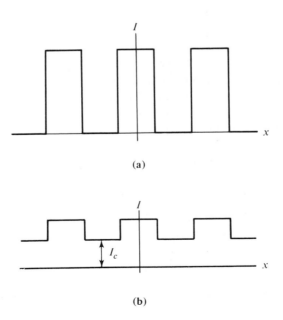

(a)

(b)

Figure 10-34

10-9f Phase Contrast

The aperture function of the grating we considered at the beginning of this discussion affects the amplitude of the transmitted wave. We now wish to consider a grating which modifies the phase. For example, if we have a periodic array composed of transparent apertures alternating with thin films producing a 90° phase shift, the aperture function is (for the initial section)

$$A(x) = 1 \quad \text{for} \quad 0 \leq x \leq \frac{d}{4}, \frac{3d}{4} \leq x \leq d$$

$$A(x) = i \quad \text{for} \quad \frac{d}{4} \leq x \leq \frac{3d}{4}$$

This grating then has a period of d and each section has a width equal to $d/2$. The Fourier series corresponding to such a function is easily seen to have the form

$$\frac{A}{A_0} = \frac{1}{2}(1 + i) + \frac{2}{\pi}(1 + i) \sum_m a_m \cos \frac{2\pi mx}{d} \qquad (10\text{-}86c)$$

for the real part is like the square wave of Eq. (A-8) and the imaginary part will have the value of zero when the real part equals unity and the value i when the real part vanishes. The term for $m = 0$ is $1 + i$, which is a complex number making an angle of 45° to the real axis. The rest of the series in Eq. (10-86c) is multiplied by $(1 - i)$, representing an angle of $-45°$, and the two terms are 90° out of phase. Two complex numbers at right angles have no components in each other's direction and hence there can be no interference between the zero order term and any of the others.

Let us next consider the effect of a 270° phase shift on the term $1 + i$. This is accomplished with a thin film at the center of the transform plane and multiplies $(1 + i)$ by $-i$, converting Eq. (10-86) to

$$\frac{A}{A_0} = (1 - i)\left[\frac{1}{2} + \frac{2}{\pi} \sum_m a_m \cos \frac{2\pi mx}{d}\right] \qquad (10\text{-}86d)$$

Since the term $(1 - i)$ has a magnitude of $\sqrt{2}$, we have essentially the same result as we obtained for the original grating; i.e., the use of a 270° phase shifting filter has converted phase variations into amplitude variations. This is the basis of the *phase contrast microscope*, invented by Frits Zernike in 1935.

Small transparent organisms are difficult to see with an ordinary microscope. It is possible, however, to take advantage of the fact that their index of refraction generally differs from that of the surrounding medium and this results in a phase difference in the transmitted light. A phase shifting film will then produce an amplitude variation which is readily visible.

10-9g Processing the Output of a Laser

Consider the expanded output beam from a typical helium-neon laser (Fig. 10-35). The pattern is irregular because dust particles and imperfections on optical surfaces in the beam produce diffraction patterns. Since the diffraction patterns are objectionable when the beam is used for holography or in photographic work, we would like to eliminate them. Think of the beam as consisting of two parts—the gaussian component (low spatial frequency) and diffraction pattern (higher spatial frequencies). When the beam is transformed by a lens, the gaussian component is located near the center of the transform while the high frequency terms are located away from the center, so that a pinhole placed at the center of the transform plane will transmit only the low frequency component. A second lens of longer focal length then collimates the beam, which is both expanded and smoothed by the assembly (Fig.

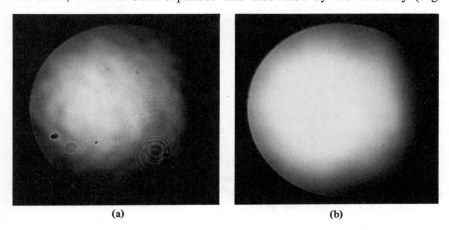

(a)   (b)

Figure 10-35 (a) Expanded output beam of laser before filtering, and (b) expanded beam after filtering.

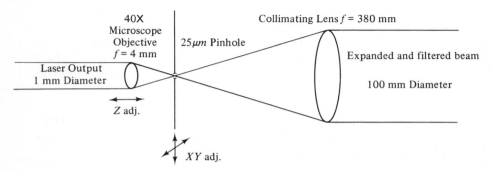

Figure 10-36

10-36). Thus, the principle of spatial filtering is used to obtain a uniform beam which then may be used for a variety of applications in the laboratory.

10-10 The Optical Transfer Function

When we considered the Fourier transform properties of lenses in Section 10-8, we first assumed that the only effect of the lens on a wave passing through it is to introduce a phase shift which depends on the separation of the two glass surfaces. Later we found that in the process of forming an image, a lens cuts out high spatial frequency components of an object. These high spatial frequency components account for fine detail. This was the basic idea in Abbe's theory of the microscope. We might expect a real lens not only to cut out certain spatial frequencies because of diffraction, but also to attenuate others because of aberrations. This suggests that the image-forming quality of a lens could be specified by a function which describes how much each spatial frequency component in an object is attenuated by the lens as it forms an image. As we shall see in this section, there is such a function, called the *optical transfer function*, which specifies lens quality. This approach is in wide use today.

10-10a Derivation of the Optical Amplitude Transfer Function

To derive the optical transfer function, we return to Eq. (10-59) and for simplicity assume:

$$x_1 = x_2 = x_L, \qquad y_1 = y_2 = y_L$$

We include the pupil function $P(x_L, y_L)$ as we did in Eq. (10-44) to explicitly include the effects of the lens on amplitude; in particular, the effects of its finite size on the spatial frequency properties of the transmitted wave. Our integrals will extend from $-$ to $+$. This will facilitate Fourier-transform operations. Equation (10-59) thus becomes

$$A(x', y') = C \iint A(x, y)P(x_L, y_L)e^{-ik(x_L{}^2 + y_L{}^2)/2f'}e^{ik[(x_L - x)^2 + (y_L - y)^2]/2t_1}$$

$$\cdot\, e^{ik[(x' - x_L)^2 + (y' - y_L)^2]/2t'_2}\, dS_L\, dS \tag{10-87a}$$

As Eq. (10-60) indicates, the quadratic term involv*ing* x_L^2 will reduce to unity when we impose the condition that

$$\frac{1}{t_1} + \frac{1}{t'_2} = \frac{1}{f'} \tag{1-6}$$

which we recognize as the Gauss lens equation. For a spherical object producing a spherical image, the quadratic terms in x^2 and x'^2 are constant. For a plane object which is not too large, this condition is approximately true and absorbing these terms into C, Eq. (10-87a) simplifies to

$$A(x', y') = C \iint A(x, y)P(x_L, y_L)e^{-ikx_L[(x/t_1) + (x'/t_2')]}$$

$$\cdot\, e^{-iky_L[(y/t_1) + (y'/t_2')]}\, dS_L\, dS \qquad (10\text{-}87b)$$

Using the expression

$$\beta = -\frac{t_2'}{t_1} \qquad (1\text{-}75)$$

for the magnification (the negative sign follows the conventions of Chapter 1) converts this to

$$A(x', y') = C \iint A(x, y)P(x_L, y_L)e^{-i(k/t_2')[x_L(x'-\beta x) + y_L(y'-\beta y)]}\, dS_L\, dS \qquad (10\text{-}88)$$

As Goodman[10-21] points out, this relation indicates that the image for $P = 1$ is the transform of the object, or the Fraunhofer pattern of the object as limited by the lens aperture, but centered on image coordinates $x' = \beta x$, $y' = \beta y$. This is what we would expect intuitively; the image has its size determined by paraxial optics, and it lies on the plane towards which the spherical waves originating at (x, y) are being made to converge by the lens. Further, for $P = 1$ over a finite distance the image represents the Airy disc plus the secondary rings of the diffraction pattern obtained by integrating over the lens opening for each point on the object and then integrating over the object to find the total contribution to each image point.

Next, we change variables as specified by

$$\bar{x} = \beta x, \qquad \bar{y} = \beta y \qquad (10\text{-}89)$$

and define spatial frequencies, following Eq. (10-45), as

$$\left.\begin{aligned} v_x &= -\frac{kx_L}{t_2'} \\[2mm] v_y &= -\frac{ky_L}{t_2'} \end{aligned}\right\} \qquad (10\text{-}90)$$

where the negative sign is used for convenience. Equation (10-88) becomes

$$A(x', y') = C \iiiint A\!\left(\frac{\bar{x}}{\beta}, \frac{\bar{y}}{\beta}\right)P\!\left(-\frac{v_x t_2'}{k}, -\frac{v_y t_2'}{k}\right)$$

$$\cdot\, e^{i[v_x(x'-\bar{x}) + v_y(y'-\bar{y})]}\, d\bar{x}\, d\bar{y}\, dv_x\, dv_y \qquad (10\text{-}91)$$

where all constants continue to be absorbed in C. The integration over the lens opening can be expressed in terms of the *point spread function* h defined by

$$h(x', y') = \iint P\!\left(-\frac{v_x t_2'}{k} - \frac{v_y t_2'}{k}\right)e^{i(v_x x' + v_y y')}\, dv_x\, dv_y \qquad (10\text{-}92)$$

which we recognize as the Fourier transform of the pupil function. If we use

Eq. (10-92), Eq. (10-91) becomes

$$A(x', y') = C \int A\left(\frac{\bar{x}}{\beta}, \frac{\bar{y}}{\beta}\right) h(x' - \bar{x}, y' - \bar{y}) \, d\bar{S} \qquad (10\text{-}93)$$

which, by Eq. (A-38a), is the convolution of h and the object. Thus, image formation—which we have already shown to be a Fourier transform computation—may also be interpreted as the convolution of the object with the transmission properties of the lens. Lipson and Lipson,[10-4] for example, show the convolution of a two-dimensional object (a photograph) with an aperture such as a slit to obtain a somewhat blurry picture. If just the object and the aperture were used, the result would be a diffraction pattern. To obtain the convolution, they use the aperture to mask the lens of an out-of-focus camera. This essentially performs the integration over the lens opening and gives a convolution rather than a transform.

10-10b Analogy with Electrical Circuits

The function h, which is defined by Eq. (10-92) and used in Eq. (10-93), has an important physical interpretation. As shown in the Appendix, the usual equation governing the behavior of a circuit containing a resistance R and an inductance L in series, by Eq. (A-40), is

$$v(t) = \left(L\frac{d}{dt} + R\right)i(t) \qquad (10\text{-}94)$$

and by taking the Fourier transform of both sides of this equation, it may be written as

$$V(\omega) = (i\omega L + R)I(\omega) \qquad (10\text{-}95)$$

or

$$I(\omega) = V(\omega)Y(\omega)$$

where

$$Y(\omega) = \frac{1}{i\omega L + R}$$

In the time domain, $v(t)$ is the excitation and $i(t)$ the response, whereas in the frequency domain, $Y(\omega)$ is known as either the complex admittance or the transfer function. In the special situation where $v(t)$ is a delta function, Eq. (A-59) gives $\mathfrak{F}[\delta(t)] = 1 = V(\omega)$ and $I(\omega)$ is identical to $Y(\omega)$. That is, the transfer function—which we originally identified in terms of the admittance (or impedance) of the circuit—is also seen to be the response to unit voltage or excitation. Furthermore, if $g(t) = v(t) = \delta(t)$ in Eq. (A-26), then Eq. (A-68) with $a = 0$ gives

$$f(t)*g(t) = f(t)*\delta(t) = f(t) = h(t)$$

This shows why $h(t)$, whose Fourier transform is

$$H(\omega) = Y(\omega) = I(\omega) \tag{10-96}$$

is called the *impulse response*.

The analogous situation in optics is the response to a point source of light. By Eq. (10-92), $h(x', y')$ represents the diffraction pattern at the image plane generated by a point source at (x, y), which is why it is called the point spread function. We further see from the Appendix that the solution of Eq. (10-94) is

$$i(t) = \int_{-\infty}^{\infty} g(\alpha)h(t - \alpha)\, d\alpha = v(t){*}y(t)$$

[see Eq. (A-38a)] which is analogous to Eq. (10-93); therefore $h(x', y')$—a function of the image coordinates—is the optical equivalent of the time domain admittance $y(t)$, i.e., it is the optical impulse response. Actually, we compute $Y(\omega)$ when we do circuit problems; thus in optics we should correspondingly work with $H(v_x, v_y)$, the transform of $h(x', y')$. Note that in circuit theory we go from the time domain to the frequency domain, whereas in optics we go from two-dimensional space to the spatial frequency domain.

10-10c The Optical Intensity Transfer Function

The function $H(v_x, v_y)$, which is the transform of $h(x', y')$, should thus be called the *optical transfer function* (or optical admittance). But h is the transform of P, so that combining Eq. (10-93) with the definition of H gives

$$H(v_x, v_y) = \mathcal{F}[h(x', y')] = \mathcal{F}\left[\mathcal{F}\left[P\left(-\frac{v_x t_2'}{k}, -\frac{v_y t_2'}{k}\right)\right]\right] \tag{10-97}$$

and since the transform of a Fourier transform is the original function, this reduces to

$$H(v_x, v_y) = P\left(-\frac{v_x t_2'}{k}, -\frac{v_y t_2'}{k}\right) \tag{10-98}$$

This transfer function $H(v_x, v_y)$ specifies the effect of a lens on the amplitude of spatial frequencies passing through, just as $Y(\omega)$ performs a similar function for the angular frequencies in a signal. However, we are usually interested in intensity rather than amplitude, and we define the *optical intensity transfer function* $F(v_x, v_y)$ by the Fourier transform relation

$$F(v_x, v_y) = \frac{\displaystyle\int_{-\infty}^{\infty}\int_{-\infty}^{\infty} h(x', y')h^*(x', y')e^{-i(v_x x + v_y y)}\, dx'\, dy'}{\displaystyle\int_{-\infty}^{\infty}\int_{-\infty}^{\infty} h(x', y')h^*(x', y')\, dx'\, dy'} \tag{10-99}$$

where the denominator is included for normalization purposes. This quantity, rather than H, is what is known as the optical transfer function (*OTF*). Equation (10-99) may be abbreviated as

$$F(v_x, v_y) = \frac{\mathfrak{F}[|h|^2]}{\mathfrak{F}[|h|^2]|_{v_x, v_y = 0}} \tag{10-100}$$

To use this result, we recast it by noting (as explained in the Appendix) that if three functions are related as

$$F(\omega) = G(\omega)H(\omega)$$

then their inverse transforms are connected by the convolution equation

$$f(t) = g(t) * h(t) \tag{A-38b}$$

as shown in the Appendix. Letting

$$g(t) = h^*(t)$$

gives

$$\mathfrak{F}[h * h^*] = HH^*$$

or in two-dimensional form

$$\mathfrak{F}\left[\iint h(\xi, \eta)h^*(\xi - v_x, \eta - v_y)\, d\xi\, d\eta\right] = H(v_x, v_y)H^*(v_x, v_y) \tag{10-101}$$

where ξ and η are dummy variables. Equation (10-101) has an inverse

$$\mathfrak{F}[h(\xi, \eta)h^*(\xi, \eta)] = \iint H(\xi, \eta)H^*(\xi + v_x, \eta + v_y)\, d\xi\, d\eta \tag{10-102}$$

as may be verified by writing out the integrals. Combining this with Eq. (10-100) converts the definition of the *OTF* into

$$F(v_x, v_y) = \frac{\iint H(\xi, \eta)H^*(\xi + v_x, \eta + v_y)\, d\xi\, d\eta}{\iint |H(\xi, \eta)|^2\, d\xi\, d\eta} \tag{10-103}$$

Letting

$$\xi' = \xi + \frac{v_x}{2}, \qquad \eta' = \eta + \frac{v_y}{2} \tag{10-104}$$

gives

$$F(v_x, v_y) = \frac{\iint H(\xi' - v_x/2, \eta' - v_y/2)H^*(\xi + v_x/2, \eta' + v_y/2)\, d\xi'\, d\eta'}{\iint |H(\xi', \eta')|^2\, d\xi'\, d\eta'} \tag{10-105}$$

Then we drop the primes on ξ' and η' and use Eq. (10-98) to write this as

$$F(v_x, v_y) = \iint P\left(\xi + \frac{t_2' v_x}{2k}, \eta + \frac{t_2' v_y}{2k}\right)P^*\left(\xi - \frac{t_2' v_x}{2k}, \eta - \frac{t_2' v_y}{2k}\right) d\xi\, d\eta \tag{10-106}$$

The denominator has also been dropped, since we can do the normalization with respect to some arbitrary value of *F*. This extremely important result indicates that, for an aberration-free lens, the *OTF* is simply the area of overlap of two displaced pupil functions, one centered about $(t_2'v_x/2k, t_2'v_y/2k)$ and the other about the diametrically opposite point $(-t_2'v_x/2k, -t_2'v_y/2k)$.

Example: A Circular Aperture

The most obvious, although not the simplest, example to which Eq. (10-106) should be applied is the circular pupil. For convenience, we will take the centers of the two displaced circles as shown in Fig. 10-37(a). There is no loss in generality, since the overlap depends only on the spacing between centers. The overlap area (shown shaded) is four times the segment labeled S

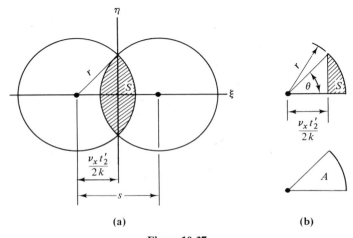

(a)   (b)

Figure 10-37

in Fig. 10-37(a). To find S, we compute the area A for the sector of angle θ and radius r, which is

$$A = \left(\frac{\theta}{2\pi}\right)(\pi r^2) = \frac{\theta r^2}{2}$$

where

$$\theta = \text{arc cos}\left(\frac{s}{2r}\right)$$

and where

$$s = \frac{v_x t_2'}{k}$$

is the spacing between centers. Then, as shown by Fig. 10-37(b),

$$S = A - \frac{1}{2}\left(\frac{s}{2}\right)\sqrt{r^2 - \left(\frac{s}{2}\right)^2}$$

By Eq. (10-106)

$$F(v_x, 0) = \frac{4S}{\pi r^2}$$

$$= \frac{2}{\pi}\left[\arccos\left(\frac{s}{2r}\right) - \left(\frac{s}{2r}\right)\sqrt{1 - \left(\frac{s}{2r}\right)^2}\right] \quad (10\text{-}107)$$

for

$$0 \le \frac{s}{2r} \le 1$$

Furthermore, $I(v_x, 0) = 0$ when $s \ge 2r$, since this means there is no overlap.

The value v_{xo} of v_x when the circles just touch is called the *cutoff frequency*. It represents the maximum spatial frequency passed by the lens and is given by

$$s = 2r$$

or

$$v_{xo} = \frac{2rk}{t_2'} = \frac{4\pi r}{\lambda t_2'}$$

Plotting Eq. (10-107), we obtain Fig. 10-38(a) (lower curve), noting that $F(0, 0) = 1$. The two-dimensional function, $F(v_x, v_y)$, is shown in Fig. 10-38(b).

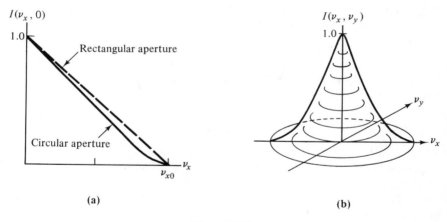

(a)

(b)

Figure 10-38

Problem 10-15

Find the transfer function for a square aperture of side $d = 2r$, where r is the radius of the circular aperture of Fig. 10-35(a). Verify the upper curve in this figure and the fact that v_{xo} is the same in both cases. ∎

10-10d Resolving Power

In order to consider the significance of the *OTF* to the lens designer and user, let us give some further thought to the concept of *resolution* or *resolving power* ℜ discussed in a qualitative way in connection with Fig. 10-8. For simplicity, we shall consider a source of light at infinity passing through a narrow slit of length l and width w. If a lens is adjacent to the slit, the diffraction pattern will be focused on a screen at a distance F'. Suppose the object (at infinity) consists of regularly spaced black and white stripes (Fig. 10-39),

Figure 10-39

and the spacing is specified in lines/mm. This quantity is the spatial analogue of the frequency of a wave (in hertz or cycles/second) and hence is the spatial frequency v_x associated with the periodic object. We therefore take it as a measure of the resolving power, since the closer the spacing, the greater the demand on the optical system. Problem 10-15 for a rectangular opening of width w gives a cutoff frequency, for a distant object, of magnitude

$$v_{xo} = \frac{kw}{f'}$$

Using a lens whose diameter d exactly matches the slit width, we then have a maximum resolution given by

$$\mathfrak{R} = \frac{k}{f'/w} \tag{10-108}$$

where f'/w is the f-number of Problem 10-3. A perfect $f/2.8$ camera lens in green light (555.0 nm) has a resolving power of about 4000 lines/mm.

Returning to the interpretation of the *OTF*, we realize that Fig. 10-38(a) shows the resolving power of two kinds of apertures used in conjunction with perfect lenses. In fact, Eq. (10-108) indicates that this quantity is directly proportional to the edge of the square (or the diameter of the circle). The graph also shows how the intensity of the diffracted image decreases with closer spacing of the black bars in Fig. 10-39. We recall, however, that we have defined two types of optical transfer function; the amplitude function H is specified by Eq. (10-97) and the intensity function F is given in Eq. (10-99). Another way of distinguishing these two quantities is in terms of coherence. Although we considered this subject at several places in Part II, let us take a very intuitive approach here and stipulate that a coherent plane wave is one for which the phase throughout the wavefront is a known function of position. We can find the intensity received at a screen by combining amplitudes and phases in the manner of Eq. (5-17) and squaring the result. On the other hand, if the wave is incoherent and the phases are random, then we must work directly with the intensities. This means that the transfer function H can be applied to a completely coherent wave and only to such a wave. If we have an aberration-free lens, the pupil function $P(x_L, y_L)$ has a value of unity when r_L is less than or equal to the lens radius R and it vanishes outside the lens. This condition can be expressed in terms of the *circle function*, defined as

$$\text{circ}\left(\sqrt{x_2 + y_2}\right) = \begin{cases} 1 \text{ when } \sqrt{x^2 + y^2} \le 1 \\ 0 \text{ otherwise} \end{cases} \tag{10-109}$$

so that the pupil function becomes

$$P(x_L, y_L) = \text{circ}\left(\frac{\sqrt{x_L^2 + y_L^2}}{R}\right) \tag{10-110}$$

A lens conforming to Eq. (10-110) is said to be *diffraction limited*; its ability to produce a perfect image is restricted only by the diffraction introduced at the edge, which converts each mathematical point in the object to an Airy disc of finite diameter.

The explicit expression for P given by Eq. (10-110) enables us to evaluate the coherent transfer function H through the use of Eq. (10-97), obtaining

$$H(v_x, v_y) = \text{circ}\left(\frac{\sqrt{x_L^2 + y_L^2}}{Rk/t_2'}\right) \tag{10-111}$$

The definition of the circle function means that the cutoff frequency will be

$$v_{co} = \frac{Rk}{t_2'} = \frac{2\pi R}{\lambda t_2'} \tag{10-112}$$

We can compare this result with the Airy disc radius r', as obtained from Eq. (10-16). Since θ is approximately given by r'/t_2', we have that

$$r' = 0.72 \frac{t_2' \lambda}{R} \tag{10-113}$$

Let us define a cutoff radius or distance r_{co} as the reciprocal of the cutoff frequency (in analogy with the relation between frequency and period for time-varying functions) so that Eq. (10-112) gives

$$r_{co} = \frac{t_2' \lambda}{2\pi R} = 0.16 \frac{t_2' \lambda}{R} \tag{10-114}$$

We thus see that there is fairly good agreement between the size of the Airy disc, as determined by elementary diffraction theory and the resolution limit of the lens specified by the frequency for which the OTF vanishes. This is not surprising, because the former is what determines the Rayleigh criterion for the minimum possible resolution.

Returning to the situation for incoherent light, let r of Eq. (10-107) have a maximum value equal to the lens radius R, so that the incoherent cutoff frequency is

$$v_{co} = \frac{4\pi R}{\lambda t'_2} \qquad (10\text{-}115)$$

which we recognize as twice the value given in Eq. (10-112). We have therefore shown that a lens passing coherent light will have a constant OTF right up to the maximum spatial frequency it can pass and beyond this value, the transfer function vanishes. For incoherent light, on the other hand, the OTF falls monotonically to a cutoff frequency which is half that for coherent light.

The next thing we must worry about is the way in which lens aberrations modify these ideal results. This requires that we have a way of expressing the effect of aberrations on a light wave, and this is the subject of the following section.

10-11   Wave Theory of Aberrations

Consider a group of parallel rays leaving an object and passing through a perfectly corrected lens. These rays will converge to a common image point P' (Fig. 10-40), and they may be regarded as originating from a spherical

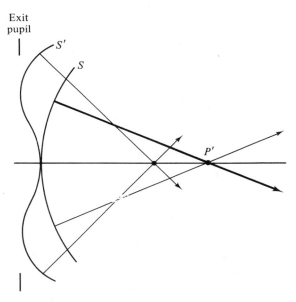

Figure 10-40

surface S which is bounded by the exit pupil. This sphere is the *wave front* producing the image.

10-11a The Wave Aberration

If the lens system producing this spherical wave front has aberrations, then the image may fall at some other point, and it may even be spread out to form a patch of light rather than a sharp point. We then realize that the sphere S is not generating the image; the image may be due to some other sphere and possibly to a distorted surface S', as shown in Fig. 10-40. Aberrations are small for small angles; therefore S' is similar to S near the axis, and the rays from this portion of the distorted wave front meet at the paraxial focus P'. However, rays from the more extreme parts of the wave front meet all along a line rather than forming a sharp image point.

To express the distortion of S into S' in a quantitative way, it is customary to use the distance w (Fig. 10-41) representing the variable spacing between

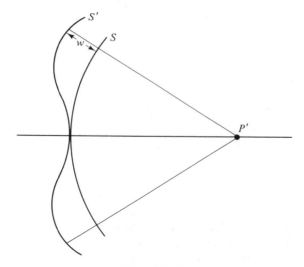

Figure 10-41

the two surfaces as measured along the radius of the sphere. This quantity is called the *wave aberration*. We shall find for spherical aberration, coma, and astigmatism that the wave aberration corresponds to a distorted surface. For curvature of the field and distortion, however, the sphere is unaltered, but displaced; the rays simply meet at the wrong point.

To study this problem quantitatively, consider the situation in Fig. 10-42. The undistorted wave front, called the *reference sphere*, has its center at the arbitrary point $P(x_0, y_0, z_0)$, and it passes through the origin (this is an

arbitrary choice on our part). We choose another point $P(x, y, 0)$ on the
distorted wave front S' which lies in the x-y plane. Then the wave front
aberration w is

$$w = r - \overline{PP_0}$$
$$= [x_0^2 + y_0^2 + z_0^2]^{1/2} - [(x_0 - x)^2 + (y_0 - y)^2 + z_0^2]^{1/2} \quad (10\text{-}116)$$

We write this as

$$w = z_0\left[\left\{1 + \frac{x_0^2 + y_0^2}{z_0^2}\right\}^{1/2} + \left\{1 + \frac{(x_0 - x)^2 + (y_0 - y)^2}{z_0^2}\right\}^{1/2}\right] \quad (10\text{-}117)$$

Assuming that

$$z_0^2 \gg x_0^2 + y_0^2$$

we may expand in a binomial series to obtain

$$w = z_0\left[1 + \frac{x_0^2 + y_0^2}{2z_0^2} - \frac{x_0^4 + 2x_0^2 y_0^2 + y_0^4}{8z_0^4}\right]$$
$$- z_0\left[1 + \frac{x_0^2 - 2xx_0 + x^2 + y_0^2 - 2yy_0 + y^2}{2z_0^2}\right.$$
$$- \frac{1}{8z_0^4}\{x_0^4 + 4x^2 x_0^2 + x^4 + y_0^4 + 4y^2 y_0^2 + y^4$$
$$+ 2(x_0^2 x^2 + x_0^2 y_0^2 + x_0^2 y^2 + x^2 y_0^2 + x^2 y^2 + y_0^2 y^2)$$
$$+ 2(4xx_0 yy_0 - 2xx_0^3 - 2x^3 x_0 - 2xx_0 y_0^2$$
$$\left.- 2xx_0 y^2 - 2yy_0 x_0^2 - 2yy_0 x^2 - 2yy_0^3 - 2y^3 y_0)\}\right]$$

or

$$w = \left\{\frac{2xx_0 - x^2 + 2yy_0 - y^2}{2z_0}\right\} + \frac{1}{8z_0^3}[x^4 + y^4 + 2x^2 y^2 + 4x^3 x_0$$
$$+ 4y^3 y_0 + 4x^2 yy_0 + 4xx_0 y^2 - 6x^2 x_0^2 - 6y^2 y_0^2 - 2x^2 y_0^2$$
$$- 2y^2 x_0^2 - 8xx_0 yy_0 + 4xx_0^3 + 4yy_0^3 + 4xx_0 y_0^2 + 4yy_0 x_0^2] \quad (10\text{-}118)$$

Although this analytical expression for the value of w appears complicated, its interpretation is quite straightforward. Following a development of
Rayces,[10-22] we show in Fig. 10-43 a point P on a distorted wave front and
a point S on the reference sphere with radius r and center Q. The two surfaces
have a common point C as they emerge from the optical system. The coordinate system is chosen so that XOY is the image plane and XOZ is the meridional plane.

Let a ray PT, normal to the actual wave front at P, strike the image plane
at T. Then $\overline{QT}$ measures the deviation of the actual image from the center Q
of the reference sphere; we therefore call it the *ray aberration*. The wave
aberration w on the other hand is specified by the distance

$$w = \overline{QS} - \overline{QP}$$
$$= r - v \quad (10\text{-}119)$$

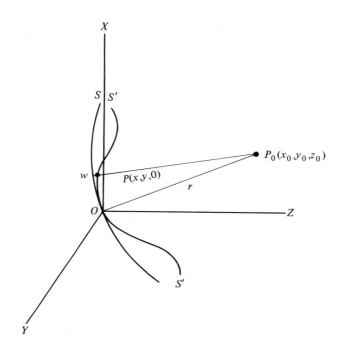

Figure 10-42

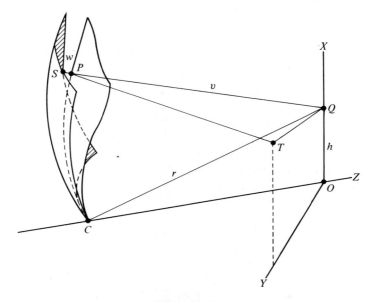

Figure 10-43

where we denote $\overline{QP}$ as v. If P has coordinates x, y, z, then the distance $v = \overline{PQ}$ may be expressed as

$$(x - h)^2 + y^2 + z^2 = v^2 \tag{10-120}$$

which may be written in the alternate form

$$G(x, y, z, v) = 0 \tag{10-121}$$

The surface containing P and C may also be specified by the equation

$$F(x, y, z) = 0 \tag{10-122}$$

or

$$z = f(x, y) \tag{10-123}$$

Further, we may eliminate z from Eqs. (10-121) and (10-122) to obtain

$$v = g(x, y) \tag{10-124}$$

From Eq. (10-123), we have

$$dz = \frac{\partial z}{\partial x} dx + \frac{\partial z}{\partial y} dy \tag{10-125}$$

and from Eq. (10-124)

$$dv = \frac{\partial v}{\partial x} dx + \frac{\partial v}{\partial y} dy \tag{10-126}$$

Differentiating the implicit function in Eq. (10-122) gives

$$\frac{\partial F}{\partial x} dx + \frac{\partial F}{\partial y} dy + \frac{\partial F}{\partial z} dz = 0 \tag{10-127}$$

and doing the same to Eq. (10-121)

$$\frac{\partial G}{\partial x} dx + \frac{\partial G}{\partial y} dy + \frac{\partial G}{\partial z} dz + \frac{\partial G}{\partial v} dv = 0 \tag{10-128}$$

If we use Eqs. (10-125) and (10-126), these two equations become

$$\left(\frac{\partial F}{\partial x} + \frac{\partial F}{\partial z} \frac{\partial z}{\partial x} \right) dz + \left(\frac{\partial F}{\partial y} + \frac{\partial F}{\partial z} \frac{\partial z}{\partial y} \right) dy = 0 \tag{10-129}$$

and

$$\left(\frac{\partial G}{\partial x} + \frac{\partial G}{\partial z} \frac{\partial z}{\partial x} + \frac{\partial G}{\partial v} \frac{\partial v}{\partial x} \right) dx + \left(\frac{\partial G}{\partial y} + \frac{\partial G}{\partial z} \frac{\partial z}{\partial y} + \frac{\partial G}{\partial v} \frac{\partial v}{\partial y} \right) dy = 0 \tag{10-130}$$

Since dx and dy are independent variables, the last two equations are valid only if the individual coefficients of dx and dy vanish. Hence,

$$\frac{\partial F}{\partial x} + \frac{\partial F}{\partial z} \frac{\partial z}{\partial x} = 0$$

$$\frac{\partial F}{\partial y} + \frac{\partial F}{\partial z} \frac{\partial z}{\partial y} = 0$$

$$\frac{\partial G}{\partial x} + \frac{\partial G}{\partial z} \frac{\partial z}{\partial x} + \frac{\partial G}{\partial v} \frac{\partial v}{\partial x} = 0$$

$$\frac{\partial G}{\partial y} + \frac{\partial G}{\partial z} \frac{\partial z}{\partial y} + \frac{\partial G}{\partial v} \frac{\partial v}{\partial y} = 0$$

from which

$$\frac{\partial v}{\partial x} = \frac{1}{\partial G/\partial v}\left(\frac{\partial G}{\partial z}\frac{\partial F/\partial x}{\partial F/\partial z} - \frac{\partial G}{\partial x}\right)\Bigg\} $$
$$\frac{\partial v}{\partial y} = \frac{1}{\partial G/\partial v}\left(\frac{\partial G}{\partial z}\frac{\partial F/\partial y}{\partial F/\partial z} - \frac{\partial G}{\partial y}\right)\Bigg\}$$

$$(10\text{-}131)$$

The partial derivatives of G, from Eq. (10-120), are

$$\frac{\partial G}{\partial x} = \frac{\partial}{\partial x}[(x - h)^2 + y^2 + z^2 - v^2] = 2(x - h)$$

$$\frac{\partial G}{\partial y} = 2y$$

$$\frac{\partial G}{\partial z} = 2z$$

$$\frac{\partial G}{\partial v} = -2v$$

and Eq. (10-130) becomes

$$\frac{\partial v}{\partial x} = \frac{(x - h) - z\dfrac{\partial F/\partial x}{\partial F/\partial z}}{v} \Bigg\}$$
$$\frac{\partial v}{\partial y} = \frac{y - z\dfrac{\partial F/\partial y}{\partial F/\partial z}}{v} \Bigg\}$$

$$(10\text{-}132)$$

We know from vector analysis that the gradient of a function F, as in Eq. (10-122), specifies the direction of the normal, and the three components of the gradient are proportional to the direction-cosines α, β, γ of this normal with respect to the axes. Hence

$$\frac{\partial F}{\partial x} : \frac{\partial F}{\partial y} : \frac{\partial F}{\partial z} = \alpha : \beta : \gamma \qquad (10\text{-}133)$$

Since $\overline{PT}$ in the figure is such a normal, we shall designate its direction-cosines α, β, γ, as used in Eq. (10-133). Also, its end points will have coordinates x', y'. It can be shown that

$$\frac{\alpha}{\gamma} = \frac{(x - h) - x'}{z} \Bigg\}$$
$$\frac{\beta}{\gamma} = \frac{y - y'}{z} \Bigg\}$$

$$(10\text{-}134)$$

for if we have a vector $\mathbf{r}$ through the origin, then its direction-cosines α, β, γ are related to the components by

$$\alpha = \frac{x}{r}, \qquad \beta = \frac{y}{r}, \qquad \gamma = \frac{z}{r}$$

so that

$$\frac{\alpha}{\beta} = \frac{x}{y}$$

and so on; Eq. (10-134) is simply a generalization of this.

Combining Eqs. (10-133) and (10-134) converts Eq. (10-132) into

$$\frac{\partial v}{\partial x} = \frac{x'}{v}, \qquad \frac{\partial v}{\partial y} = \frac{y'}{v} \qquad (10\text{-}135)$$

Finally, by Eq. (10-118)

$$\frac{\partial v}{\partial x} = -\frac{\partial w}{\partial x}, \qquad \frac{\partial v}{\partial y} = -\frac{\partial w}{\partial y}$$

and

$$\frac{\partial w}{\partial x} = \frac{x'}{r - w}, \qquad \frac{\partial w}{\partial y} = \frac{v'}{r - w} \qquad (10\text{-}136)$$

These equations relate the wave aberration to the ray aberration. They are usually used in the approximate form

$$\frac{\partial w}{\partial x} = -\frac{x'}{r}, \qquad \frac{\partial w}{\partial y} = -\frac{y'}{r} \qquad (10\text{-}137)$$

Returning to Eq. (10-118), we consider the first three terms of the final expression on the right. These are

$$\frac{x^4 + y^4 + 2x^2y^2}{8z_0^3} = \frac{(x^2 + y^2)^2}{8z_0^3} \qquad (10\text{-}138)$$

We convert to polar coordinates in the plane of the exit pupil by

$$x = \rho \cos \theta$$
$$y = \rho \sin \theta \qquad (10\text{-}139)$$

Equation (10-138) is then

$$\frac{(x^2 + y^2)^2}{8z_0^3} = C_1 \rho^4 \qquad (10\text{-}140)$$

where $C_1 = 1/8z_0^3$ is a constant as far as the distorted sphere is concerned. Let us assume that this term represents the only contribution to w. Then

$$w = C_1 \rho^4$$

and by Eq. (10-137), ignoring minus signs,

$$x' = rC_1 \left[4\rho^3 \frac{\partial \rho}{\partial x} \right]$$

$$= 4rC_1 \rho^3 \frac{x}{\rho} = 4rC_1 x \rho^2$$

and

$$y' = 4rC_1 y \rho^2$$

We then see that the radius ρ' of the image is

$$\rho' = \sqrt{x'^2 + y'^2} = 4rC_1\rho^2\sqrt{x^2 + y^2} = 4rC_1\rho^3 \qquad (10\text{-}141)$$

so that the image diameter is proportional to the cube of the exit pupil diameter. This is just the behavior we found in Chapter 2 for spherical aberration when we are beyond the paraxial range but not yet into the fifth-power region.

The next four terms, if acting alone, would give a wave aberration

$$
\begin{aligned}
w &= \frac{4(x^3 x_0 + y_1^3 y_0 + x^2 y y_0 + x x_0 y^2)}{8z_0^3} \\
&= \frac{4\rho^3(x_0 \cos^3\theta + y_0 \sin^3\theta + y_0 \cos^2\theta \sin\theta + x_0 \cos\theta \sin^2\theta)}{8z_0^3} \\
&= \frac{\rho^3(x_0 \cos\theta + y_0 \sin\theta)}{2z_0^3} = \frac{(xx_0 + yy_0)\rho^2}{2z_0^3} \qquad (10\text{-}142)
\end{aligned}
$$

It is convenient at this point to choose the image plane coordinates so that $x_0 = 0$. This is simply a matter of rotating the system about the z axis of Fig. 10-42 until P_0 lies in the y-z plane and causes no loss in generality. Then Eq. (10-142) becomes

$$w = C_2\rho^2 y y_0 = C_2(x^2 + y^2)yy_0 \qquad (10\text{-}143)$$

where $C_2 = \frac{1}{2}z_0^3$ and by Eq. (10-137)

$$
\begin{aligned}
x' &= 2y_0 C_2 xy \\
y' &= y_0 C_2 (x^2 + 3y^2)
\end{aligned} \qquad (10\text{-}144)
$$

or converting back to polar coordinates

$$
\begin{aligned}
x' &= rhC_2\rho^2 \sin 2\theta \\
y' &= rhC_2\rho^2 (2 - \cos 2\theta)
\end{aligned} \qquad (10\text{-}145)
$$

where we have written y_0 as h, the paraxial image height.

Equations (10-145) have the form of the well-known parametric equations for a circle, which are

$$x = R\cos\varphi$$

$$y = R\sin\varphi$$

Hence, to each zone ($\rho = $ constant) in the plane of the exit pupil, there is a circle in the image plane of radius $rhC_2\rho^2$ and with center on the y axis at the point $(0, -2rhC_2\rho^2)$. As ρ changes, so does the radius and the position of the image circle, which we recognize as a description of coma as the sole aberration.

Jumping to the last four terms in Eq. (10-118), we have

$$w = \frac{(x_0^3 + x_0 y_0^2)x + (y_0^3 + y_0 x_0^2)y}{2z_0} = \rho_0^2(xx_0 + yy_0) \qquad (10\text{-}146)$$

Then

$$x' = \frac{\rho_0^2 x_0 r}{2z_0}$$

$$y' = \frac{\rho_0^2 y_0 r}{2z_0}$$

(10-147)

or

$$\rho' = \frac{r\rho_0^3}{2z_0}$$

(10-148)

Thus, the position of the image point depends on the cube of the distance of the paraxial image point from the axis. This leads to what we have previously identified as distortion.

Problem 10-16

Identify the remaining terms in the coefficient of $(1/8z_0^3)$ with astigmatism and curvature of the field.   ■

10-11b   The OTF for a Lens with Aberrations

Now that we have an analytical expression for the wave aberration which shows its connection with the third-order geometric aberrations, we can consider the *OTF* for a real lens. Actually, a simple example to start with involves the phenomenon of defocusing, which is not a true aberration. The quadratic terms in the first expression on the right of Eq. (10-118) give a contribution

$$w = \frac{x^2 + y^2}{2z_0} = \frac{\rho^2}{2z_0}$$

(10-149)

Then by Eq. (10-137)

$$x' = \frac{xr}{z_0}, \qquad y' = \frac{yr}{z_0}$$

and

$$\rho' = \frac{r}{z_0}$$

(10-150)

Hence, each ring of radius ρ in the exit pupil corresponds to a ring of radius ρ' at the image plane. However, recall that we obtained Eq. (10-87b) through the process of eliminating quadratic terms in Eq. (10-62) by satisfying the Gauss equation. In the general case, these terms involve $\exp(x_L^2 + y_L^2)$, so that in our present notation, they correspond to $(x^2 + y^2)$, as measured in the exit pupil. Thus, we associate the quadratic expression in Eq. (10-149) with defocusing, and all we need do to eliminate this error is shift the image plane.

The effect of the wave aberration w on the light passing through a lens is to change the path length of each ray, as indicated in Fig. 10-41. This corresponds to a phase shift kw, which we incorporate into the diffraction integrals and the definition of the *OTF* by the exponential term exp (ikw). Thus the pupil function $P(x, y)$, which is unity for an ideal lens, becomes

$$P(x, y) = e^{ikw(x,y)}$$

and the *OTF* integral in Eq. (10-106) is now

$$F(v_x, v_y) = \int_{-\infty}^{\infty} \int_{-\infty}^{\infty} e^{ikw[\xi + (t_2'v_x/2k), \eta + (t_2'v_y/2k)]} e^{-ikw[\xi + (t_2'v_x/2k), \eta - (t_2'v_y/2k)]} \, d\xi \, d\eta$$

$$(10\text{-}151)$$

In the most general case, w will be the power series of Eq. (10-118), but it is easier to consider each aberration individually. Also, it is customary to write Eq. (10-118) in a different form when using it in *OTF* calculations. Let the polar coordinates of Eq. (10-139) be converted into reduced form by the relation

$$\rho' = \frac{\rho}{R}$$

where R is the radius of the exit pupil. We then drop the prime and let ρ designate a relative displacement. In the same way, $h = y_0$ of Fig. 10-41 will be treated as a relative value. If we then substitute cos θ, sin θ, 0, and h for x, y, x_0, and y_0, respectively, in Eq. (10-118), and consolidate all the constants into a set C_{ijk}, we obtain

$$w(h, \rho, \theta) = (C_{020}\rho^2 + C_{111}h\rho \cos \theta) + (C_{040}\rho^4 + C_{131}h\rho^3 \cos \theta$$
$$+ C_{222}h^2\rho^2 \cos^2 \theta + C_{220}h^2\rho^2 + C_{311}h^3\rho \cos \theta) \quad (10\text{-}152)$$

where the subscripts on the C_{ijk} designate the exponents i, j, k in the expression $C_{ijk}h^i\rho^j \cos^k \theta$, and the two groups of terms constitute the defocusing effects and the third-order aberrations, respectively.

It has been shown by Longhurst[10-1] that this result, which we have laboriously calculated, is almost intuitively obvious. For example, having located the meridional plane by letting $x_0 = 0$, we see that $w(h, \rho, \theta)$ can depend on θ only through cos θ and not sin θ, since the aberrations are symmetric about this plane. For those aberrations which do not depend on θ— that is, they are symmetric with respect to the axis of the wave front—the terms involving h and ρ must be of the form h^2, ρ^2, or $h^2\rho^2$, since they must be independent of sign.

Problem 10-17

Explain the remaining terms in Eq. (10-152) in similar fashion. ■

As an example of an actual determination of the *OTF*, Hopkins[10-23] has calculated the frequency response of a defocused lens with either a square or a circular exit pupil. Using Eq. (10-151) with only $C_{020} \neq 0$ in Eq. (10-152) and simplifying to a one-dimensional problem gives

$$F(v_x, 0) = \int \int e^{ikC_{020}[(\zeta + s/2)^2 + \eta^2]} e^{-ikC_{020}[(\zeta - s/2)^2 + \eta^2]} \, d\xi \, d\eta \qquad (10\text{-}153)$$

where $s = v_x t_2'/k$, as before. Using relative coordinates, the overlap of two round openings has a maximum value of π (the area of a unit circle); thus Eq. (10-153), when normalized, is

$$F(v_x, 0) = \frac{1}{\pi} \int \int e^{ia\xi} \, d\xi \, d\eta \qquad (10\text{-}154)$$

where

$$a = 2kC_{020}s$$

When the inclination of r in Fig. 10-37 is smaller than θ, we have

$$\left(\frac{s}{2} + \xi\right)^2 + \eta^2 = r^2 = 1$$

so that the limits for the integration over ξ in one quadrant are 0 and $\sqrt{1 - \eta^2} - s/2$. Then

$$F(v_x, 0) = \frac{4}{\pi a} \int_0^{\sqrt{1 - (s/2)^2}} \sin\left[a\left\{\sqrt{1 - \eta^2} - \frac{s}{2}\right\}\right] d\eta \qquad (10\text{-}155)$$

This integral can be expressed as a power series in Bessel functions, but it is simpler to do it numerically.

Problem 10-18

Show that Eq. (10-155) reduces to the correct form when $C_{020} = 0$. ■

The behavior of the *OTF* as a function of the normalized spatial frequency v_x/v_{xo} is shown in Fig. 10-44. These plots were generated by letting

$$C_{020} = \frac{n\lambda}{\pi}$$

where $n = 0, 1, 2, 3$, and 4. Then a takes on values

$$a = 2ks \frac{n\lambda}{\pi}$$

$$= 4ns$$

and s goes from 0 to 2, the maximum value permitted in the upper limit.

Problem 10-19

Verify the integration numerically for $C_{020} = n\lambda/4$, $n = 0$ to 4. ■

A very significant feature of the curves for $n = 3$ or 4 is that the *OTF* can actually go negative. This implies a phase reversal in the image, as illus-

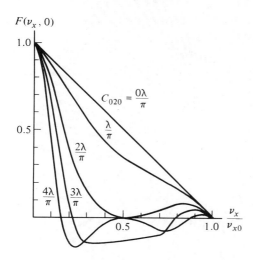

Figure 10-44

trated by the effect on the radial object of Fig. 10-45(a). The image shows an interchange of black and white areas corresponding to the places where $F(v_x, 0)$ crosses or touches the axis (Fig. 10-45(b)). This is reminiscent of the diffraction pattern due to a circular aperture (Fig. 10-6), and in fact, it is possible to confuse the Airy disc of an aberration-free lens with the first zero in the *OTF* curve of a lens for which the focusing error is about 1 wavelength. This is the well-known effect called *spurious resolution*.

Problem 10-20

(a) Evaluate Eq. (10-151) for a unit square aperture, showing that

$$F(v_x, 0) = \frac{2}{a} \sin\left[\frac{(1-s)a}{2}\right] \tag{10-156}$$

(b) Show that when the focusing error is large, the cutoff frequency becomes

$$v_{ox} = \frac{\pi}{f'} C_{020}$$

The third-order aberrations can be treated in a similar fashion. For example, the spherical aberration term in Eq. (10-152) is $C_{040}\rho^4$. Using this in Eq. (10-151) gives

$$F(v_x, 0) = \iint e^{ikC_{040}[(\xi + s/2)^2 + \eta^2]^2} e^{-ikC_{040}[(\xi - s/2)^2 + \eta^2]^2} \, d\xi \, d\eta$$

$$= \iint e^{ikC_{040}(4\xi^3 s + \xi s^3 + 4\xi s\eta^3)} \, d\xi \, d\eta \tag{10-157}$$

This integral must be evaluated numerically for both variables, and the results for small values of C_{040} are similar to those for $C_{020} = \lambda/\pi$ in Fig. 10-42.

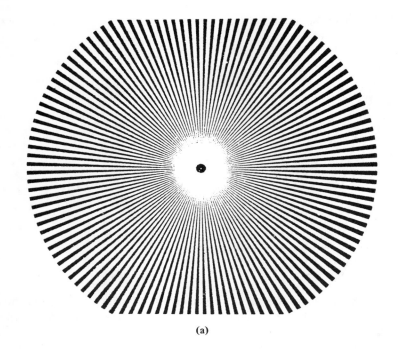

(a)

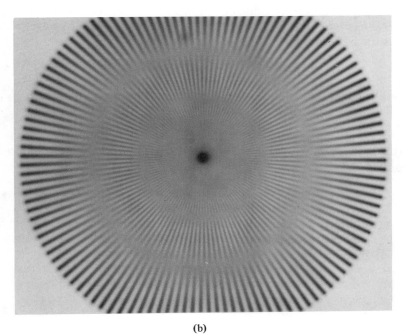

(b)

Figure 10-45

10-11c Image Formation as a Convolution Process

The discussion of the *OTF* just given indicated that the diffraction of light by a given optical system implies that there is an upper limit to the ability of the system to resolve a bar target such as in Fig. 10-39. There is another vital piece of information, however, which comes from *OTF* calculations or measurements, and this we shall consider now. Following a development of Francon,[10-24] we shall show in a rather simple way that image formation is a convolution process representing a combination of the energy output of the object and the diffraction properties of the lens. Figure 10-46 shows a flat

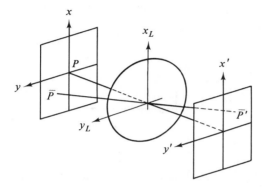

Figure 10-46

object in the *x-y* plane, the exit pupil of the lens in the x_L-y_L plane, and the image in the *x'-y'* plane. Let $E(x', y')$ be the illuminance of the image, each point of which is an Airy disc, and let $B(x, y)$ be the luminance (or brightness) of the corresponding point on the object. If we shift the object point P to a to a new position $\bar{P}$, the Airy disc at P' also moves, but the nature of the diffraction effect—except for the associated magnification β—is unchanged. Hence, if we normalize the image plane coordinates with respect to β, we can say that the illumination E anywhere on the image plane is a function only of the relative position of the image point with respect to the object point and may be written as $E(x' - x, y' - y)$. If we also normalize B and E so that we need not worry about units, then the luminous flux or relative intensity I received at the image plane is the integral of all the individual contributions, or

$$I(x', y') = \iint E(x' - x, y' - y)B(x, y) \, dx \, dy \qquad (10\text{-}158)$$

As seen from the Appendix, this is a convolution relation. It is shown there, in fact, that a convolution arises when two independent physical phenomena are jointly contributing to a particular result. Equation (10-158) indicates that the brightness B of the object, when combined with the diffraction pattern E of each point [this is what we previously identified as the point spread function or impulse response $h(x', y')$]—determines the nature of the image.

An interesting demonstration of optical convolution, which brings out the meaning of the mathematical processes involved, has been devised by Haskell.[10-25] He takes as an example the convolution of a two-slit pattern with a three-slit pattern (Fig. 10-47(a) and (b)). As described in connection with Fig. A-4 of the Appendix, the convolution is formed by reversing one of the patterns, say $h(\alpha)$, to obtain Fig. 10-47(c). Then we shift $h(-\alpha)$ by an amount t, chosen in this case as $t = -2$ (Fig. 10-47(d)). The corresponding values of $g(\alpha)$ and $h(t - \alpha)$ are then multiplied to obtain the product of Fig. 10-47(e). The total shaded area amounts to 2.0 units, and this is indicated in the convolution $f(t)$ of Fig. 10-47(f).

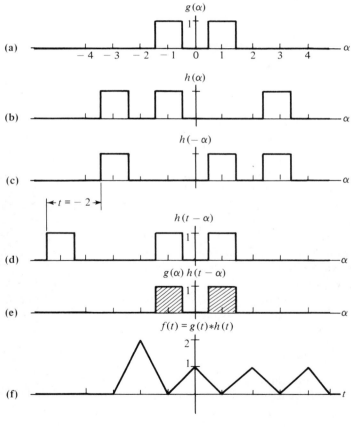

Figure 10-47

Problem 10-21

Justify the remainder of Fig. 10-47(f). ▄

The experimental arrangement is shown in Fig. 10-48(a). A laser beam is expanded and forms an image of a three-slit pattern $h(\alpha)$ at the plane where

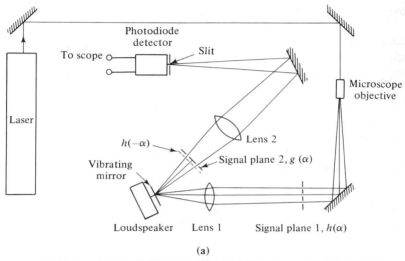

(a)

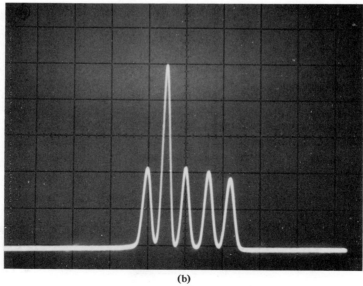

(b)

Figure 10-48

the two-slit pattern $g(\alpha)$ is located. Lens 1 gives an inverted image, which is equivalent to generating $h(-\alpha)$. The vibrating mirror sweeps $h(-\alpha)$ across $g(\alpha)$, thus performing the integration, and the received signal is displayed on an oscilloscope swept by the same signal that drives the speaker. The resulting trace (Fig. 10-48(b)) agrees very well with Fig. 10-47(f).

10-11d The Modulation Transfer Function

Returning to Eq. (10-158), our intuitive look at the meaning of the convolution process should indicate that image formation represents the combination of two independent functions: the variable energy output of the object and the effect of the diffraction process on each point in the object. To talk about the resulting resolution of the lens, we should use the black-and-white bar target as the object. It is convenient, however, to regard the target as alternate regions of high and low (but not zero) luminance and then decompose this pulse function into its Fourier series components. Let a typical target have a luminance of the form

$$B(x) = b_0 + b_1 \cos kx \qquad (10\text{-}159)$$

It is shown in Fig. 10-49(a). Note that b_0 represents the average value of this component of the total output, and that the maximum and minimum values

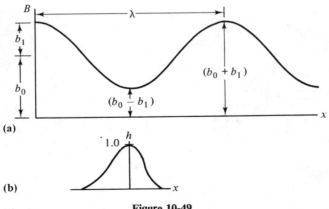

Figure 10-49

are, respectively, $(b_0 + b_1)$ and $(b_0 - b_1)$. Since the intensity I of the radiation is proportional to B, we can define the *contrast* or *modulation* by the relation

$$M = \frac{I_{max} - I_{min}}{I_{max} + I_{min}} \qquad (10\text{-}160)$$

which we recognize as identical in form to Eq. (6-14). The quantity M refers to variations in the illuminance received on a screen, whereas V specifies the corresponding behavior of a fringe system observed by the eye. Using Eq. (10-159), we obtain

$$M_0 = \frac{b_1}{b_0} \qquad (10\text{-}161)$$

where the subscript refers to the object function.

Let us next postulate some sort of arbitrary diffraction behavior represented by the impulse response curve of Fig. 10-49(b); the skew shape represents the distortion of sinc x by aberrations. We then perform the convolution of h with the cosine curve, but we shall do it analytically rather than graphically. The quantity h is a measure of the illuminance at each point of the image so that this convolution will determine the image intensity I. Equation (10-58) in one dimension, using Eq. (A-46), becomes

$$
\begin{aligned}
I(x') &= \int h(x)\{b_0 + b_1 \cos k(x' - x)\}\, dx \\
&= b_0 \int h(x)\, dx + b_1 \int h(x) \cos kx' \cos kx\, dx \\
&\quad + b_1 \int h(x) \sin kx' \sin kx\, dx
\end{aligned} \qquad (10\text{-}162)
$$

Normalizing $I(x')$ with respect to $\int h(x)\, dx$ (and retaining the same symbol) gives

$$
\begin{aligned}
I(x') &= b_0 + b_1 \frac{\int h(x) \cos kx\, dx}{\int h(x)\, dx} \cos kx' + b_1 \frac{\int h(x) \sin kx\, dx}{\int h(x)\, dx} \sin kx' \\
&= b_0 + b_1 h_R(k) \cos kx' + b_1 h_I(k) \sin kx'
\end{aligned} \qquad (10\text{-}163)
$$

where $k_R(k)$ and $h_I(k)$ are the real and imaginary parts, respectively, of the complex quantity

$$h(k) = \frac{\int h(x) e^{ikx}\, dx}{\int h(x)\, dx} \qquad (10\text{-}164)$$

We may write Eq. (10-163) as

$$I(x') = b_0 + b_1 |h(k)| \cos (kx' - \varphi) \qquad (10\text{-}165)$$

where

$$|h(k)| = \sqrt{h_R^2(k) + h_I^2(k)}$$

and

$$\varphi = \arctan \frac{h_I(k)}{h_R(k)}$$

Then the modulation M_i of the image, from Eqs. (10-160) and (10-165), is

$$
\begin{aligned}
M_i &= \frac{b_1 |h(k)|}{b_0} \\
&= M_0 |h(k)|
\end{aligned} \qquad (10\text{-}166)
$$

Thus, we see that the ratio of the image modulation to the object modulation is given by the magnitude of the impulse response. Since we have used $h(x)$ as a measure of intensity, we can say that $|h|$—known as the *modulation transfer function (MTF)*—is in fact the absolute magnitude of the spread or impulse response function we have previously used. The places where either the *OTF* or the *MTF* goes to zero represent a complete reversal in contrast since F in Eq. (10-99) depends on h.

The theory based on Eqs. (10-151) and (10-152) can be applied only when we know the aberration coefficients C_{ijk}. Since these are difficult to determine, another approach to the computation of the *MTF* is via the *spot diagram*, as described by Smith.[3-1] Consider a plane object and a group of points distributed uniformly and concentrically about the axis. The places where a parallel set of rays from each of these points strikes the paraxial image plane shown in Fig. 10-50(a) constitute the spot diagram. It may not reproduce the original distribution because of aberrations, and we obtain the spread function by simply combining the number of spots per unit width of the pattern. (Strictly speaking, this is a *line spread function*, which is appropriate for the image of a line in a cylindrical lens, but we are doing a one-dimensional problem in any case.) The function $h(x)$ obtained in this way is normalized with respect to its maximum value (Fig. 10-50(b)) and used to evaluate the integrals in Eq. (10-163), which are approximated by summations. It is also possible to measure the *MTF* directly by using cosine-wave targets and evaluating the integrals with data-processing techniques. A complete survey of the various methods is given by Rosenhauer and Rosenbruch.[10-26]

The significance of the results we have obtained here has been summarized by Smith in the following terms: excellent reproduction (serifs are distin-

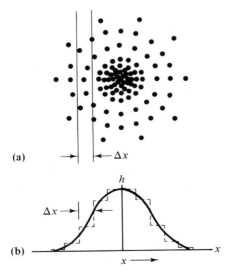

(a)

(b)

Figure 10-50

guishable) of a printed page requires the ability to resolve eight line pairs (one black and one white bar of identical width) per height of a lowercase e; legible reproduction requires five line pairs; and barely decipherable copy corresponds to three line pairs.

Problem 10-22

By regarding a plane wave as a series of wave fronts spaced a distance equal to the wavelength λ, show that the Bragg law of X-ray diffraction (see Halliday and Resnick[10-2]) may be obtained from Huyghens' principle.

Problem 10-23

Show that the diffraction pattern of an equilateral triangle is a six-pointed star, each arm of which is very similar to the pattern for a single slit. This result can be verified very easily by forming such an opening from three razor blades taped together with edges of about 1 mm. The aperture is then placed in front of a laser.

Problem 10-24

Figure 10-51(a) shows a pattern and Fig. 10-51(b) indicates its appearance after being spatially filtered. Describe the shape of the filter and explain how it produced the filtered image.

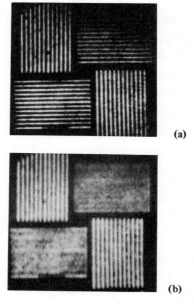

(a)

(b)

Figure 10-51

References

10-1 R. S. Longhurst, *Geometrical and Physical Optics*, 2nd ed., John Wiley & Sons (1967).

10-2 D. Halliday and R. Resnick, *Phyics*, John Wiley and Sons, New York (1966).

10-3 M. Abramowitz and I. A. Stegun, eds., *Handbook of Mathematical Functions*, U.S. Government Printing Office, Washington, D.C. (1964).

10-4 S. G. Lipson and H. Lipson, *Optical Physics*, Cambridge University Press (1969).

10-5 S. George, *Phys. Educ.* **6**, 349 (1971).

10-6 G. R. Graham, *Phys. Educ.* **6**, 352 (1971).

10-7 R. Boyer and E. Fortin, *Am. J. Phys.* **40**, 74 (1972).

10-8 A. Sommerfeld, *Optics*, Academic Press, New York (1954).

10-9 H. M. Smith, *Principles of Holography*, John Wiley & Sons, New York (1969).

10-10 I. S. Gradshteyn and I. M. Ryzhik, No. 3693. *Tables of Integrals, Series, and Products*, 4th ed., Academic Press (1965).

10-11 E. Abbe, *Archiv. Mikros. Anat.* **9**, 413 (1873).

10-12 A. B. Porter, *Phil. Mag.* **11**, 154 (1906).

10-13 Judith C. Brown, Am. J. Phys. **39**, 797 (1971).

10-14 A. Gerrard, *Am. J. Phys.* **31**, 723 (1963).

10-15 A. R. Shulman, *Optical Data Processing*, John Wiley & Sons, N.Y. (1970).

10-16 L. N. Peckham, M. O. Hagler and M. Kristiansen, *IEEE Trans., Educ.* **EP-13**, 60 (1970).

10-17 R. A. Phillips, *Am. J. Phys.* **37**, 536 (1969).

10-18 L. S. Watkins, *Proc. IEEE* **57**, 1634 (1969).

10-19 R.B. Hoover, Am. J. Phys. **37**, 871 (1969).

10-20 J. M. Stone, *Radiation and Optics*, McGraw-Hill, New York (1963)

10-21 J. W. Goodman, *Introduction to Fourier Optics*, McGraw-Hill, New York (1968).

10-22 J. L. Rayces, *Optica Acta* **11**, 85 (1964).

10-23 H. H. Hopkins, *Proc. Roy. Soc.* **A231**, 91 (1955).

10-24 M. Francon, *Modern Applications of Physical Optics*, Wiley-Interscience (1963).

10-25 R. E. Haskell, *IEEE Trans. Educ.* **E-14**, 110 (1971).

10-26 K. Rosenhauer and K. Rosenbruch, *Repts. Prog. Phys.* **30**, 1 (1962).

11

HOLOGRAPHY

11-1 Introduction

The word *hologram* comes from the Greek word *hólos* meaning the whole. In a hologram the entire wave front including amplitude and phase (which is very important) is recorded. As we have seen previously, phase variations in an object greatly affect its diffraction pattern. In the ordinary photographic process, the film records the intensity I, or E^2, throwing away the phase information. But in the holographic process, the film records both amplitude and phase. When a hologram is properly illuminated, an exact replica of the original wave front is reconstructed. This means that information is recorded in three dimensions in a hologram, as compared to two dimensions by an ordinary photographic camera.

The waves used to record holograms must be spatially and temporally coherent, thus requiring a laser as a source. We shall see that there are many different ways in which holograms are used. One is for information storage devices; a very high density can be stored. Another feature of holograms is that multiple exposures can be made on the same film. When two exposures are made of the same object, the reconstructed waves interfere, and this can be used to detect small changes in the dimension of an object. Holograms are also used in vibration analysis and in contouring.

The hologram is analogous to a modulated carrier wave in communications. When the modulated wave is mixed with a reference wave at the detector, it is demodulated, and this leaves the pure signal. Both the phase and amplitude of the signal are recovered in this process.

We are already familiar with methods that permit recording and display of phase information. In the phase contrast microscope (Chapter 10), variations in the phase of an object are converted into variations of amplitude. This is done by shifting the phase of the carrier (dc term). The general method then of recording phase information is to combine a wave with a uniform coherent reference wave.

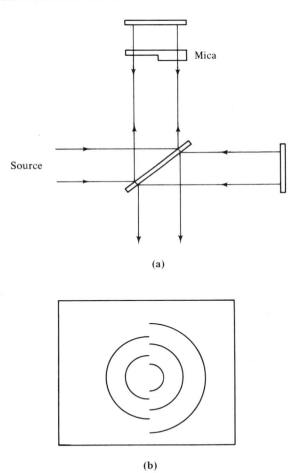

(a)

(b)

Figure 11-1 A piece of mica with a step in the surface is inserted in one arm of a Michelson interferometer. (a) Phase changes in the wave passing through the step are recorded by combining the wave with a reference wave from the other arm (b).

The procedure will be illustrated by a Michelson interferometer with a piece of mica in one arm. (See Section 7-8.) The mica is assumed to have a step across one surface. Since mica is transparent, it introduces phase variations in the wave passing through it. There will be a discontinuity in the phase of the wave front across the step. These phase variations are recorded when the wave is mixed with the reference wave from the other arm (Fig. 11-1).

11-2 A Plane-Wave Hologram

We will first consider the simplest possible hologram, one produced by recording the interference pattern of two plane waves (Fig. 11-2). One wave is designated the *reference wave* and the other is designated the *object wave*.

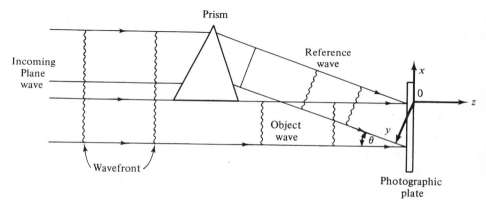

Figure 11-2 The arrangement for recording a hologram. The reference and object are plane waves.

These two plane waves set up a standing wave pattern which is recorded by the photographic plate. The object wave consists of a plane wave propagating in the z direction and is given (Chapter 5) by

$$E = E_1 e^{i(kz-\omega t)} \tag{11-1}$$

The reference wave propagates at an angle θ with respect to the z axis. This wave is described by the propagation vector $\mathbf{k}$, and the factor $\mathbf{k} \cdot \mathbf{r}$ is

$$\mathbf{k} \cdot \mathbf{r} = k_x x + k_z z \tag{11-2}$$

For small values of the angle θ, k_x and k_z are

$$k_x = k \sin \theta \sim k\theta \tag{11-3a}$$

$$k_z = k \cos \theta \sim k \tag{11-3b}$$

The reference wave then is described by

$$E_r = E_0 e^{i(k\theta x + kz - \omega t)} \tag{11-4}$$

and the combined field of the two waves is

$$E_T = E + E_r = e^{i(kz-\omega t)}[E_1 + E_0 e^{ik\theta x}] \tag{11-5}$$

The intensity I is obtained from Eq. (5-15) but we shall ignore the factor of $\frac{1}{2}$ since we are interested only in relative values. Thus

$$
\begin{aligned}
E_T^2 &= E_1^2 + E_0^2 + E_1 E_0 e^{ik\theta x} + E_1 E_0 e^{-ik\theta x} \\
&= E_1^2 + E_0^2 + 2E_1 E_0 \cos(k\theta x)
\end{aligned}
\tag{11-6}
$$

Next we must consider the photographic process. The density of exposed grains d in the film is given by

$$d = \gamma \log_{10} E_T^2 \tag{11-7}$$

where γ is a characteristic of the film that depends on E_T^2. After the film is developed and illuminated by a uniform plane wave, the transmission is described by the amplitude transmission factor t, related to d through the equation

$$t = 10^{-d/2} \tag{11-8}$$

Substituting the expression for d into Eq. (11-8) yields

$$t = (E_T^2)^{-\gamma/2} \tag{11-9a}$$

or

$$t = [E_0^2 + E_1^2 + 2E_0 E_1 \cos(k\theta x)]^{-\gamma/2} \tag{11-9b}$$

If we assume $E_0^2 \gg E_1^2$, this can be expanded to

$$
\begin{aligned}
t &= (E_0^2)^{-\gamma/2}\left[1 - \frac{\gamma}{2}\frac{E_1^2}{E_0^2} - \gamma\frac{E_1}{E_0}\cos(k\theta x) + \cdots\right] \\
&= C[E_0^2 - \gamma E_1^2 - 2\gamma E_0 E_1 \cos(k\theta x)]
\end{aligned}
\tag{11-9c}
$$

where C is a constant. If a plane wave of amplitude E_3 illuminates the plate, the transmitted amplitude is given by

$$E_{\text{tran}} = E_3 t \tag{11-10}$$

Assuming that E_0^2 is much greater than E_1^2, Eqs. (11-9c) and (11-10) show that, to within some constant factors, the transmitted wave is identical to the object wave captured on film.

Suppose we illuminate the film by a plane wave and ask. "What does the transmitted wave look like to an observer?" The transmitted wave is simply the diffraction pattern of the hologram. As shown in the previous chapter, the field distribution at a great distance from the film can be obtained by placing the film in front of a lens and looking for the field distribution on the focal plane of the lens. Also, the field distribution across the focal plane is the Fourier transform of the field distribution across the aperture.

The aperture function is

$$P(x) = E_3 t$$

$$= a + b \cos(k\theta x) \qquad (11\text{-}11)$$

and the transform (Eq. (A-22)) in terms of the spatial frequency $v = ku/f$
(Eq. (10-45)) is

$$E(u) = \int_{-\infty}^{\infty} (a + b \cos k\theta x) e^{ivx} \, dx \qquad (11\text{-}12)$$

$$= a\delta(0) + b\delta\left(\theta \pm \frac{u}{f}\right) \qquad (11\text{-}13)$$

There will be three spots on the focal plane behind the lens. The central one
at $u = 0$ (the dc term) represents the undeviated beam. There are two side
spots, one at $u = f\theta$ and the other at $u = -f\theta$. Writing the cosine in
exponential form, we see that one spot comes from $e^{+ik\theta x}$ and the other from
$e^{-ik\theta x}$. These are the first-order diffraction maxima (Fig. 11-3). Note that
only the first order is present in the diffraction pattern because there is only
one Fourier component in the aperture function. Without the lens, we would
have three plane waves leaving the hologram; an undeviated beam ($\theta = 0$)
and two beams deviated $\pm \theta$. It is instructive to look at this result employing
the terminology of holography; we previously considered it from another
point of view in Section 10-9.

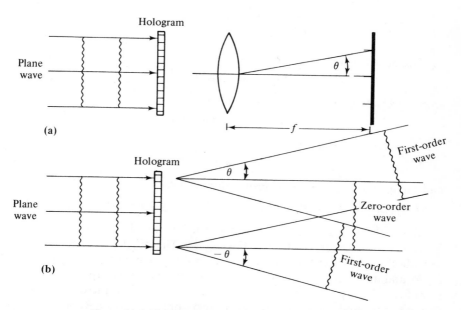

Figure 11-3 The reconstruction of the object wave front.

11-3   A Hologram of a Point Source

Next, let us place a screen having a pinhole in the object beam (Fig. 11-4). Light will be transmitted through the pinhole, so that the effective object will be a point source, and a spherical wave will be generated. The phase of the spherical wave at the center of the photographic plate is taken as kz, and at a distance x from the center of the plate the phase difference is $kx^2/2l$ (Section 10-8).

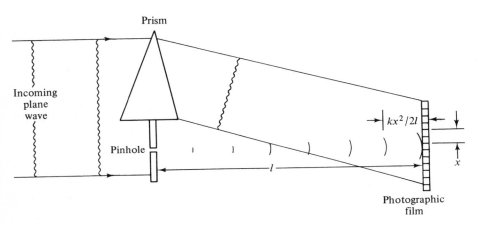

Figure 11-4 The reference wave is a plane wave; the object is a spherical wave.

The object wave is

$$E = E_1 e^{i[(kx^2/2l) + kz - \omega t]} \tag{11-14}$$

and the reference wave, as before, is

$$E_r = E_0 e^{i(k\theta x + kz - \omega t)}$$

The total field E_T is given by

$$E_T = e^{i(kz - \omega t)}[E_0 e^{ik\theta x} + E_1 e^{ikx^2/2l}] \tag{11-15}$$

and the intensity at the film is proportional to E_0^2

$$E_T^2 = E_T E_T^* = E_0^2 + E_1^2 + E_0 E_1[e^{i[(kx^2/2l) - k\theta x]} + e^{-i[(kx^2/2l) - k\theta x]}] \tag{11-16}$$

When the film is illuminated by a plane wave, the quantity $E_0^2 + E_1^2$ leads to a direct or dc beam (Fig. 11-5). The term $E_0 E_1 \, e^{i[(kx^2/2l) - k\theta x]}$ describes a wave traveling in the $-\theta$ direction, as indicated by the factor $e^{-ik\theta x}$, and it is spherical due to the factor $e^{ikx^2/2l}$. This wave appears to come from behind the plate and is diverging, i.e., identical with the original wave that was captured by the hologram. It therefore produces a virtual image. The term $e^{-i[(kx^2/2l) - k\theta x]}$ represents a wave traveling in the $+\theta$ direction;

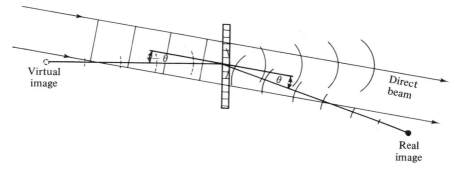

Figure 11-5 When the hologram of Fig. 11-4 is illuminated by the reference wave, three beams are diffracted. In the $-\theta$ direction, there is a virtual image. In the $+\theta$ direction, there is a real image. The direct beam is undeviated by the hologram.

it is a spherical wave converging to a point at a distance l from the plate, and it forms a real image in front of the plate.

Problem 11-1

Plot the phase as a function of x for $f(x) = e^{i[(kx^2/2l)-k\theta x]}$. When a hologram having transmission $t = f(x)$ is illuminated by a plane wave, $E = E_0 e^{ikz}$, the transmitted field is given by $E_T = E_t$. Plot the phase of E_T as a function of x at a given position z. Show that this is identical to the distribution produced by a point source located a distance l behind the hologram and a distance $x' = l\theta$ above the z axis. ∎

11-4 A Hologram of a General Object

Any arbitrary object can be represented by a series of points, so that when the object is illuminated, each point on it will become a source. The analysis of a hologram produced by an arbitrary object is a straightforward generalization of the previous analysis for a single point source.

Figure 11-6 shows the arrangement Leith and Upatnieks[11-1] used to record holograms. The object wave is given by

$$E = E(x, y)e^{i[kz-\omega t+\varphi(x,y)]} \tag{11-17}$$

where x and y are the coordinates on the photographic plate. The amplitude information is contained in $E(x, y)$ and the phase information in $\varphi(x, y)$. The reference wave is given by Eq. (11-4). Then

$$E_T^2 = E_0^2 + E^2(x, y) + E_0 E(x, y)[e^{i[k\theta x-\varphi(x,y)]} + e^{-i[k\theta x-\varphi(x,y)]}] \tag{11-18}$$

and the recording on the film is proportional to each of the terms in Eq. (11-18). The developed plate bears no resemblence to the original object;

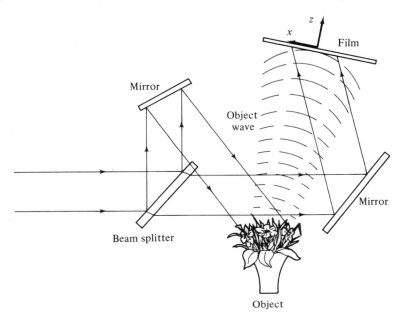

Figure 11-6 Lieth's and Upatnieks' arrangement for recording a hologram.

the interference pattern looks like a jumble of wavy lines. A photomicrograph
of a hologram is shown in Fig. 11-7.

Upon reconstruction using the reference wave E_0, we again obtain three
waves: the direct wave is $[E_0^2 + E^2 (x, y)]$; the wave diffracted in the $-\theta$
direction, $E_0 \, E(x, y) \, e^{i\varphi(x,y)}$, forms a virtual image; and the wave diffracted
in the $+\theta$ direction, $E_0 \, E(x \, y) \, e^{i\varphi(x,y)}$, forms a real image.

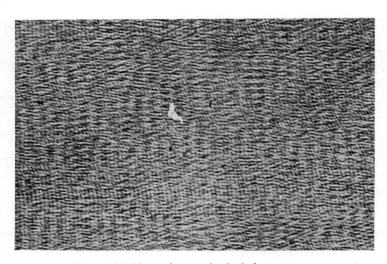

Figure 11-7 Photomicrograph of a hologram.

Earlier we found that the reconstructed wave is an exact copy of the original object wave. As a consequence of this, a hologram records three-dimensional information. Figure 11-8(a) shows the positions of the plate and object during the recording. At point *A* the object has a different perspective than at point *B*. In the reconstruction process, when an observer looks through the plate from *A* (Fig. 11-8(b)), he sees the perspective that

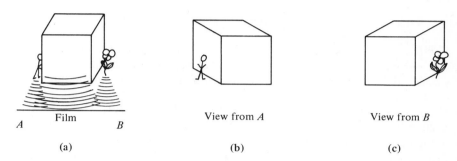

Film	View from *A*	View from *B*
A *B*		
(a)	(b)	(c)

Figure 11-8 Three-dimensional information is recorded on a hologram in (a). The reconstructed wave has one perspective (b) when viewed from the plate at *A* and another perspective (c) when viewed from the plate at *B*.

he would have seen originally at *A*. As he moves his eye to the other end of the plate (Fig. 11-8(c)), he sees a different perspective. In this manner, one can look around corners of a building.

Another striking demonstration that the reconstruction (including phase) is faithful involves a hologram of a scene containing a magnifying lens focused on a scale, for example. When an observer examines the reconstructed image and moves his eye in relation to the holographic plate, he sees different portions of the scale magnified. Furthermore, the reconstructed wave can be photographed with a camera just as if it were the original wave emanating from the object.

Although a laser is needed to record a hologram, it is not required to reconstruct the wave front. A monochromatic source such as a spectral lamp works almost as well as a laser. The difference in the quality of the reconstruction is in the resolution, since edges are slightly rounded when nonlaser light is used.

11-5 Experimental Considerations

Holograms record an interference pattern between two waves, and the three-dimensional effect is most pronounced when a wide field of view is recorded. We shall now calculate the fringe spacing on the hologram, and then use it to determine the resolution requirements for the film. The x component of **k** for an oblique wave is given by Eq. (11-3a) as

$$k_x = k \sin \theta$$

and the maxima of the interference pattern occur when (see Eq. (11-6))

$$\cos (kx \sin \theta) = 1$$

or at separations of

$$\Delta x = \frac{\lambda}{\sin \theta} \qquad (11\text{-}19)$$

For $\theta = 30°$ the fringe spacing is 2λ, which sets the requirement on the film resolution since it must handle a pair of lines at this spacing. Only very special, high-resolution films meet this requirement; the two commonly used are Kodak 649F and Afga-Gevaert 128.

High-resolution films have very fine grain and are therefore slow. Using a 20-milliwatt helium-neon laser as a source, typical exposure times for holograms are on the order of seconds. Since an interference pattern is recorded, both the object and reference mirror must be stationary throughout the recording time. Movement of either by a fraction of a wavelength during this time will shift the fringes, making the pattern wash out. Therefore holograms are recorded on specially constructed tables that eliminate vibrations. The tables are massive and often mounted on air cushions for isolation. Stability is usually the most critical requirement in obtaining good quality holograms.

Problem 11-2

A medium-resolution film can resolve up to 150 lines/mm. If the film is used to record a hologram, what is the maximum permissible angle between the reference and object beams?   ███

11-6 Four-Dimensional Holograms; Interference Between the Object and the Reconstructed Waves

After recording a hologram as shown in Fig. 11-6, developing the film, and returning it to the original position, an interesting experiment can be performed. The laser is turned on, the reference beam illuminates the plate, and the direct beam illuminates the scene. First, the beam illuminating the scene is blocked off with a card. The reference beam passing through the hologram reconstructs the wave front and a viewer sees a bright scene. Then, the position of the card is changed so that the reference beam is blocked off. The viewer sees the illuminated scene through the plate. If the card alternately blocks off one beam and then the other, the viewer cannot see any difference between the real scene and the faithfully reconstructed one.

This suggests that the direct and reconstructed wave fronts can interfere with each other. If the plate is returned precisely to its original position, the two wave fronts will be in phase everywhere, and any small change in the position or size of the object will produce regions of destructive interference between the two wave fronts. This will appear to the viewer as dark bands superimposed on the object. To observe this interference effect, the film emulsion must be prohibited from shrinking during development and drying, usually by mounting on glass plates.

Another way that holographic interference can be used is by doubly exposing a plate. A hologram of the original scene is made, and the object is given a slight displacement which can arise, for example, from heating, strain, or growth if it is a living object. Then the second hologram is made on the same plate in its original position and is developed. When illuminated by the reference beam, both objects will be reconstructed. Interference fringes will be seen for a displacement of $\lambda/4$ (or odd multiples) between the objects (Fig. 11-9).

An interesting application of the double-exposure idea was developed by T. Jeong and is shown in Fig. 11-10. In this configuration, the cellulose based photographic film is placed in a hollow cylindrical holder and an object is placed inside the cylinder. The output beam from a laser passes through a very short focal length lens which is mounted on the axis of the cylinder (a microscope objective is well suited for this application). The beam comes

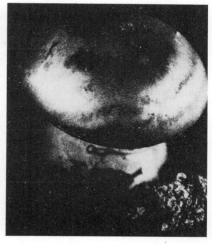

(a)   (b)

Figure 11-9 Holograms of a growing object. A singly-exposed hologram (a) of a mushroom is shown. A doubly-exposed hologram (b) shows interference fringes superimposed on the plant. A dark fringe can be seen for displacements between exposures which are odd multiples of $\lambda/4$. (*Courtesy of R. Wuerker, TRW Inc.*)

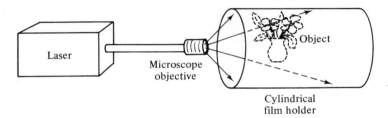

Figure 11-10. An arrangement for recording a hologram having a 360° field of view.

to a point focus and diverges into a cone. The part of the beam which intercepts the film forms the reference beam, while light scattered from the object forms the scene beam. After the hologram has been recorded, the film is reversed, a second object replaces the first, and another hologram is taken on the same film. The object is then removed, the doubly exposed film is developed, and replaced in the holder, and the laser (and hence the reference beam) is then turned on. The laser beam passes through the lens and illuminates the hologram. In one orientation of the film, the first object is seen, and when the film is reversed, the second object is in its place. A particularly interesting feature of this system is that a 360° view of the object is reconstructed: as the cylinder is rotated, each side of the object comes into view.

11-7   Other Applications of Holography

Holograms of several different scenes can be recorded on the same plate. These multiple exposures can be separated if each is recorded with the reference beam oriented at a different angle with respect to the normal to the film. Each reconstructed scene is observed at its own orientation of the plate with respect to the reference beam, and as many as 200 nonoverlapping exposures have been made on one plate.

Holography can also be used to analyze the modes of a vibrating object. Standing waves are set up when a bound object such as the top of a drum vibrates. The standing waves have nodes, that is, positions at which the displacement is zero (see Fig. 5-2). When a hologram is taken of a vibrating object, the nodes and sides of the object are recorded in the usual manner. However, the antinodes have displacements greater than a fraction of a wavelength, and the information on the hologram corresponding to these points is washed out. The reconstructed image of the object has black bands at each of the antinode regions (Fig. 11-11).

In addition, the motion of an object can be frozen by taking the hologram with a pulsed ruby laser instead of a continuously operating laser. A *Q*-switched ruby laser (see Chapter 15) generates a pulse of light of 10^{-8} second

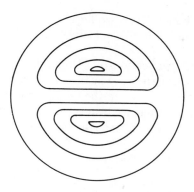

Figure 11-11 The reconstructed image of a vibrating drum head. The dark fringes indicate the position of the antinodes.

duration. During this time, slowly moving objects have very little displacement, and a hologram is recorded of their instantaneous position.

In a particularly interesting application of this idea, Wuerker, Brooks, and Heflinger[11-2] made a hologram of an airborne fruit fly. The reconstructed wave was magnified by a lens. The position of the lens was adjusted until it focused on the fruit fly. The result was a magnified view of a small object in motion. This could not have been accomplished by conventional photography, for a lens must have a short focal length for large magnification and a practical working distance between object and image. A lens with a short focal length has a short depth of field. A flying insect would be impossible to photograph (witness how difficult it is to even catch a fruit fly!). By means of holography and a pulsed laser, a recording with an infinite depth of field was made and the object located and magnified after the fact.

Problem 11-3

A hologram of a moving object is to be recorded on a film having a resolution of 2000 lines/mm using a pulsed ruby laser. The pulse of light emitted by the laser has a duration of 10^{-8} second and wavelength 694.3 nm (6943 Å). What is the maximum velocity toward the film that the object can have? How does this compare to the velocity of a fruit fly? ■

Problem 11-4

Now consider a point object moving parallel to the film. When the object is 50 cm from the film, the curvature of the wave front arriving at the film will not be very large, and the fringe pattern recorded on the film will be widely spaced. Calculate the spacing if the reference wave is a plane wave incident normal to the film. If the point moves a distance d parallel to the plate, what will be the displacement of the fringe system? How far can the point move in 10^{-8} second without washing out the pattern recorded on the film? Find the maximum velocity parallel to the film. How does it compare to the velocity of a 22-caliber bullet? How would the apparent angular size of the object affect the maximum allowable velocity? ◼

Contouring is another novel application of holography. To contour an object, the laser must operate simultaneously at two wavelengths. λ_1 and λ_2. We will assume that both are in phase when they illuminate the object and when they arrive on the film in the reference beam. Let us consider one point on the object which will be taken at a distance l from a point on the film. If

$$(k_1 - k_2)l = 2\pi m \tag{11-20}$$

where m is an integer, then both waves from the object are in phase on the film, and an interference fringe is formed. When the waves cancel and

$$(k_1 - k_2)l = (2m + 1)\pi \tag{11-21}$$

the fringe has zero visibility.

An actual object has regions of varying distances from the film. Some satisfy Eq. (11-20) and others Eq. (11-21), so that the reconstructed image has bands or contours which correspond to a constant distance. Let us calculate the spacing dl/dm of the contours. Rearranging Eq. (11-20)

$$l = \frac{2\pi m}{k_1 - k_2} = \frac{m}{(1/\lambda_1) - (1/\lambda_2)} \sim \frac{m\lambda^2}{\Delta\lambda} \tag{11-22}$$

where $\lambda_1 \sim \lambda_2 = \lambda$ and $\Delta\lambda = \lambda_2 - \lambda_1$. Differentiating Eq. (11-22) yields

$$\frac{dl}{dm} = \frac{\lambda^2}{\Delta\lambda} \tag{11-23}$$

For a ruby laser ($\lambda = 693.2$ nm) operating in two longitudinal cavity modes 0.1 nm apart, the contoured interval is 5 mm.

Problem 11-5

Could a laser operating at two wavelengths in the visible be used to holographically contour a sheet of mica having a step 20 Å high so that the height of the step could be measured? ◼

Finally we consider color holograms, which are unique because they provide a color image from black and white film. The hologram is recorded at three different wavelengths—usually in the blue, green, and red—and these are sufficient to give a complete color rendition of an object to the human eye. In recording this hologram, the reference beam enters the emulsion from the back of the plate. The reference and object waves create longitudinal standing waves rather than the transverse standing waves across the emulsion previously described. Antinodes are spaced every half wavelength, forming layers of developed grains (Fig. 11-12). When the developed plate

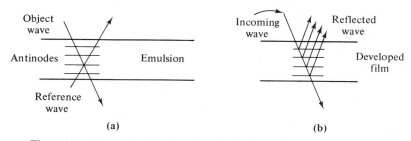

Figure 11-12 In a color hologram longitudinal standing waves are set up in the emulsion (a). When illuminated by white light (b), the developed plate reflects the wavelength corresponding to the one used to make the recording.

is illuminated by white light, those wavelengths that match the spacing of the recorded layers undergo constructive interference and are reflected. With three layers, three color images are reflected.

It should be emphasized that monochromatic light is not required for the reconstruction. Incoherent white light is satisfactory because the hologram selects the wavelength to be reflected out of the continuous spectrum. These holograms can be mounted in frames and hung on the wall like pictures. Although from a distance they look like grey sheets, when a viewer steps up to one that is suitably positioned next to a white light, he sees the scene apparently located deep inside the wall!

References

11-1 E. N. Leith and J. Upatnieks, *J. Opt. Soc. Am.* **52,** 1123 (1962); **54,** 1295 (1964).

11-2 R. E. Brooks, L. O. Heflinger, R. F. Wuerker, and R. A. Briones, *Appl. Phys. Letters,* **7,** 92 (1965)

General References

E. N. Leith and J. Upatnieks, "Laser Photography," *Scientific American*, p. 24 (June 1965).

H. M. Smith, *Holography*, John Wiley & Sons, New York (1969).

J. DeVelis and G. O. Reynolds, *Theory and Applications of Holography*, Addison Wesley, Reading, Mass. (1967).

M. Francon, *Optical Interferometry*, Academic Press, New York (1966).

G. Stroke, *Introduction to Coherent Optics and Holography*, Academic Press, New York (1966).

PART

IV

THE INTERACTION
OF LIGHT AND MATTER

The subject of physical optics covers all the phenomena associated with light as a wave: interference, diffraction, and polarization. We have already considered interference and diffraction. The latter led us to Fourier optics in Chapter 10 and holography in Chapter 11. In this final portion of the book, we shall deal with polarization, which involves consideration of the interaction of light and matter. This will lead us to absorption and emission of light, including the laser and its applications, and detectors.

12

ABSORPTION AND DISPERSION

12-1 Introduction

Objects are colored because they absorb some of the components of white light and transmit or reflect others. When the spectrum of light transmitted by a solution containing a dye is analyzed, dark bands are seen. These bands correspond to the wavelengths that were absorbed. Black materials such as carbon absorb at every wavelength. Some powders such as table salt appear white, but single crystals of these materials are transparent, absorbing very little. Because of the multitude of surfaces in a powder, most of the incident light of each color is reflected.

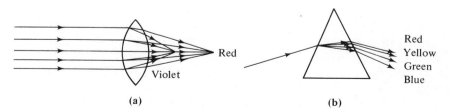

Figure 12-1 The effects of dispersion. In (a), shorter wavelengths (violet) are brought to a focus closer to the lens than longer wavelengths (red). In (b), rays of different colors are deviated by different amounts.

Dispersion (see Section 4-6) is the variation of index of refraction with wavelength. It is the cause of chromatic aberration of a lens and the familiar decomposition of white light by a prism (Fig. 12-1).

Although not immediately obvious, dispersion and absorption are intimately connected. In this chapter each phenomenon will be described separately, and then the relationship between them will be established.

12-2 Absorption

When light passes through an absorbing material, a constant fraction of the light is absorbed or transmitted in a given thickness of the material. For example, if one millimeter transmits one third of the incident energy, then two millimeters will transmit one ninth. This experimentally observed behavior can be described by the equation

$$dI = -\alpha I \, dl \tag{12-1}$$

where I is the incident intensity, dI is the change in intensity, dl is the thickness of the material, and α is the absorption coefficient. The negative sign indicates a reduction in intensity. Integrating Eq. (12-1) and evaluating the constant of integration by taking $I = I_0$ at $l = 0$ yields the law of absorption

$$I = I_0 e^{-\alpha l} \tag{12-2}$$

The values of the absorption coefficient α vary over many orders of magnitude, depending on the material. For example, in glass the absorption is very low—about one part in 10^5 per cm, or less. This gives $\alpha = 10^{-5}$ cm^{-1}, using the approximation that $\exp(-x) = 1 - x$ when x is small. However, in saturated copper sulfate ($CuSO_4$) solution, red light is attenuated 18 orders of magnitude in 1.5 inches, giving $\alpha = 10.9/\text{in}$. Furthermore, materials can have a high absorption at one wavelength and a low absorption at another, and copper sulfate, a strong absorber at 700 nm, is transparent at 350 nm. This can be understood from an energy-level diagram for the material, as discussed in Section 16-4.

12-2a Electronic Transitions

Several different types of transitions between energy levels take place in materials. In a gas composed of atoms, *electronic transitions* involve a change in the orbit of an electron around the nucleus, and the energies correspond to frequencies in the ultraviolet or visible region. An example is gaseous sodium, whose energy-level diagram (Fig. 12-2) shows several transitions. A characteristic of this type of transition is that the lines are narrow, generally less than 0.1nm wide. In an emission spectrum, more lines appear than in an absorption spectrum, and these may terminate in an excited state. Not all of the emission lines are seen in absorption because of the Boltzmann dis-

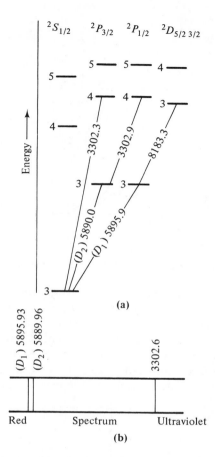

Figure 12-2 The energy-level diagram for sodium, and the absorption spectrum.

tribution. At room temperature, almost all of the atoms are in the ground state, but when a gas is heated or subjected to a powerful electric current, the upper levels are populated.

In some complex molecules such as dyes, the electrons do not belong to a single atom but rather are shared by several atoms in the molecule. These molecular orbits are weakly bound and their transitions are often in the visible and near infrared regions. The bands are broad, frequently about 100.0 nm.

Some additives can be incorporated into glass, providing a permanent and durable carrier. Such glass can be used as absorption filters to selectively remove part of the spectrum of a source. Figure 12-3 shows the transmission curves for a number of these filters.

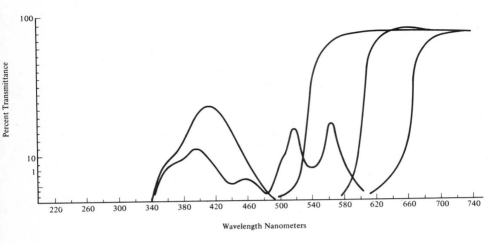

Figure 12-3 Transmission curves for some absorption glass filters.

The theory discussed in Section 16-4 explains why the absorption properties of a crystal are characterized by bands. When the frequency of the incoming wave is equal to or greater than an energy known as the *gap width*, it is absorbed, but when it is less than this amount, it is transmitted. (See Fig. 12-4.) Some crystals have several absorption bands. Table 12-1 lists a number of crystals and the short wavelength edge of their absorption bands.

Further out in the infrared region or at longer wavelengths, light can be

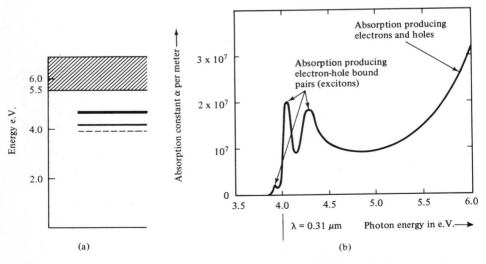

Figure 12-4 The energy-level diagram of a solid (a), and its absorption spectrum (b).

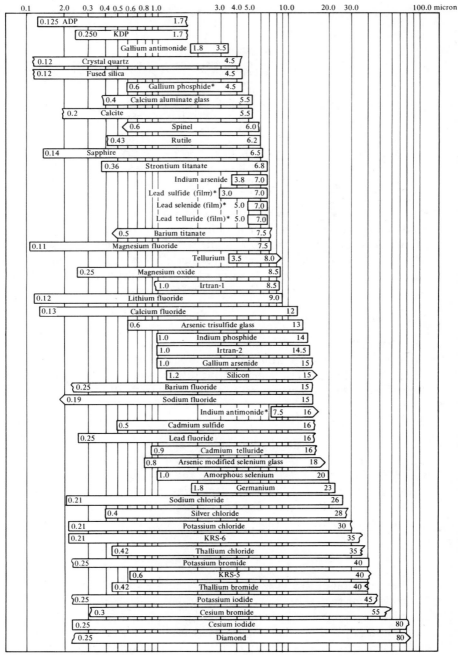

*Maximum external transmittance of less than 10%

Figure 12-5 Transmission regions for various materials. Key: Sharp cutoff,
⊐ , gradual cutoff, ⊐∫ , transmits at least as far as ⊐▷.

Table 12-1
Some Crystals and Their Absorption Band Edges

Material	Band Edge (nm)	Material	Band Edge (nm)
Quartz (SiO$_2$)	120	TiO$_2$	414
Sapphire (Al$_2$O$_3$)	140	AgI	444
Calcite (CaCO$_3$)	200	ZnSe	480
NaCl	210	CdS	510
Diamond	232	GaP	550
ZnS	345	Cu$_2$O	590
ZnO	390	GaAs	885
AgCl	390	PbS	3550

absorbed by being converted into elastic waves (phonons) involving the motion of the nuclei of the lattice, rather than the electrons. The absorption of materials therefore increases at some wavelength in the infrared. Knowledge of the transmission properties of materials is particularly important in selecting window materials for infrared or ultraviolet applications. Figure 12-5 illustrates the transmission regions of some common materials.

12-2b Molecular Transitions

The atoms of a molecule can vibrate back and forth or rotate around the center of mass. Transitions from one vibrational energy level to another are observed in the infrared region, typically between 5–50 μm. Rotational transitions involve less energy and are generally from 100 to 500 μm.

12-2c Electron Spin Transitions

Electrons have an internal motion: spin. In some materials there is an energy difference in the spin-up and spin-down states. When an electron undergoes a transition in its spin state, a photon in the microwave region must be absorbed or emitted, the frequency depending on the material. Microwave spectroscopy, used to study these transitions, will identify substances and help to understand their properties.

12-3 Dispersion

The variation of the index of refraction with wavelength is usually small in transparent materials. For instance, in crown glass $n = 1.514$ at 656.3 nm and $n = 1.528$ at 434.0 nm. Nevertheless, dispersion of this order of magnitude can be important. White light passing through a prism made of such material is broken up into its different components, and the deviation depends on wavelength.

The index of refraction can be measured by finding the angle of minimum deviation ψ_{min} as defined in Fig. 12-6. When the deviation is ψ_{min}, the incoming

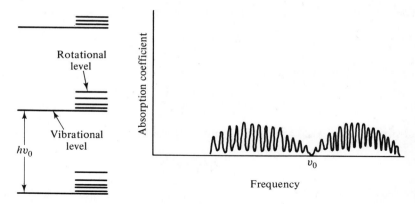

Figure 12-6 Energy-level diagram and absorption spectrum of a molecular gas. Selection rules require that the rotational level change by one step. The rotational levels are not evenly spaced.

and emerging rays are symmetric with respect to a plane which bisects the apex angle A of the prism. If this symmetry did not exist, we could argue that reversing the direction of the light would imply that there are two angles of incidence for minimum deviation. From Fig. 12-7

$$\psi = 2(\theta - \varphi)$$

(since the total deviation is twice that occurring at each surface) or

$$\theta = \frac{\psi}{2} + \varphi \tag{12-3}$$

Also

$$\beta + \varphi = \frac{\pi}{2} \tag{12-4}$$

and

$$\beta + \frac{A}{2} + \frac{\pi}{2} = \pi \tag{12-5}$$

since the sum of the angles in a triangle is π. Eliminating the angle β from Eqs. (12-4) and (12-5) yields

$$\varphi = \frac{A}{2} \tag{12-6}$$

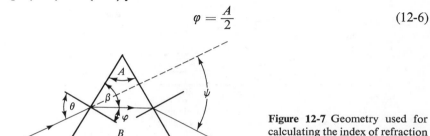

Figure 12-7 Geometry used for calculating the index of refraction from the angle of minimum deviation.

When this expression for φ is substituted into Eq. (12-3), we obtain

$$\theta = \frac{\psi + A}{2} \tag{12-7}$$

Snells law applied at the surface is

$$n = \frac{\sin \theta}{\sin \varphi} \tag{12-8}$$

Substituting Eqs. (12-6) and (12-7) into (12-8) yields

$$n = \frac{\sin \left[(\psi + A)/2\right]}{\sin (A/2)} \tag{12-9}$$

With a given prism the procedure is to first measure the apex angle A. Then the angle of minimum deviation is measured at several wavelengths, and the dispersion curve for the material is calculated from Eq. (12-9).

The variation of index with wavelength, $dn/d\lambda$, is not constant. Generally for transparent materials, it is larger for shorter wavelengths (i.e., in the blue and violet regions). Both the value of the index and the dispersion depend on the material; typical dispersion curves are shown in Fig. 12-8.

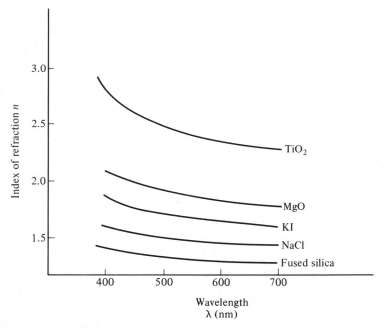

Figure 12-8 Dispersion curves for various substances.

Problem 12-1

Crown glass has $n = 1.5319$ at 4046 Å (404.6 nm) and $n = 1.5141$ at 6708 Å (670.8 nm). What is the minimum angular spread of the visible region produced by a 60° prism of this material? ■

When a prism is used as the dispersing element in a spectrometer, it is desirable that the dispersion be as large as possible, so the spectrum will be spread out and the resolution will be high.

The *chromatic resolving power* $\Delta\lambda/\lambda$ of a prism spectrometer is determined from the diffraction limit of the effective aperture of the prism. In Fig. 12-9,

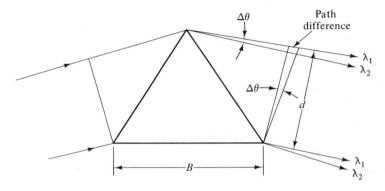

Figure 12-9 Two wavelengths, λ_1 and λ_2, are refracted through angles which differ by $\Delta\theta$.

the base of the prism is B, the effective aperture is d, and the index of refraction at λ_1 is n. Consider the path of the two rays shown for λ_1. Since the wave front is plane, the optical paths of the top and bottom rays must be the same. A ray for a second wavelength λ_2 emerges from the prism at an angle $\Delta\theta$ with respect to the first, and at the path difference of the top ray is $d\Delta\theta$. This must equal the path difference of the bottom ray (caused by dispersion), i. e.

$$B\Delta n$$

Equating these two yields

$$d\Delta\theta = B\Delta n \tag{12-10}$$

If we assume that the smallest angle that can be resolved is set by the diffraction limit (see Section 10-3), then

$$\Delta\theta = \frac{\lambda}{d} \tag{12-11}$$

Substituting this expression for $\Delta\theta$ into Eq. (12-10) gives us

$$\lambda = B\Delta n$$

Dividing by $\Delta\lambda$ then gives

$$\frac{\lambda}{\Delta\lambda} = B\frac{\Delta n}{\Delta\lambda} \tag{12-12}$$

Thus the resolving power of a prism is equal to the product of the base and the dispersion.

Problem 12-2

What is the average chromatic resolving power of a prism made of crown glass in Problem 12-1 if the base is one inch? (The resolving power is actually greater in the violet region than in the red region.) Can a spectrometer employing the above prism as the dispersing element resolve two components of the Hg green line at 5461 Å that are 0.1 cm⁻¹ apart? What is the smallest prism that can resolve them?   ■

12-4   The Lorentz Model of the Atom

The connection between absorption and dispersion can be shown using the *Lorentz model* for an atom. Lorentz assumed that the atom is made up of positive and negative charges which are attracted to each other, and when the charges are displaced from their equilibrium positions, a restoring force is set up. The conclusions of the theory are independent of the detailed structure of the atom, so they are not invalidated by refinements in our understanding of atomic physics or quantum mechanics. The beauty of this theory is that a simple and easily understood picture of the atom explains so many things.

The Lorentz model assumes the time-averaged centers of positive and negative charges coincide (Fig. 12-10). The precise expression for the restoring force $F(x)$ need not be known, but the displacement x from equilibrium is

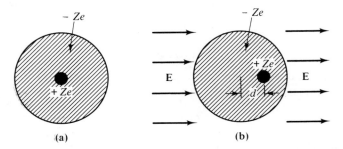

(a)   (b)

Figure 12-10 The equilibrium centers of positive and negative charges coincide in the atom at rest. Under external influences, the centers are displaced setting up powerful electromagnetic restoring forces.

small and the force can be expressed in a Taylor series as

$$F(x) = F(0) + \left(\frac{\partial F}{\partial x}\right)_0 x + \dots$$

The force must be zero when the displacement is zero or else the system would not be equilibrium; hence $F(0) = 0$. The unknown quantity $\partial F/\partial x)_0$ can be represented by $-k$, the negative sign indicating that it is a restoring force. Hence

$$F(x) = -kx \qquad (12\text{-}13)$$

In addition, the charges do not continue to oscillate indefinitely; the motion is damped and gradually dies out. To account for this damping, Lorentz postulated an additional force F_d, proportional to the velocity, of the form

$$F_d = -m\gamma\frac{dx}{dt} \qquad (12\text{-}14)$$

When an electric field of strength E interacts with the atom, the charges are separated by the Coulomb force

$$F_C = eE \qquad (12\text{-}15)$$

The equation of motion for the atom, using Newton's second law, is

$$\sum F = m\frac{d^2x}{dt^2} \qquad (12\text{-}16)$$

or

$$m\frac{d^2x}{dt^2} + m\gamma\frac{dx}{dt} + kx = eE \qquad (12\text{-}17)$$

If the external field arises from a light wave of angular frequency ω and amplitude E_0, it can be represented by

$$E = E_0 e^{i\omega t} \qquad (12\text{-}18)$$

Equation (12-17) can be solved if we assume that the displacement x follows the applied field, or

$$x = x_0 e^{i\omega t} \qquad (12\text{-}19)$$

Substituting Eqs. (12-18) and (12-19) into Eq. (12-17) and replacing k/m by ω_0^2 yields

$$(-m\omega^2 x_0 + i\gamma m\omega x_0 + kx_0)e^{i\omega t} = eE_0 e^{i\omega t} \qquad (12\text{-}20)$$

Solving for x_0 then gives

$$x_0 = \frac{eE_0/m}{\omega_0^2 - \omega^2 + i\gamma\omega} \qquad (12\text{-}21)$$

Problem 12-3

For small displacements from the equilibrium position, the restoring force F on an object can be expanded in a power series in x, the displacement.

The lowest term in the expansion is linear, $F = -kx$. The next term is quadratic, $F = k'x^2$. How does this term in combination with the linear term affect the frequency response of the atom? (*Hint*: Try different expressions for x in Eq. (12-20)).   ■

12-4a   Polarization and Index of Refraction

The equation of motion using the Lorentz model has been solved for x_0, the amplitude of the displacement of the center of the positive charge with respect to the center of the negative charge. Next we need to relate the displacement to the index of refraction. The dipole moment[5-2] p of the atom is defined as the charge e times the displacement x between the centers of the positive and negative charges.

$$p = ex \tag{12-22}$$

It is known that the displacement and hence the dipole moment is proportional to the applied field. Therefore

$$p = \alpha E \tag{12-23}$$

where α is the polarizability of the atom. This parameter α expresses the ease with which the charges can be separated. Combining Eqs. (12-22) and (12-23), we obtain

$$\alpha = \frac{ex}{E} \tag{12-24}$$

Our calculations have been carried out on an atomic scale, but measurements in the laboratory involve bulk properties of materials; thus we introduce the bulk polarization P as

$$P = Np \tag{12-25}$$

where N is the number of atoms per unit volume, and P is the dipole moment per unit volume. On a macroscopic scale

$$P = \chi E \tag{12-26}$$

where χ is the susceptibility. Combining Eqs. (12-23), (12-25), and (12-26) yields χ in terms of atomic parameters as

$$\chi = N\alpha \tag{12-27}$$

It can be shown from Maxwell's equations[5-2] that the velocity v of propagation of a wave in a medium whose permittivity is ϵ and whose permeability is μ will be given by the relation

$$v = \frac{1}{\sqrt{\epsilon\mu}}$$

The quantity ϵ is the product of ϵ_o, the permittivity of free space, and ϵ_R, the relative permittivity or dielectric constant of the medium. Similarly,

$\mu = \mu_o\mu_R$. In a vacuum, the corresponding velocity is $c = 1/\sqrt{\epsilon_o\mu_o}$ and the index of refraction is then

$$n = \frac{c}{v} = \sqrt{\epsilon_R\mu_R}$$

For the majority of the media we shall be considering, the relative permeability is very close to unity and therefore we have the simple relation

$$n^2 = \epsilon_R = 1 + \chi \tag{12-28}$$

Then for the Lorentz atom

$$n^2 - 1 = \frac{(Ne^2/m)}{\omega_0^2 - \omega^2 + i\gamma\omega} \tag{12-29}$$

12-4b   The Real and Imaginary Parts of the Index of Refraction

The above expression for the index of refraction contains real and imaginary parts. We would like to know the meaning of this complex representation of n; i.e., what physical phenomenon is described by the imaginary part. We therefore write n as the sum of a real part n' and an imaginary part n'', so that

$$n = n' + in'' \tag{12-30}$$

The electric field associated with a wave is written as

$$E = E_0 e^{i(kx - \omega t)}$$
$$= E_0 e^{i[(2\pi/\lambda)nx - \omega t]} \tag{12-31}$$

Substituting the expression for n from Eq. (12-30) into Eq. (12-31) gives

$$E = E_0 e^{i[(2\pi n'x/\lambda) - \omega t] - (2\pi n''x/\lambda)}$$

or

$$E = E_0 e^{-(2\pi n''x)/\lambda} e^{i[(2\pi n'x/\lambda) - \omega t]} \tag{12-32}$$

The imaginary part n'' is contained in the term $e^{-(2\pi n''x)/\lambda}$ which corresponds to an exponentially decreasing amplitude with distance (in other words, absorption) and $4\pi n''x/\lambda$ is simply the absorption coefficient; therefore

$$\alpha = \frac{4\pi n''}{\lambda} \tag{12-33}$$

On the other hand, the real part n' describes the velocity of propagation, as Eq. (12-32) indicates.

Each side of Eq. (12-29) can also be separated into real and imaginary parts. With the aid of the relation

$$\frac{1}{A + iB} = \frac{A}{A^2 + B^2} - i\frac{B}{A^2 + B^2} \tag{12-34}$$

applied to the right-hand side, it becomes

$$n^2 - 1 = \frac{(Ne^2/m)(\omega_0^2 - \omega^2)}{(\omega_0^2 - \omega^2)^2 + \gamma^2\omega^2} - i\frac{(Ne^2/m)\gamma\omega}{(\omega_0^2 - \omega^2)^2 + \gamma^2\omega^2} \qquad (12\text{-}35)$$

Now that we have separated n^2 into real and imaginary parts, we must find a way to express n the same way.

12-4c   Gases

The separation of the complex index of refraction into real and imaginary parts proceeds along two different paths depending on the state of material considered, i.e., gas or solid. The simplest case is for a gas, where $n \sim 1$. The identity

$$n^2 - 1 = (n + 1)(n - 1) \qquad (12\text{-}36)$$

can be approximated by

$$n^2 - 1 = 2(n - 1) \qquad (12\text{-}37)$$

Combining Eqs. (12-30), (12-35) and (12-37) and equating the real and imaginary parts, we obtain

$$n' = 1 + \frac{(Ne^2/2m)(\omega_0^2 - \omega^2)}{(\omega_0^2 - \omega^2)^2 + \gamma^2\omega^2} \qquad (12\text{-}38)$$

and

$$n'' = \frac{-(Ne^2/2m)\gamma\omega}{(\omega_0^2 - \omega^2)^2 + \gamma^2\omega^2} \qquad (12\text{-}39)$$

These two functions are plotted in Fig. 12-11. The variation of n' with frequency (or wavelength) is called dispersion.

   The dispersion curve first rises as the frequency increases. This is what is normally encountered in transparent materials and is labeled *normal dispersion* in Fig. 12-11. In most glasses, for example, n', the index of refraction, is larger in the violet region than in the red region of the spectrum. As the frequency continues to increase and approaches the center of the absorption band, the index decreases. Since this is contrary to what is usually observed, this part of the curve is called the region of *anomalous dispersion*. The center of the anomalous dispersion region almost coincides with the peak in absorption, and the width of the absorption region is γ. *This is the physical interpretation of the damping factor in Eq. (12-17).* For large amounts of damping, the absorption curve is broad and not very high; the changes in n with frequency are small. For small values of γ, the absorption curve is narrow and high, and the index of refraction undergoes larger variations near the absorption line.

   Now we clearly see that absorption and dispersion are intimately connected: if there were no absorption, there would be no dispersion. Note that each curve is centered very near ω_0, the natural frequency of the atom.

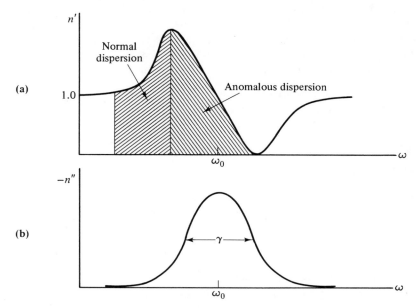

Figure 12-11 The real and imaginary parts of the index of refraction is shown.

12-4d Liquids and Solids

In condensed matter, the approximation $n \sim 1$ of Eq. (12-36) is not valid. Instead, we must work with Eq. (12-30) and obtain

$$n^2 = n'^2 - n''^2 + 2in'n'' \qquad (12\text{-}40)$$

This leads to curves like those displayed in Fig. 12-11 for n' and n'', but their calculation is considerably more complicated and no new ideas are introduced. Equations (12-38) and (12-39) satisfactorily describe liquids and solids if high accuracy is not required.

Problem 12-4

Assume that for a given material the absorption is small $n'' \ll 1$. Using Eq. (12-35), solve explicitly for n' and n''. ■

12-4e Anomalous Dispersion

In the case of transparent materials the absorption frequency ω_0 lies in the ultraviolet. For fused silica ω_0 is well into the ultraviolet; for lead glass, it is close to the visible. Hence the index of refraction of lead glass is larger at a given visible wavelength than for fused silica, and its transmission in the

UV cuts off at a wavelength closer to the visible. Organic dyes, on the other hand, have absorption frequencies in the visible region. Fuschin is one common dye; its absorption and dispersion curves are shown in Fig. 12-12. In white light, an alcohol solution of the dye appears magenta, but when the light transmitted by the solution is analyzed with a spectrometer, it is seen

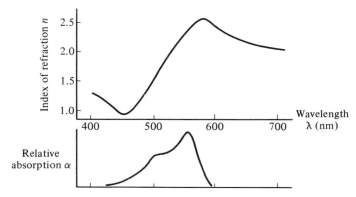

Figure 12-12 The dispersion and absorption curves for Fuschin.

that all of the colors are present except the greens. Hence the center ω_0 of the absorption band lies in the green region, and the dye looks magenta because its complementary color, green, is missing. Table 12-2 lists the color that the eye sees when a band of one color is removed from white light.

Table 12-2

Color Removed from White Light	Color Seen by the Eye
red	blue-green
orange	light blue
yellow	deep blue
green	magenta
light blue	orange
dark blue	yellow-orange
violet	yellow

Since fuschin is a strong absorber, a prism of the material does not transmit much light, and it is difficult to measure n by the method of minimum deviation. An approach for such materials is to measure Brewster's angle for different wavelengths, for then the reflected light is completely polarized. A glass substrate can be coated with a thick film (0.1 mm) of dye (Fig. 12-13) by vacuum deposition. The surface of the film, following the contour of the glass substrate, will be flat, enabling accurate measurements of Brewster's angle to be made.

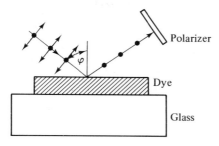

Figure 12-13 The index of refraction of an absorbing material can be calculated from Brewster's angle, ϕ.

12-5 Group and Phase Velocity

In Figs. 12-12 and 12-13, the index of refraction drops below 1.0 in the absorbing region. Since $v = c/n$, this means that the velocity of light in the material is greater than c. We need to reconcile this with the postulate that energy cannot travel with a velocity greater than c. We start by introducing the *phase velocity*, which is the velocity of a pure monochromatic infinitely long wave. In practice, however, energy travels in wave packets or pulses which consist of more than one wavelength, and the spectrum of the pulse is given by the Fourier transform of the energy as a function of time (See the Appendix.). Thus a short (in time) pulse must have many wavelengths in its spectrum (see Fig. 12-14). Let us consider the propagation of just two components having equal amplitude E_0. Their frequencies are designated ω and $\omega + \Delta\omega$, and the propagation constants are k and $k + \Delta k$. The total field is

$$E = E_0[e^{i(kx-\omega t)} + e^{i[(k+\Delta k)x-(\omega+\Delta\omega)t]}]$$

$$= 2E_0 e^{i\left[\left(k+\frac{\Delta k}{2}\right)x+\left(\omega+\frac{\Delta\omega}{2}\right)t\right]} \cos\left(\frac{x\Delta k}{2} - \frac{t\Delta\omega}{2}\right) \tag{12-41}$$

This equation describes an amplitude-modulated wave with frequency equal to the average frequency of the components, and with a sinusoidal envelope (Fig. 12-15) with frequency equal to the difference frequency.

The velocity v of a monochromatic wave is described by the first term on the right-hand side of Eq. (12-41).

$$v \simeq \frac{\omega}{k} = \frac{c}{n} \tag{12-42}$$

This is the velocity of a point on the wave front of constant phase and is referred to as the *phase velocity*. The velocity μ of a pulse (the velocity of the envelope) is called the *group velocity* and is given by the second term on the right-hand side of Eq. (12-41).

$$\mu = \frac{\Delta\omega}{\Delta k}$$

For small variations this can be written as

$$\mu = \frac{d\omega}{dk} \tag{12-43}$$

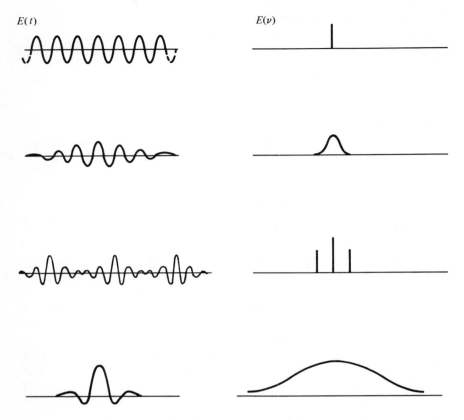

Figure 12-14 A number of wave trains and their spectra. The spectrum $E(\nu)$ of a wave train is the Fourier transform of the $E(t)$.

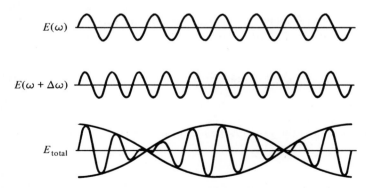

Figure 12-15 The summation of two waves of differing frequency is a wave having an envelope with frequency equal to the difference frequency.

Michelson was the first to measure the difference between phase velocity and group velocity. He measured $n = 1.64$ in CS_2 with a refractometer, giving $v = c/1.64$, and he measured the velocity of pulses to be $\mu = c/1.76$. Lord Rayleigh correctly explained the difference between v and μ along the above lines.

To understand what happens in the region where $n < 1$, we must relate $d\omega/dk$ to the dispersion curve. This can be accomplished if we start with

$$\omega = \frac{c}{n} k \tag{12-44}$$

and differentiate to obtain

$$d\omega = c\left(\frac{dk}{n} - \frac{k}{n^2} dn\right) \tag{12-45}$$

Rearranging terms gives

$$\frac{dk}{d\omega} = \frac{1}{c}\left[n + \omega \frac{dn}{d\omega}\right]$$

and

$$\frac{d\omega}{dk} = \frac{c}{n + \omega(dn/d\omega)} \tag{12-46}$$

There is a region for large values of ω in Fig. 12-11(a) where $n' < 1$ and $dn/d\omega$ is positive. The denominator $n + \omega(dn/d\omega)$ is then greater than unity and $\mu < c$.

In the region of anomalous dispersion, the slope $dn/d\omega$ is negative, and in part of this region $n \leq 1$. The two terms in the denominator are less than unity, and $\mu > c$. This was verified by Faxvog and Carruthers,[12-2] who found that both v and μ for a laser pulse in neon exceed c by one part in 10^4. This can be explained if we note that near ω_0, the absorption is highest. The trailing edge of the pulse is preferentially absorbed, causing the peak to be displaced forward, so that the energy in the leading edge of the pulse travels at a velocity less than c but the peak of the pulse travels faster than c.

Problem 12-5

Find the dispersion $dn/d\lambda$ in CS_2 using Michelson's measured data (above), and compare it to the value in glass in Problem 12-1. ■

For a discussion of pulse propagation in a laser, see Chap. XII of Ref. 12-3.

12-6 Dispersion and Pulse Spreading

A pulse that propagates in a dispersive medium is broadened. The various spectral components that make up the pulse get out of step and the pulse spreads out. We can see how this happens if a pulse is represented by a series of waves of the form $E_m \cos(k_m x - \omega_m t)$

$$E = \sum_m |E_m| \cos(k_m x - \omega_m t) \tag{12-47}$$

Then at $t = 0$ and $x = 0$, the initial conditions, each component is a maximum and E is a maximum. At any other x, each component will be a maximum for

$$t = \frac{k_m x}{\omega_m} = \frac{n_m x}{c} \tag{12-48}$$

Because of dispersion, the value of n for the different components is different and the time t at which the maxima occur (Eq. (12-48)) is different for each component, producing a broadening of the pulse. Furthermore, conservation of energy requires that the peak be reduced as it is broadened (Fig. 12-16).

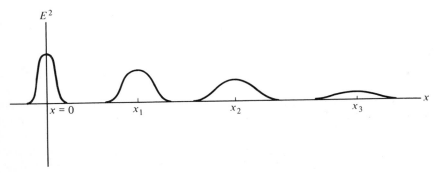

Figure 12-16 When a pulse propagates in a dispersive medium, it is broadened.

References

12-1 *Handbook of Military Infrared Technology*, ed. Wm. L. Wolfe, U. S. Govt. Printing Office, Washington, D.C. (1965).

12-2 F. R. Faxvog et al, *Appl. Phys. Letters* (1970).

12-3 A. M. Ratner, *Spectral, Spatial, and Temporal Properties of Lasers*, R. A. Phillips, ed., Plenum Press, N. Y. (1972).

General References

J. Stone, *Radiation and Optics*, McGraw-Hill, New York (1963).

F. Jenkins and H. White, *Fundamentals of Optics*, McGraw-Hill, New York (1957).

M. Garbuny, *Optical Physics*, Academic Press, New York (1965).

M. Lorentz, *Theory of Electrons*, Dover, New York (1952).

V. Weisskopf, "How Light Interacts with Matter," *Scientific American*, Sept. (1968).

13

CRYSTAL OPTICS AND THE
POLARIZATION OF LIGHT

13-1 Introduction

In treatments of absorption and dispersion it is frequently assumed that the material interacting with the light wave is *isotropic*, i.e., it looks the same from all directions. This is a valid assumption for gases and liquids but not for solids. Most molecules and unit cells of crystals are *anisotropic*, i.e., their physical properties depend on direction (see Fig. 13-1). For example, some molecules are shaped like cigars. Clearly, a physical property measured along the axis of such a molecule can be different from when measured across the waist. In the gas or liquid state, particles are randomly oriented, and measurements of bulk properties involve averages over all orientations, so that the anisotropy of the individual particle is not detected. In a solid, however, atoms or molecules are placed in a regular array. The directional dependence is inherent in the lattice, and it becomes part of the large-scale properties of the crystal. Some of the most interesting and useful applications of light depend on the anisotropy of materials.

In the Lorentz model (Section 12-4), anisotropy can be taken into account by letting the spring constant depend on direction (Fig 13-2). There will be three principal values for the constant: k_x, k_y, and k_z. In the most general case, each of these is different, but it is also possible for two—or even all

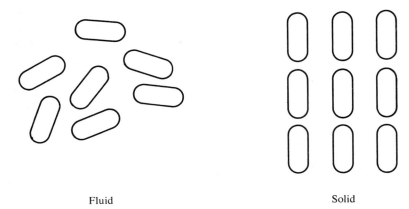

Fluid   Solid

Figure 13-1 A fluid with random orientations of anisotropic molecules appears isotropic when bulk properties are averaged over large distances. In a solid, the molecular arrangement is orderly and the directional dependence of the individual molecules or unit cells is preserved.

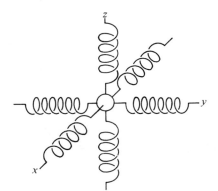

Figure 13-2 Anisotropy in the Lorentz model of the atom is taken into account by letting k depend on direction.

three—to be identical. Whenever k and γ (the damping coefficient) depend on direction, n and α do, also.

The polarization of light and the anisotropy of materials play an important role in the undertanding of the interaction of light and matter. Light is a transverse wave, i.e., the amplitude is perpendicular to the direction of propagation. Therefore, for each direction of propagation, there are two orthogonal, or perpendicular, polarizations. The easiest type to visualize is *linear polarization*. If propagation is in the z direction, then the amplitude of the wave can be along the x or y direction.

(If light were a longitudinal wave, its interaction with matter would still depend on the direction of propagation, but it could not display polarization properties.)

13-2 Polarization of Light

Light usually interacts most strongly with materials via its electric field. The magnetic interaction is proportional to v/c, where v is the velocity of the electrons in the material and c is the velocity of light. Hence, it is small.

An early demonstration of the polarization of electromagnetic waves was performed by Hertz. His demonstration can be repeated using a set of parallel wires. When the wires are oriented along the direction of the electric field of a radio wave, as determined by the orientation of the transmitting antenna, currents are set up along their length. By conservation of energy, the wave must be absorbed in the process. When the wires are rotated 90° so that they are perpendicular to the direction of the electric field, the wave is not absorbed (Fig. 13-3(a)).

This experiment was repeated in 1963 by Bird and Parrish[13-1] using light waves. They deposited gold strips onto a plastic diffraction grating having 50,000 lines/inch. The gold film was evaporated onto the grating from the side. Viewed from the top, the gold looked like wires, each less than a wavelength wide. When incident light was polarized parallel to the gold

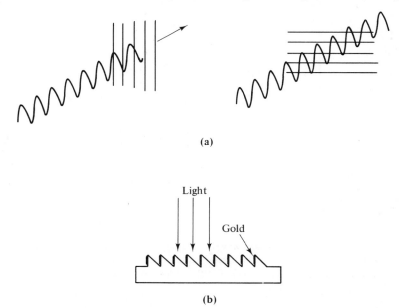

(a)

(b)

Figure 13-3 Two experiments illustrating that electromagnetic waves are polarized and interact with materials via the electric field. In Hertz's experiment (a), wires parallel to E absorbed a radio wave, whereas those perpendicular to E transmitted it. Using a diffraction grating (b), the same results were obtained in visible light.

"wires," it was absorbed, and when polarized in the perpendicular direction, it was transmitted (Fig. 13-3(b)).

It should be pointed out that the usual analogy of polarized light with waves on a string and a picket fence is off by 90° because of the orientiation of the electric vector **E**. When the amplitude of string waves is polarized parallel to the openings of a picket fence, the waves are transmitted through the fence. But electromagnetic waves polarized parallel to wires are absorbed.

The amplitude of a light wave is a vector—it has both magnitude and direction—and the simplest type of polarization is linear polarization. In this case, the amplitude vector points in a fixed direction normal to the direction of propagation (**k** is also a vector). This is also known as *plane-polarized light*, because the amplitude and the wave vector define a plane.

The state of polarization of a wave can be described by a subscript. For example, in Fig. 13-4(a) it is

$$\mathbf{E} = \mathbf{E}_z e^{i(kx - \omega t)} \qquad (13\text{-}1)$$

and the wave in Fig. 13.4(b) is expressed as

$$\mathbf{E} = \mathbf{E}_y e^{i(kx - \omega t)} \qquad (13\text{-}2)$$

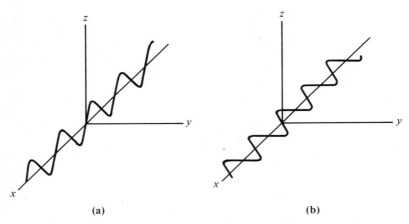

(a)   (b)

Figure 13-4 Plane-polarized waves. The magnitude of the wave as a function of position is shown at a given instant. In (a), the wave is polarized in the z direction and is labeled $\mathbf{E}_z$. In (b), the wave is polarized in the y direction and is labeled $\mathbf{E}_y$.

The state of polarization may also be along any arbitrary direction in the y-z plane, say at 45° with respect to z. Such a wave is described as

$$\mathbf{E} = (\mathbf{E}_z \pm \mathbf{E}_y)e^{i(kx - \omega t)} \qquad (13\text{-}3)$$

where $|\mathbf{E}_z| = |\mathbf{E}_y|$.

Light can also be *circularly polarized*. The wave has two components $\mathbf{E}_x$

and $\mathbf{E}_y$ with a 90° phase difference, and it may therefore be expressed as

$$\mathbf{E} = \mathbf{E}_z e^{i(kx-\omega t)} + \mathbf{E}_y e^{i(kx-\omega t+\pi/2)} \qquad (13\text{-}4)$$

We will take $x = 0$ for simplicity. Then at $t = 0$, $\mathbf{E} = \mathbf{E}_z$; at $t = \pi/2\omega$, $\mathbf{E} = \mathbf{E}_y$; at π/ω, $\mathbf{E} = -\mathbf{E}_z$; and at $3\pi/\omega$, $\mathbf{E} = -\mathbf{E}_y$. The amplitude thus rotates. If the phase difference is $\pi/2$, it rotates in the clockwise sense and is *right circularly polarized;* if the phase difference is $3\pi/2$ it rotates counterclockwise and is left circularly polarized. Figure 13-5 shows the direction of the amplitude at a specific time and at a series of different locations for a circularly polarized wave.

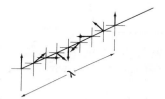

Figure 13-5 The amplitude of a circularly polarized wave rotates in a clockwise manner tracing out a helix in space. One wavelength is pictured.

Elliptical polarization is the most general state. The pattern traced out in space by an elliptically polarized wave resembles a flattened helix, and when viewed from behind, the helix looks like an ellipse. The ratio of the major axis to the minor axis of the ellipse can take on any value between unity and infinity. We see that linear polarization is a special case of elliptical polarization with the ratio of ∞, and circular polarization is a special case with the ratio unity. Several polarization states are shown in Fig. 13-6.

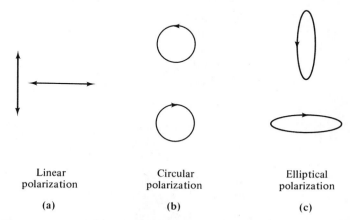

Linear polarization	Circular polarization	Elliptical polarization
(a)	(b)	(c)

Figure 13-6 The path that the (electric) amplitude vector traces out for different states of polarization. The wave is viewed end-on, that is, k is perpendicular to the plane of the page. In (c), the ratio of the major to minor axes of the ellipse is 5.

13-3 Double Refraction

13-3a Uniaxial Crystals

Double refraction was discovered in 1669 by Bartholinus, who observed two images when he looked at an object through a crystal of Iceland spar (calcite, $CaCO_3$) (see Fig. 13-7). The effect was explained by assuming that the crystal

Figure 13-7 Double refraction of a calcite crystal.

had two different indices of refraction. Thus, light could be refracted into two different directions as it entered the crystal, explaining the name, double refraction.

Calcite has several cleavage planes. That is, when the edge of a knife or safety razor blade is placed on the crystal along one of three specific directions and struck with a small hammer, the crystal splits apart with very smooth and planar faces. A slab of crystal can be cleaved so that two opposite faces are parallel. When reading through a plate with cleaved faces, very little distortion is introduced by the surfaces and a sharp image results. Each face of a perfect calcite crystal is a parallelogram with angles of 71° and 109° (Fig 13-8).

The indices for this double refraction are referred to as *ordinary* (n_o) and *extraordinary* (n_e) (Fig. 13-9). We then have two Snell's laws, namely

$$\frac{\sin \theta_1}{\sin \varphi_o} = n_o \tag{13-5a}$$

and

$$\frac{\sin \theta_2}{\sin \varphi_e} = n_e(\varphi_e) \tag{13-5b}$$

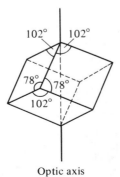

Optic axis

Figure 13-8 A cleaved calcite crystal.

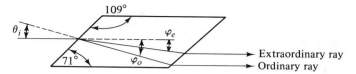

Figure 13-9 Double refraction in a calcite prism.

The extraordinary index depends on the direction of propagation φ_e, as will be explained in Section 13-15.

Even when $\theta = 0$, two images are seen. For the ordinary index n_o, the directions of the rays and energy are the same, but for the other index n_e, the direction of energy is not the same as the direction of the rays (as will be considered in Section 14-4). However, when light emerges from the crystal, both beams are parallel, one image being displaced with respect to the other.

13-3b The Optic Axis

There is one direction of propagation in calcite for which double refraction does not occur. (In a cleaved crystal, it is the direction connecting the opposite corners having three obtuse angles. See Fig. 13-7.) This is the direction of the *optic axis*, and the index of refraction for both polarizations is n_0 for propagation along this axis.

13-3c Polarization and Double Refraction

The phenomena described above in Sect. 13-3b can be explained in terms of the anisotropy of the unit cell of calcite and the polarization of light (Fig. 13-10). There is one direction in calcite, the z direction, that is unique. When the amplitude of a light wave is polarized in this direction, i.e., $E = E_z$, the wave has the extraordinary index of refraction, n_e, but when the wave is polarized perpendicularly such that $E = E_x$ or E_y, it has the ordinary index n_o.

When the propagation direction is perpendicular to z, say $k = k_x$, the wave can be polarized either along or perpendicular to z. Then E_z corresponds to n_e and E_y to n_o; in this direction, double refraction occurs. But when the propagation direction is along z, $k = k_z$, and the amplitude lies in the x-y plane. Since light is a transverse wave, there is no component along z, and both of the polarization states—whatever they may be (E_x and E_y, or $E_x + E_y$ and $E_x - E_y$)—now have the index n_o. Double refraction vanishes and the propagation is along the optic axis.

Cubic crystals have the same dimensions in all three principal directions, and they are generally isotropic. However, most crystals are anisotropic. Those in which one dimension is different from the other two are *uniaxial*.

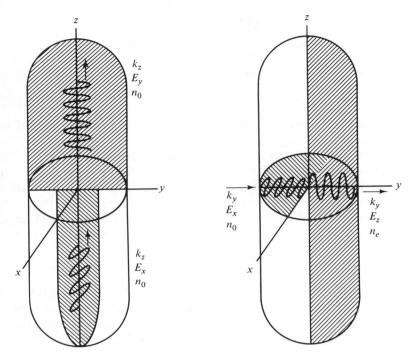

Figure 13-10 The relationship between anisotropy, polarization of light, and index of refraction for a uniaxial material. The x and y dimensions of the molecule or unit cell are the same, but the z dimension is different. In (a), the direction of propagation is down the z axis, k_z, and both polarizations experience n_o. In (b), the direction of propagation is perpendicular to the axis. The wave polarized in the x direction, E_x, has index n_o, but the wave polarized in the z direction, E_z, has index n_e.

If $n_e > n_o$ the crystal is called positive; otherwise, it is negative. Those crystals in which all three dimensions are different are *biaxial*, and will be treated later. Table 13-1 lists some common examples.

13-3d Double-Refracting Prisms

A prism made of a double-refracting crystal produces two spectra of an unpolarized source. If the double refraction is greater than the dispersion, the spectra will be separated, but if the double-refraction is weak, the spectra will overlap. For a source polarized along one of the principal directions of the crystal (e.g., in Fig. 13-11, perpendicular to the page), then only the spectrum that corresponds to that polarization will be seen (the extraordinary one) in Fig. 13-11(a)).

Table 13-1

Symmetry and Index of Refraction at 589 nm of Some Common Crystals

Cubic		n		
Sodium chloride		1.544		
Sodium chlorate		1.513		
Diamond		2.417		
Uniaxial		n_o	n_e	
Quartz		1.544	1.553	pos.
Ice		1.309	1.310	pos.
Rutile		2.616	2.903	pos.
Zircon		1.923	1.968	pos.
Calcite		1.568	1.486	neg.
Tourmaline		1.669	1.638	neg.
Biaxial	n_a	n_b	n_c	
Mica	1.552	1.582	1.588	
Cordierite	1.540	1.549	1.553	
Sodium nitrite	1.355	1.413	1.655	

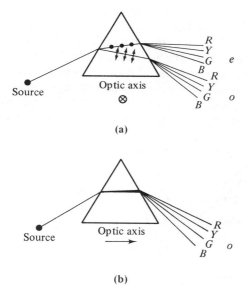

Figure 13-11 A prism made of double-refracting material produces two spectra. One is polarized parallel to the optic axis and the other perpendicular to the axis. For propagation down the axis, only one spectrum is produced.

13-4 Polarizing Prisms

Double refraction can be used to polarize light and analyze its state of polarization. Often this is done by making a prism such that light of one polarization suffers total internal refraction while light of the other polarization is transmitted.

The *Nicol prism* (invented in 1828) was the first device that operated on this principle. To fabricate a Nicol prism, a calcite crystal is cleaved so that it is three times as long as it is wide. The entrance and exit faces on the side are cut at an angle of 3° with respect to the cleavage plane and polished flat, making the acute angle between surfaces equal to 68° instead of 71°. The prism is then cut along the diagonal, the surfaces polished flat, and the two pieces cemented together with Canada balsam (Fig. 13-12). Under these con-

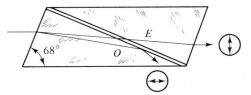

Figure 13-12 The Nicol prism.

ditions, the extraordinary ray is transmitted across the joint, but the ordinary ray suffers total internal refraction. Although the incident light can be unpolarized, the light emerging from the prism will have the polarization of the extraordinary ray.

The incident light must be parallel to the base of the prism to within about ±11°. The emerging beam is displaced laterally with respect to the incident beam. If the polarizer is rotated, the image will rotate with it, which is a disadvantage in many optical applications. The advantage of the Nicol prism is that very little of the parent crystal is wasted, so it is relatively inexpensive.

The Canada balsam cement used in a Nicol prism absorbs ultraviolet light at about 350 nm, limiting the range of its usefulness. In the *Foucault prism* (Fig. 13-13), the cement is replaced by an air gap and the transmission extends

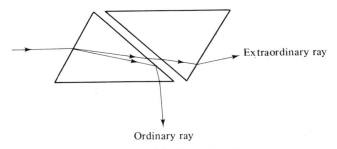

Extraordinary ray

Ordinary ray

Figure 13-13. The Foucault prism.

down to 215 nm, where calcite begins to absorb. This prism is shorter, with a length-to-aperture ratio of unity, but the usable acceptance angle is ±4° with respect to a normal to the face. It also has the undesirable feature that the exit beam is displaced with respect to the entrance beam; in addition, multiple reflections from the air-calcite surfaces can be troublesome.

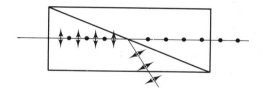

Figure 13-14 The Glan-Thompson prism.

Glan-Thompson prisms (Fig. 13-14) have the entrance and exit faces perpendicular to the incident beam, so that the exit beam is undisplaced. In making the faces perpendicular, more calcite is wasted and a larger crystal must be used in the fabrication. Large high-quality crystals of calcite are rare and expensive.

The cutoff frequency for ultraviolet transmission depends on the thickness of the cement layer; therefore shorter prisms, which have a diagonal more nearly vertical, have thinner effective cement layers and can be used further out in the ultraviolet region. A Glan-Thompson prism with a length-to-aperture ratio of 3:1 transmits to 360 nm with an angular acceptance of $\pm 12.5°$. A prism with a ratio of 2:1 transmits to about 300 nm and has an angular acceptance of $\pm 7.5°$. Glan-Thompson prisms can also be made with an air space; these transmit to 214 nm and are useful for high-intensity light where the cement would be degraded by the light. Fig. 13-15 shows the transmission properties of a number of polarizing prisms.

The *Rochon prism* (Fig. 13-16) separates a light beam into two polarized beams which emerge at slightly different angles. It is useful when small differ-

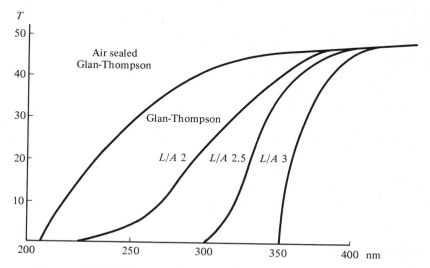

Figure 13-15 The percent transmission vs wavelength for several polarizing prisms. When the incident light is unpolarized, 50 percent is the maximum transmission.

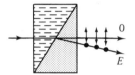

Figure 13-16 The Rochon prism.

ences in the strength of polarized components are to be measured. Rochon prisms are often made of quartz (SiO_2); a crystal with less birefringence, but one that is more readily available than calcite.

Problem 13-1

An air-spaced Glan-Thompson prism is to be made of quartz. Determine the maximum angle that the diagonal face can have with respect to the entrance face. If Canada balsam with $n = 1.530$ is used to cement the faces of the prism together, what can the maximum angle be? ■

13-5   The Law of Malus

Malus, a French army engineer, was fascinated by the optical properties of calcite. He frequently carried a piece of it with him and demonstrated its properties to his friends. One afternoon in 1808, while visiting the palace of Luxembourg, he looked through his crystal at the sunlight reflected from the windows of the palace and noticed that the two images varied in intensity as he rotated the crystal. He did not offer an explanation of the observation, but he had actually found that light was partially polarized by reflection.

According to Malus' law, the intensity of light transmitted by a polarizer varies as the square of the cosine of the angle between the plane of polarization and the plane of transmission of the polarizer. This can be understood if we consider the amplitude E_0 of a wave (Fig. 13-17) which makes an angle of θ with respect to the plane of transmission of a polarizer. The component of E_0 along this direction is $E_0 \cos \theta$. The polarizer transmits this component and rejects the perpendicular component $E_0 \sin \theta$, so that the intensity trans-

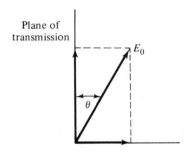

Plane of
transmission

E_0

θ

Figure 13-17 E_0 is inclined at an angle θ with respect to the plane of transmission of a polarizer.

mitted, which is proportional to E^2, is then

$$I = E_0^2 \cos^2 \theta = I_0 \cos^2 \theta \tag{13-6}$$

The initial plane of polarization of the light wave can be set by another polarizer with its axis of transmission at an angle θ. Malus' law can then be regarded as specifying the fraction of light transmitted between two polarizers with an angle θ between their planes of transmission.

13-6 The Extinction Ratio for Crossed Polarizers

According to Eq. (13-6) the intensity transmitted by crossed polarizers can be reduced to zero if $\theta = 90°$. In practice this cannot be attained, since the orientation of the plane of polarization of the beam emerging from a polarizer depends in part on the quality of calcite used in the prism. Bubbles, striae, inclusions, or small variations in the direction of the optic axis affect the plane of polarization.

The figure of merit for polarizers is the *extinction ratio*, which is the ratio of minimum transmission ($\theta = 90°$) to maximum transmission ($\theta = 0°$). Low ratios require a high grade of calcite in the fabrication of prisms.

Table 13-2 lists the extinction ratio of a prism produced from various grades of calcite; for comparison, the ratio for polarizing sheet is included.

Table 13-2
The Extinction Ratio for a Pair of Crossed Polarizers

Material	Grade	Ratio
Calcite	ellipsometer grade	10^{-7}
"	grade A	10^{-6}
"	grade B	10^{-4}
Polarizing sheet	sunglasses	10^{-3}
Tourmaline		10^{-2}

For grade-A calcite the extinction ratio is about 10^{-6}. This sets a limit on the definition of the direction of polarization that can be obtained by a given polarizer. If we let $\Delta\theta$ be the uncertainty in the orientation of the direction of polarization, then when two polarizers are crossed

$$10^{-6} = \cos^2 (\pi/2 - \Delta\theta) = \sin^2 (\Delta\theta) \sim (\Delta\theta)^2$$

or

$$\Delta\theta = 10^{-3}$$

This represents an uncertainty in the orientation of about 3 minutes of arc.

13-7 The Half-Wave Plate

Birefringent materials can be used as optical elements; an example is the *half-wave plate* (Fig. 13-18). Actually, it is much thicker than a half wave-

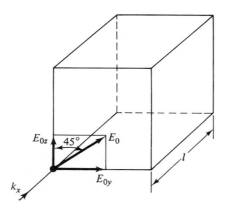

Figure 13-18 A half-wave plate.

length; the name means that the optical path difference $\Delta(nl)$ for the two polarizations is a half wavelength. The optic axis is in the z direction, with the incoming wave E_0 polarized at 45° with respect to z. The direction of propagation as specified by k_x is along the $-x$ axis.

The amplitude can be separated into components parallel and perpendicular to the optic axis, written as

$$
\left.\begin{aligned}
E_{0y} &= \frac{E_0}{\sqrt{2}} \\[2mm]
E_{0z} &= \frac{E_0}{\sqrt{2}}
\end{aligned}\right\}
\tag{13-7}
$$

and at the exit face, the path difference for these components is

$$
\Delta = (n_e - n_o)l
\tag{13-8}
$$

Since $\Delta = \lambda/2$, we have

$$
l = \frac{\lambda}{2(n_e - n_o)}
\tag{13-9}
$$

As the components emerge from the plate, the phase difference is then

$$
\delta = \frac{2\pi}{\lambda}\Delta = \pi
\tag{13-10}
$$

Taking the phase of the z component E_{lz} at the exit face as reference, we see that E_{ly} is

$$
E_{ly} = \left(\frac{E_0}{\sqrt{2}}\right)e^{i\delta} = -\frac{E_0}{\sqrt{2}}
\tag{13-11}
$$

where

$$
E_{lz} = \frac{E_0}{\sqrt{2}}
$$

Thus the wave plate has changed the sign of the y component. When the y

and z components are added, this produces a rotation of the plane of polarization by 90° (Fig. 13-19).

Figure 13-19 A relative phase shift of π for E_y rotates the plane of polarization of E by 90°.

Problem 13-2

A half-wave plate for $\lambda = 435.8$ nm is to be made of quartz. How thick should it be? If it is for $\lambda = 694.3$ nm? If it is to be made of calcite?   ■

13-8 The Quarter-Wave Plate and Circular Polarization

If the thickness of a wave plate satisfies $\Delta(nl) = \lambda/4$, it is a *quarter-wave plate*, which shifts the phase of E_y by 90° or by $e^{i\pi/2} = i$. A wave with components E_y and E_z which are 90° out of phase is circularly polarized as explained in Section 13-2.

13-9 The Jones Calculus

The state of polarization of a wave and the action of optical elements can be described in a notation invented by R. Clark Jones. In this notation, the polarization is described by a column vector, with the three entries in the column representing the amplitude and phase of each component. For example

$$\mathbf{E} = \begin{bmatrix} E_x \\ E_y \\ E_z \end{bmatrix} \tag{13-12}$$

A simplification results when we choose one coordinate as the direction of propagation (say x). Then in free space, the component in this direction vanishes ($E_x = 0$) and the column is reduced to just two elements. Consider the half-wave plate as an example. The linearly polarized wave entering it is represented by

$$\mathbf{E}_{in} = \frac{E_0}{\sqrt{2}} \begin{bmatrix} 1 \\ 1 \end{bmatrix} \tag{13-13}$$

and the emerging wave E_{out} by

$$\mathbf{E}_{out} = \frac{E_0}{\sqrt{2}} \begin{bmatrix} -1 \\ 1 \end{bmatrix} \tag{13-14}$$

where the column matrix of Eq. (13-12) is written with only two elements. In complex notation, a phase difference of $\pi/2$ is represented by $e^{i\pi/2} = i$. The two states of circular polarization, right-handed and left-handed, are then represented by

$$\mathbf{E}_r = \frac{E_0}{\sqrt{2}} \begin{bmatrix} 1 \\ i \end{bmatrix}, \qquad \mathbf{E}_l = \frac{E_0}{\sqrt{2}} \begin{bmatrix} 1 \\ -i \end{bmatrix} \tag{13-15}$$

13-9a Orthogonality

The condition that the states of polarization of two waves $\mathbf{E}_1$ and $\mathbf{E}_2$ be *orthogonal* (perpendicular) is $\mathbf{E}_1^* \mathbf{E}_2 = 0$. In matrix notation, $\mathbf{E}_1^*$ is obtained by converting the column vector to a row and replacing each element by its complex conjugate. If $\mathbf{E}_1 = \begin{bmatrix} A \\ B \end{bmatrix}$, then $\mathbf{E}_1^* = [A^* \ \ B^*]$. If we let $\mathbf{E}_2 = \begin{bmatrix} C \\ D \end{bmatrix}$, the product $\mathbf{E}_1^* \mathbf{E}_2$ is

$$\mathbf{E}_1^* \mathbf{E}_2 = (A^*C + B^*D) \tag{13-16}$$

which is a scalar quantity.

As a simple example, we shall show that two states of linear polarization $\mathbf{E}_1 = E_0 \begin{bmatrix} 1 \\ 0 \end{bmatrix}$ and $\mathbf{E}_2 = E_0 \begin{bmatrix} 0 \\ 1 \end{bmatrix}$ are orthogonal. $\mathbf{E}_1^* \mathbf{E}_2$ is then

$$\mathbf{E}_1^* \mathbf{E}_2 = E_0^2(1 \cdot 0 + 0 \cdot 1) = 0 \tag{13-17}$$

As another example, we shall also show that right- and left-handed circularly polarized states are also orthogonal. Using

$$\mathbf{E}_r = E_0 \begin{bmatrix} 1 \\ i \end{bmatrix}$$

and

$$\mathbf{E}_l = E_0 \begin{bmatrix} 1 \\ -i \end{bmatrix}$$

gives

$$\mathbf{E}_r^* \mathbf{E}_l = E_0^2 [1 \ \ -i] \begin{bmatrix} 1 \\ -i \end{bmatrix} = 0$$

There are an infinite number of pairs of orthogonal states of polarization, of which linear and circular are perhaps the most familiar. The general state of polarization of a wave is elliptical. For example, a set of elliptical states may be given by $\mathbf{E}_1 = \begin{bmatrix} 5 \\ i \end{bmatrix}$ and $\mathbf{E}_2 = \begin{bmatrix} 1 \\ -5i \end{bmatrix}$. We note that $\mathbf{E}_1^* \mathbf{E}_2 = 0$, indicating that the two waves are orthogonal. However, E_1 can be viewed as the superposition of $\begin{bmatrix} 4 \\ 0 \end{bmatrix}$ and $\begin{bmatrix} 1 \\ i \end{bmatrix}$, i.e., a linear and circular wave, and $\mathbf{E}_2$

can be viewed as $\begin{bmatrix} 1 \\ -i \end{bmatrix}$ and $\begin{bmatrix} 0 \\ -4i \end{bmatrix}$. Figure 13-6 illustrates the behavior of the amplitude for the polarization states $\begin{bmatrix} 1 \\ 0 \end{bmatrix} \begin{bmatrix} 0 \\ 1 \end{bmatrix}$, $\begin{bmatrix} 1 \\ i \end{bmatrix} \begin{bmatrix} 1 \\ -i \end{bmatrix}$, and $\begin{bmatrix} 5 \\ i \end{bmatrix} \begin{bmatrix} 1 \\ -5i \end{bmatrix}$. At a given instant, the amplitude is plotted at different locations along the direction of propagation.

13-9b Normalization

It is often convenient to deal with vectors of unit length, and a vector **E** can be *normalized* if it is divided by its magnitude $(E_x^2 + E_y^2 + E_z^2)^{1/2}$. For example, if $\mathbf{E} = E_0 \begin{bmatrix} 1 \\ 1 \end{bmatrix}$, the normalization factor is $E_0/\sqrt{2}$ and the normalized vector is $\mathbf{E} = \dfrac{1}{\sqrt{2}} \begin{bmatrix} 1 \\ 1 \end{bmatrix}$.

13-9c Optical Elements

An optical element is represented by an *operator* Ω which converts an input wave $\mathbf{E}_{in}$ to the output $\mathbf{E}_{out}$. The process can be described by the equation

$$\mathbf{E}_{out} = \Omega \mathbf{E}_{in} \tag{13-18}$$

For example, the operator $\Omega_{\lambda/2}$ for a half-wave plate is given by

$$\Omega_{\lambda/2} = \begin{bmatrix} 1 & 0 \\ 0 & -1 \end{bmatrix} \tag{13-19}$$

Normalizing the waves of Eq. (13-13) and (13-14) gives

$$\mathbf{E}_{in} = \frac{1}{\sqrt{2}} \begin{bmatrix} 1 \\ 1 \end{bmatrix} \tag{13-20}$$

and

$$\mathbf{E}_{out} = \frac{1}{\sqrt{2}} \begin{bmatrix} 1 \\ -1 \end{bmatrix} \tag{13-21}$$

The representation of this process in operator notation [Eq. (13-18)] is

$$\frac{1}{\sqrt{2}} \begin{bmatrix} 1 & 0 \\ 0 & -1 \end{bmatrix} \begin{bmatrix} 1 \\ 1 \end{bmatrix} = \frac{1}{\sqrt{2}} \begin{bmatrix} 1 \\ -1 \end{bmatrix} \tag{13-22}$$

In the other example, we saw that a quarter-wave plate converts a plane-polarized wave

$$\mathbf{E}_{in} = \frac{1}{\sqrt{2}} \begin{bmatrix} 1 \\ 1 \end{bmatrix} \tag{13-23}$$

to a circularly polarized wave

$$\mathbf{E}_{\text{out}} = \frac{1}{\sqrt{2}} \begin{bmatrix} 1 \\ i \end{bmatrix} \tag{13-24}$$

The Jones matrix for the quarter-wave plate is

$$\Omega_{\lambda/4} = \begin{bmatrix} 1 & 0 \\ 0 & i \end{bmatrix} \tag{13-25}$$

and the operator equation becomes

$$\frac{1}{\sqrt{2}} \begin{bmatrix} 1 & 0 \\ 0 & i \end{bmatrix} \begin{bmatrix} 1 \\ 1 \end{bmatrix} = \frac{1}{\sqrt{2}} \begin{bmatrix} 1 \\ i \end{bmatrix}$$

Note that two quarter-wave plates in succession are equivalent to a half-wave plate, since

$$\Omega_{\lambda/4} \Omega_{\lambda/4} = \Omega_{\lambda/2}$$

or

$$\begin{bmatrix} 1 & 0 \\ 0 & i \end{bmatrix} \begin{bmatrix} 1 & 0 \\ 0 & i \end{bmatrix} = \begin{bmatrix} 1 & 0 \\ 0 & -1 \end{bmatrix} \tag{13-26}$$

Polarizers transmit one component and reject the other. A linear polarizer with transmission along the z direction is represented by the matrix

$$\Omega_{LP} = \begin{bmatrix} 0 & 0 \\ 0 & 1 \end{bmatrix} \tag{13-27}$$

To verify this, consider two waves $\mathbf{E}_y = \begin{bmatrix} 1 \\ 0 \end{bmatrix}$ and $\mathbf{E}_z = \begin{bmatrix} 0 \\ 1 \end{bmatrix}$. Then

$$\Omega_{LP} \mathbf{E}_z = \mathbf{E}_z$$

since

$$\begin{bmatrix} 0 & 0 \\ 0 & 1 \end{bmatrix} \begin{bmatrix} 0 \\ 1 \end{bmatrix} = \begin{bmatrix} 0 \\ 1 \end{bmatrix} \tag{13-28}$$

and

$$\Omega_{LP} \mathbf{E}_y = 0$$

because

$$\begin{bmatrix} 0 & 0 \\ 0 & 1 \end{bmatrix} \begin{bmatrix} 1 \\ 0 \end{bmatrix} = \begin{bmatrix} 0 \\ 0 \end{bmatrix} \tag{13-29}$$

This shows that the polarizer transmitted the wave of one polarization and rejected the other. Jones matrices for a number of optical elements are given in Table 13-3.

Problem 13-3

A film is made by cementing two thin optical elements together. The thickness of the combination is approximately 1.0 mm. When the sheet with one side

Table 13-3

Jones Matrices for Some Optical Elements

The horizontal direction is taken as y and the vertical as z.

Linear polarizer, transmission along z	$\begin{pmatrix} 0 & 0 \\ 0 & 1 \end{pmatrix}$
" " " along y	$\begin{pmatrix} 1 & 0 \\ 0 & 0 \end{pmatrix}$
" " " 45° from y	$\frac{1}{2}\begin{pmatrix} 1 & 1 \\ 1 & 1 \end{pmatrix}$
" " " 45° from z	$\frac{1}{2}\begin{pmatrix} 1 & -1 \\ -1 & 1 \end{pmatrix}$
Circular polarizer, right-handed	$\frac{1}{2}\begin{pmatrix} 1 & i \\ -i & 1 \end{pmatrix}$
" " left-handed	$\frac{1}{2}\begin{pmatrix} 1 & -i \\ i & 1 \end{pmatrix}$
Isotropic phase retarder	$\begin{pmatrix} e^{i\phi} & 0 \\ 0 & e^{i\phi} \end{pmatrix}$
Relative phase changer	$\begin{pmatrix} e^{i\phi} & 0 \\ 0 & e^{-i\phi} \end{pmatrix}$
Half-wave plate, optical axis along z (positive crystal)	$\begin{pmatrix} 1 & 0 \\ 0 & -1 \end{pmatrix}$
" " optical axis at 45° (negative crystal)	$\begin{pmatrix} 0 & 1 \\ 1 & 0 \end{pmatrix}$
Quarter-wave plate, optical axis along z (positive crystal)	$\begin{pmatrix} 1 & 0 \\ 0 & i \end{pmatrix}$
" " optical axis along z (negative crystal)	$\begin{pmatrix} 1 & 0 \\ 0 & -i \end{pmatrix}$
Quarter-wave plate, axis 45°	$\begin{pmatrix} 1 & -i \\ -i & 1 \end{pmatrix}$

up is laid on a dime, it appears shiny. When the sheet is turned over and laid on the dime, it appears dark. What does the film consist of? Write the Jones matrix for each element and show that the observed behavior can be explained using Jones' calculus. Does this violate the principle of reversability of light rays that is used in geometrical optics? ▬

13-10 Transformation of Coordinates

The nature of the representations of a polarization state by a column vector and an optical element by a square matrix depends on the orientation with respect to a given coordinate system, as indicated by the examples of Table 13-3.

There are some problems for which it is necessary to be able to describe the polarization state and optical element for an arbitrary orientation. Once we have a representation for a specific orientation, we can obtain it for any other orientation by applying the rules for the transformation of coordinates. Knowledge of those techniques is not only important for problems involving optical elements, but is also useful for a larger class of problems in crystal optics involving the propagation of light in anisotropic media.

We will start with a vector $\mathbf{V} = \begin{bmatrix} A \\ B \end{bmatrix}$ in a given coordinate system yz, and we wish to describe the same vector in a system rotated through the angle θ with respect to the original system (Fig. 13-20). The vector can be represented as

$$\mathbf{V} = A\hat{\mathbf{y}} + B\hat{\mathbf{z}} \tag{13-30}$$

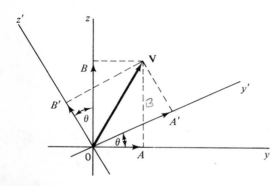

Figure 13-20 The vector **V** described in two coordinate systems. The primed system was obtained by a rotation about the x axis through an angle θ.

where $\hat{\mathbf{y}}$ and $\hat{\mathbf{z}}$ are unit vectors in the coordinate directions y and z, respectively. In the rotated system the components A' and B' are the sums of the projections of the components A and B onto the respective axes y' and z', giving

$$\mathbf{V}' = A'\hat{\mathbf{y}}' + B'\hat{\mathbf{z}}' \tag{13-31}$$

where

$$A' = A \cos \theta + B \sin \theta \tag{13-32}$$
$$B' = -A \sin \theta + B \cos \theta$$

as can be seen from the diagram. This set of equations can be described in matrix form as

$$S\begin{bmatrix} A \\ B \end{bmatrix} = \begin{bmatrix} A' \\ B' \end{bmatrix} \tag{13-33}$$

where S is the rotation matrix

$$S = \begin{bmatrix} \cos \theta & \sin \theta \\ -\sin \theta & \cos \theta \end{bmatrix} \tag{13-34}$$

Equation (13-33) can alternately be written in operator notation as

$$S\mathbf{V} = \mathbf{V}'$$

For every transformation S that we will encounter (rotations, inversions, reflections), there will be another transformation S^{-1} that returns the coordinates to their original positions. In the foregoing example, S is a rotation through an angle θ about the x axis, and the inverse transformation is a rotation through an angle $-\theta$ about the same axis. Then S^{-1} becomes

$$S^{-1} = \begin{bmatrix} \cos\theta & -\sin\theta \\ \sin\theta & \cos\theta \end{bmatrix} \tag{13-35}$$

which may be obtained by replacing θ by $-\theta$ in Eq. (13-34) and using the trigonometric identities $\sin(-\theta) = -\sin\theta$ and $\cos(-\theta) = \cos\theta$. We also note that

$$S^{-1}S = SS^{-1} = I \tag{13-36}$$

where I is the unit matrix $\begin{bmatrix} 1 & 0 \\ 0 & 1 \end{bmatrix}$.

Turning now to the transformation of matrix operators, we return to Eq. (13-18) and write it in another coordinate system designated by primed quantities as

$$\Omega' E'_{in} = E'_{out} \tag{13-37}$$

If we return to the original Eq. (13-18), and multiply both sides by S we obtain

$$S\Omega E_{in} = SE_{out} \tag{13-38}$$

Equation (13-33) can be written for the electric field vector **E** as follows

$$\boxed{SE = E'} \tag{13-39}$$

Applying this to the right-hand side of Eq. (13-38), we obtain

$$S\Omega E_{in} = E'_{out} \tag{13-40}$$

Then we insert $S^{-1}S = 1$ into the left-hand side of Eq. (13-40), which gives

$$S\Omega S^{-1} \, SE_{in} = E'_{out} \tag{13-41a}$$

or

$$S\Omega S^{-1} \, E'_{in} = E'_{out} \tag{13-41b}$$

This may also be written as

$$\Omega' E'_{in} = E'_{out} \tag{13-37}$$

Now we see by comparing Eq. (13-37) and (13-41b) that the transformed operator Ω' is given by

$$\boxed{\Omega' = S\Omega S^{-1}} \tag{13-42}$$

The two Eqs. (13-39) and (13-42) describe how operators and vectors are transformed from one coordinate system to another. As a final point, it should be stressed that matrix multiplication is not commutative and must always be performed in the order shown.

Example 13-1

Consider a polarizer specified by the operator

$$\Omega_{LP} = \begin{bmatrix} 1 & 0 \\ 0 & 0 \end{bmatrix} \tag{13-43}$$

which transmits the horizontal, or y, component of the electric field. Let the input wave be y-polarized, so that

$$\mathbf{E}_{out} = \begin{bmatrix} 1 & 0 \\ 0 & 0 \end{bmatrix}\begin{bmatrix} 1 \\ 0 \end{bmatrix} = \begin{bmatrix} 1 \\ 0 \end{bmatrix} \tag{13-44}$$

Now we rotate the coordinate system through-90°. The transformation matrix S is given by

$$S = \begin{bmatrix} 0 & -1 \\ 1 & 0 \end{bmatrix}$$

and

$$S^{-1} = \begin{bmatrix} 0 & 1 \\ -1 & 0 \end{bmatrix}$$

Applying Eq. (13-39) to E gives

$$\mathbf{E}' = E_0 \begin{bmatrix} 0 & -1 \\ 1 & 0 \end{bmatrix}\begin{bmatrix} 1 \\ 0 \end{bmatrix}$$

$$\mathbf{E}' = E_0 \begin{bmatrix} 0 \\ 1 \end{bmatrix}$$

and applying Eq. (13-42) to Ω gives

$$\Omega' = \begin{bmatrix} 0 & -1 \\ 1 & 0 \end{bmatrix}\begin{bmatrix} 1 & 0 \\ 0 & 0 \end{bmatrix}\begin{bmatrix} 0 & 1 \\ -1 & 0 \end{bmatrix}$$

$$= \begin{bmatrix} 0 & 0 \\ 0 & 1 \end{bmatrix}$$

Problem 13-4

Given the expression for the linear polarizer with axis along z, Table 13-3, derive the expression for the same element with axis oriented at 45° with respect to z. ■

13-11 Dichroism

The absorption properties of some materials are anisotropic and depend on the direction of propagation and the state of polarization of the wave. Materials in which there are two distinct kinds of absorption are called *dichroic*, which means "two-colored," since the color observed depends on orientation. Biot discovered dichroism in the mineral tourmaline in 1815. The unit cells have a different length in one direction than in the other two (Fig. 13-9). Light polarized normal to the optic axis is absorbed, while light with parallel polarization is transmitted. The absorption is so strong that a 1-mm thick plate absorbs essentially all the light of one kind of polarization and very little of the other; hence the output is highly polarized.

Polarizers are made from tourmaline by cutting the mineral in plates such that the optic axis lies in the plane of the plate. For a wave of normal incidence, there are pure ordinary and extraordinary waves inside the crystal. Polarizers of this type have large apertures and acceptance angles but their extinction ratios are not as good as those of polarizers which operate on the principle of total internal reflection.

Polaroid is a dichroic sheet polarizer invented by E. H. Land. He started work in 1928 while he was an undergraduate student. The original *J*-sheet was made by covering a sheet of transparant polyvinyl alcohol with tiny crystals of herapathite (or quinine sulfate periodide), which are needle shaped and originally oriented at random. The polyvinyl alcohol is then heated and stretched in one direction to several times its original length. The needle-shaped molecules untangle and line up with their long axes parallel to the direction of stretch. This orders the crystals of herapathite. The sheet is then attached to a thicker piece of transparent material, so it will not unstretch, and covered with a third sheet for protection.

The *H*-sheet is an improved version. The polyvinyl alcohol consists of long, tangled molecules. When it is stretched, the molecules untangle and tend to form long chains in the direction of stretch. The stretched sheet is cemented to a sheet of clear cellulose aceto butylrate to keep its shape, and it is then dipped into an iodine solution. The iodine diffuses into the polyvinyl alcohol and forms a complex within the coils of the long chains of the host. The iodine complexes absorb most of the visible light polarized parallel to the length of the chains but transmit the light polarized perpendicular to the chains and the result is an inexpensive sheet polarizer. In general, the extinction ratio of sheet polarizers is poorer than that of prism polarizers, and they do not stand up well under heat and humidity. The largest quantities of the sheet are used in sunglasses. Other applications are displays, 3-D movies, laboratory uses, and polarized headlights.

13-12 Haidinger's Brush

That the human eye can detect polarized light was discovered by Haidinger in 1844. *Haidinger's brush* can be observed by shining a 100-watt light bulb onto a piece of white paper. (See Fig. 13-21.) After an observer looks at the paper through a polarizer for about 30 seconds, he will see a small blue

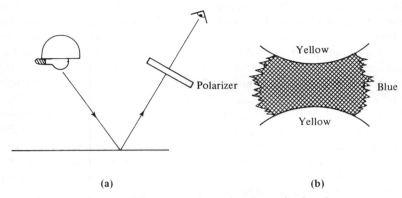

(a)   (b)

Figure 13-21 The Haidinger's brush can be observed using the arrangement in (a). The brush (b) subtends about 2°. The light-blue region lies
parallel to the direction of polarization of the light.

cylinder (the brush) subtending about 2° with yellow regions above and below
it. The axis of the blue cylinder is perpendicular to the polarization of the
light. Rotating the polarizer 90° causes the brush to rotate the same amount.
Some people can see this effect more easily than others, and a few individuals
see it while looking at the blue sky, since light scattered from molecules in
the atmosphere is partially polarized.

An explanation was given by Helmholtz: light-sensitive cones in the
fovea (the central section of the retina) are densely packed in filaments. Light-
sensitive molecules are anisotropic and line up with their axes along the axes
of the filaments. Intense incident light polarized in the horizontal direction
saturates those molecules whose axes are horizontal; the observer sees blue
light in the horizontal direction. The molecules with axes vertical are unsatu-
rated, and the observer sees a yellow color (Fig. 13-22).

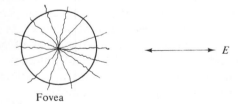

Fovea

Figure 13-22 Explanation of Haidinger's brush. A polarized wave
preferentially activates those regions of the fovea in which the
molecules are lined up parallel to
the direction of polarization.

13-13   The Relationship between Birefringence and Dichroism

The phenomena of birefringence and dichroism (double refraction and
double absorption) are closely related, just as dispersion and absorption are
closely related. They represent the real and imaginary parts of the index of

refraction. When the symmetry of a material is such that there are two possible values for the real part of n, then there are also two possible values for
the imaginary part. For example, if the lengths of the unit cell along the x
and y axis are the same but are different from the z axis length, then the
absorption and dispersion for E_x will be the same as for E_y, but the values
for E_z will be different.

In the situation shown in Fig. 13-23, the absorption frequency ω_0 is
slightly different for the ordinary and extraordinary polarizations. The
strengths of the absorption lines in these two directions need not be the same,
nor do the widths have to be the same. In fact, there are some cases in which
absorption exists only for one polarization.

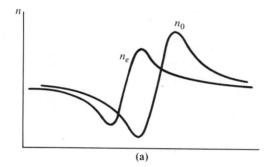

(a)

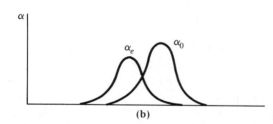

(b)

Figure 13-23 The dispersion and
absorption curves for a uniaxial
crystal showing birefringence and
dichroism.

13-14   Biaxial Crystals

In our earlier treatment of the relationship between symmetry and optical
properties of materials, we found that the symmetry of a molecule or unit cell
determines its optical properties. In cubic crystals, the cell dimensions in
the three principal directions are the same (materials of cubic symmetry are
distinct from truly isotropic materials; the latter are spherically symmetric).
In uniaxial crystals, two lengths are equal and the third one is different, but
in biaxial crystals, the length in each principal direction is different from
the others.

Most crystals are biaxial, especially those with complicated compositions. Although potentially very useful, biaxial crystals have not been studied as much as cubic and uniaxial crystals because of their greater complexity.

13-15   The Index Ellipsoid

The index of refraction for different polarizations can be represented by the distance from the center to the surface of an ellipsoid. (See Fig. 13-24). For propagation down one axis, say x, there is a value of n for E_y and one for E_z.

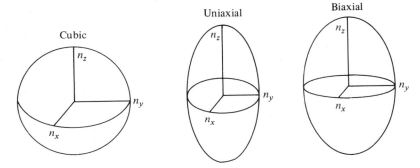

Figure 13-24 The index ellipsoid for different symmetries.

These will be plotted on the y and z axes, respectively. Since the index depends on the polarization of the wave but not on its direction of propagation, the axes of the ellipsoid refer to polarization states.

For cubic crystals, $n_x = n_y = n_z$, and the index ellipsoid degenerates to a sphere, but in uniaxial crystals, $n_x = n_y$, so that the figure representing these materials is an ellipsoid of revolution. For biaxial crystals, each value of n is different and we have a general ellipsoid.

In uniaxial materials, for propagation along z (the optic axis), the indices of refraction are represented by the x, y plane. This figure is a circle, indicating that both polarizations have the same value of n_o. For propagation along x, on the other hand, the intersection with the surface is an ellipse, giving distinct values for n_o and n_e.

As the direction of propagation changes in the xz plane, one index is always n_o but the other changes from n_e to n_o. Thus, *the extraordinary index depends on direction of propagation* and is given by the equation

$$\frac{1}{n_e(\theta)^2} = \frac{\cos^2 \theta}{n_o^2} + \frac{\sin^2 \theta}{n_e^2} \tag{13-45}$$

where θ is the angle between the optic axis and k. (This equation will be derived in Section 14-3). This is summarized in Fig. 13-25, where the two values of n are plotted for each direction of propagation. At $\theta = 0$, both indices are shown equal to n_o.

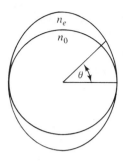

Figure 13-25 The two indices of refraction for a wave propagating in a uniaxial crystal are plotted. θ is the direction of propagation with respect to the optic axis (z).

In biaxial crystals, the middle index is designated as n_y, so that $n_x < n_y < n_z$. Let us now consider a varying value for **k** in the x-z plane. The first thing we notice is that one index will be n_y, regardless of the direction of **k**. Next, we see that the other index varies from n_x which is less than n_y to n_y which is greater (Fig. 13-26). At some angle $\beta/2$ which corresponds to an optic axis, this latter index must obviously equal n_y. By referring to Fig. 13-26, we

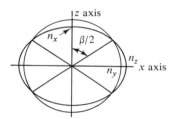

Figure 13-26 The indices of refraction plotted for direction of propagation for a biaxial crystal. There are two directions in which double refraction is zero.

see that there are two such angles, designated as $\beta/2$ and $\pi - (\beta/2)$; hence there are two optic axes. The reason these crystals are called biaxial is that there are two directions of propagation in which birefringence disappears.

The angle between optic axes is the *optic angle* β. It can be calculated from the equation

$$\tan \beta = \sqrt{\frac{n_z^2(n_y^2 - n_x^2)}{n_x^2(n_z^2 - n_y^2)}} \tag{13-46}$$

The derivation of this and Eq. (13-45) is quite involved; we shall return to them in Section 14-3.

The plane containing the two optic axes is called the *optic plane*. The angle β can vary between 0 and 90° from one material to another, and it can vary appreciably with wavelength in a given material if one or more of the indices has noticeable dispersion.

Mica, a common mineral, cleaves easily into sheets. The optic angle is 47° and the two optic axes make an angle of $23\frac{1}{2}°$ with respect to the normal to the cleavage plane. The normal can be taken as the x axis.

The indices of refraction for mica at 589 nm are 1.552, 1.582, and 1.588. For propagation perpendicular to the cleaved sheets, the last two indices are observed. Birefringence is small, only 0.006. Mica is a convenient material to use in making wave plates. A half-wave plate should be 0.1 mm thick, and it is relatively easy to cleave a sheet to this thickness.

Problem 13-5

Calculate the optic angle for mica from the index of refraction data.   ◼

13-16 Pleochroism

When a material has three different principal indices of refraction, it also has three different absorption coefficients. This is called *pleochroism* ("many-colored"). There are two particularly interesting crystals, vivianite and cordierite, that demonstrate this phenomenon.

Vivianite is a mineral which grows naturally in long thin prisms. When one looks at a white light source through the flat side of the prism at an angle of about $+45°$, the crystal appears green; if the prism is rotated to $-45°$, it changes color to red (Fig. 13-27).

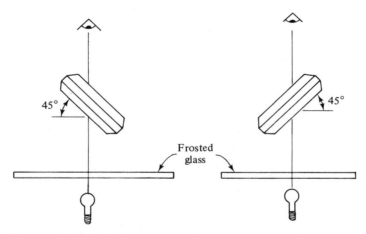

Figure 13-27 Vivianite demonstrates pleochroism. In (a), the crystal is green; in (b), it has changed color to red.

Cordierite occurs in large chunks. When light is polarized along one axis in cordierite, it is almost completely absorbed and the crystal looks dark. When it is polarized along either of the other two axes, it is transmitted, and the crystal looks transparent.

The Vikings used cordierite, which they found in Denmark, as an aid to navigation. At night, the sky in the North Atlantic was usually clear and they were able to use the stars for navigation, but during the day the sky was frequently cloudy and the sun could not be used. Without fully understanding what was happening, the Vikings made use of the fact that light scattered from air molecules in the atmosphere is polarized. When the angle between incident and scattered beams is 90°, the light is almost completely polarized. When the angle between incident and scattered beams is small and the light is scattered in the forward direction, it is unpolarized (Fig. 13-28). Looking through a crystal of cordierite at the sky in a direction about 90° from the sun, they could extinguish the light transmitted through the crystal by rotating it to the proper orientation. When they looked in the direction of the sun, even though it was obscured by clouds, the light was unpolarized and could not be extinguished. After determining the direction of the sun, they were able to navigate in the daytime. The Vikings were satisfied with the explanation that one could not extinguish the sun because it was so powerful. This was one of the earliest practical applications of the optics of biaxial crystals.

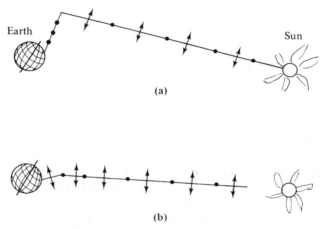

Figure 13-28 Skylight is polarized in (a) when the scattering angle is 90°. When the scattering angle is small, (b), it is unpolarized.

Reference

13-1   G. Bird and M. Parrish, *Journal of Optical Society of America*, *41*, 957 (1951).

General References

E. Wood, *Crystals and Light*, D. Van Nostrand, New York (1964).

W. Shurcliff and S. Ballard, *Polarized Light*, D. Van Nostrand, New York (1964).

E. Wahlstrom, *Optical Crystallography*, John Wiley & Sons, New York (1969).

J. Strong, *Concepts of Classical Optics*, Freeman, San Francisco (1958).

G. Fowles, *Introduction to Modern Optics*, Holt, Rinehart and Winston, New York (1968).

14

NONLINEAR OPTICS

Until this point we have used an inductive or historical approach to describe the various properties of light. The key experimental observations which formed the basis of our knowledge were described and the associated subjects were built up around them. This approach is simpler and easier to understand than the mathematical or deductive approach.

In nonlinear and advanced crystal optics, the situation is reversed. The inductive approach can be continued but it is more complicated, less elegant, and the underlying unity between various phenomena does not become clear. Here we will introduce Maxwell's equations and solve them in aniso-tropic and nonlinear media.

14-1 The Dielectric Tensor

From electromagnetic theory,* we know that in isotropic materials the permittivity ϵ relates the displacement field $\mathbf{D}$ to the electric field $\mathbf{E}$ by the equation

$$\mathbf{D} = \epsilon \mathbf{E} \tag{14-1}$$

where $\epsilon = \epsilon_R \epsilon_0$, ϵ_R is the relative dielectric constant, and ϵ_0 is the free-space value of ϵ. The polarization $\mathbf{P}$ is related to $\mathbf{E}$ through the susceptibility χ via

$$\mathbf{P} = \epsilon_0 \chi \mathbf{E} \tag{14-2}$$

*See, for example, Ref. 14-1.

with

$$\epsilon = \epsilon_0(1 + \chi) \tag{14-3}$$

The index of refraction in isotropic material is*

$$n = \sqrt{\epsilon_R} \tag{14-4}$$

The dielectric "constant" ϵ_R is not really a constant at all, but actually a function of frequency, and in Section 12-4, we calculated the frequency dependence of n using the Lorentz model of the atom.

In anisotropic media, the relation between **D** and **E** is more complicated. Returning to our cigar-shaped molecule, we would expect that it would be easier to polarize such a molecule along one axis than another. If an external field is applied at some arbitrary angle with respect to the axis, then its axial component will be more effective than the other component perpendicular to the axis. The induced polarization **P** will therefore lie along a direction closer to the axis than the field E_1 (Fig. 14-1).

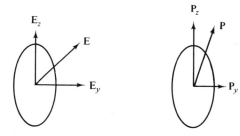

Figure 14-1 An external electric field **E** is applied to an anisotropic molecule. The molecule is easier to polarize in the z direction. Therefore the induced polarization is proportionately greater in that direction. The effect of the anisotrophy is that **P** is in a different direction than the applied field **E**.

To describe this kind of behavior, ϵ and χ must be tensors rather than scalars. The most general expression relating each of the three components of **D** to **E** is

$$D_x = \epsilon_0\epsilon_{xx}E_x + \epsilon_0\epsilon_{xy}E_y + \epsilon_0\epsilon_{xz}E_z$$
$$D_y = \epsilon_0\epsilon_{yx}E_x + \epsilon_0\epsilon_{yy}E_y + \epsilon_0\epsilon_{yz}E_z \tag{14-5}$$
$$D_z = \epsilon_0\epsilon_{zx}E_x + \epsilon_0\epsilon_{zy}E_y + \epsilon_0\epsilon_{zz}E_z$$

This may be abbreviated to

$$D_i = \epsilon_0 \sum_j \epsilon_{ij}E_j \qquad (i, j = x, y, z) \tag{14-6}$$

or in matrix form

$$\begin{bmatrix} D_x \\ D_y \\ D_z \end{bmatrix} = \epsilon_0 \begin{bmatrix} \epsilon_{xx} & \epsilon_{xy} & \epsilon_{xz} \\ \epsilon_{yx} & \epsilon_{yy} & \epsilon_{yz} \\ \epsilon_{zx} & \epsilon_{zy} & \epsilon_{zz} \end{bmatrix} \begin{bmatrix} E_x \\ E_y \\ E_z \end{bmatrix} \tag{14-7}$$

*See, for example, Ref. 14-1.

For some orientations, the dielectric tensor assumes a simpler form—only diagonal elements ϵ_{xx}, ϵ_{yy}, ϵ_{zz} are nonzero. This is called the *principal axis system*.* The forms of the dielectric tensor in the principal axis system for cubic, uniaxial, and biaxial crystal classes are shown in Table 14-1. Different crystals of the same class have different numerical values of the coefficients ϵ_{ij}.

Table 14-1
The Dielectric Tensor for Crystals

Principal axis systems		
Cubic	Uniaxial $\begin{cases}\text{trigonal}\\\text{tetragonal}\\\text{hexagonal}\end{cases}$	Biaxial $\begin{cases}\text{triclinic}\\\text{monoclinic}\\\text{orthorhombic}\end{cases}$
$\epsilon_0 \begin{bmatrix} \epsilon & 0 & 0 \\ 0 & \epsilon & 0 \\ 0 & 0 & \epsilon \end{bmatrix}$	$\epsilon_0 \begin{bmatrix} \epsilon_{xx} & 0 & 0 \\ 0 & \epsilon_{xx} & 0 \\ 0 & 0 & \epsilon_{zz} \end{bmatrix}$	$\epsilon_0 \begin{bmatrix} \epsilon_{xx} & 0 & 0 \\ 0 & \epsilon_{yy} & 0 \\ 0 & 0 & \epsilon_{zz} \end{bmatrix}$

14-2 Characteristic Waves

We will now relate the characteristics of the dielectric tensor to some of the physical properties of crystals which were discussed in previous sections. Let us start by reviewing the half-wave plate formed from a piece of uniaxial crystal. The optic axis is in the vertical (z) direction. A wave is incident normal to the plate ($-k_x$) and polarized at 45° with respect to z, giving $\mathbf{E}_{in} = E_0 \begin{bmatrix} 1 \\ 1 \end{bmatrix}$. The polarization of the emerging wave is rotated by 90° so that $\mathbf{E}_{out} = E_0 \begin{bmatrix} 1 \\ -1 \end{bmatrix}$, the crystal changing the state of polarization. (See Fig. 14-2.)

On the other hand, when $\mathbf{E}_{in} = E_0 \begin{bmatrix} 1 \\ 0 \end{bmatrix}$ or $\begin{bmatrix} 0 \\ 1 \end{bmatrix}$, the emerging wave has the same polarization as the incident wave. Evidently there are certain waves which can propagate through the crystal undisturbed. These are *characteristic waves* for the material or *eigenwaves* or *eigenvectors*. The index of refraction for the characteristic waves corresponds to the square root of the principal (or diagonal) values of the dielectric tensor; these are also called *eigenvalues*.

To understand how light of any polarization state propagates in a crystal, we must first find the eigenvectors and eigenvalues. Then the incoming wave is represented by a linear sum of eigenvectors, and since each eigenvector propagates through the material unchanged in polarization state, at the exit face they are recombined to find the emerging wave.

---

*For example, consider a cigar-shaped object. If we select one axis of a coordinate system along the axis of the cigar, it would be a principal axis coordinate system. If the coordinate axis made an arbitrary angle with respect to the axis of the cigar, we might intuitively expect that the mathematical description of phenomena would be much more complicated.

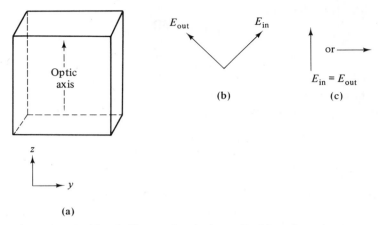

(a)

Figure 14-2 In (a), a half-wave plate is shown. In (b), an incoming wave polarized at 45° with respect to z is rotated by 90°. In (c) and (d), the state of the incoming wave is unaltered by the plate. These latter two are characteristic waves for the plate.

We will restrict our attention to those optical effects described by the linear relation between **D** and **E** in Eq. (14-6) or (14-7) (or between **P** and **E**). This is the region of *linear optics*. When a coordinate system is chosen such that the dielectric tensor is in diagonal form, the eigenvalues are just the diagonal elements of the tensor, and the eigenvectors correspond to these eigenvalues. Thus, for biaxial crystals the dielectric tensor is of the form

$$\begin{vmatrix} \epsilon_{xx} & 0 & 0 \\ 0 & \epsilon_{yy} & 0 \\ 0 & 0 & \epsilon_{zz} \end{vmatrix}$$

where the value ϵ_{xx} is for the eigenvector $\begin{bmatrix} 1 \\ 0 \\ 0 \end{bmatrix}$ or E_x, the value ϵ_{yy} is for $\begin{bmatrix} 0 \\ 1 \\ 0 \end{bmatrix}$ or E_y, etc.

14-3 The Eigenvalue Equation

A central problem in crystal optics is to find the eigenvectors and eigenvalues for a given direction of propagation. As we shall see, this involves diagonalizing a matrix. For propagation along one principal axis, the matrix is already in diagonal form, the two eigenvectors lie in the direction of the other two axes, and the eigenvalues are the diagonal elements. However, for propagation in an arbitrary direction, the problem is more involved. We start with Maxwell's equations[5-2]

$$\text{curl } \mathbf{H} = \frac{\partial \mathbf{D}}{\partial t} \qquad (14\text{-}8)$$

$$-\text{curl } \mathbf{E} = \frac{\partial \mathbf{B}}{\partial t} \qquad (14\text{-}9)$$

where, in nonmagnetic materials, $\mathbf{B} = \mu_0 \mathbf{H}$.

For plane waves of the form

$$\mathbf{E} = \mathbf{E}_0 e^{i(\mathbf{k}\cdot\mathbf{r}-\omega t)}$$

$$\mathbf{H} = \mathbf{H}_0 e^{i(\mathbf{k}\cdot\mathbf{r}-\omega t)}$$

$$\mathbf{D} = \mathbf{D}_0 e^{i(\mathbf{k}\cdot\mathbf{r}-\omega t)}$$

the operations called for in Eqs. (14-8) and (14-9) give

$$\text{curl } \mathbf{E} = i\mathbf{k} \times \mathbf{E} \qquad (14\text{-}10)$$

$$\text{curl } \mathbf{H} = i\mathbf{k} \times \mathbf{H} \qquad (14\text{-}11)$$

$$\frac{\partial \mathbf{H}}{\partial t} = -i\omega \mathbf{H} \qquad (14\text{-}12a)$$

$$\frac{\partial \mathbf{D}}{\partial t} = -i\omega \mathbf{D} \qquad (14\text{-}12b)$$

With these substitutions, Eqs. (14-8) and (14-9) become

$$\mathbf{k} \times \mathbf{H} = \omega \mathbf{D} \qquad (14\text{-}13)$$

$$\mathbf{k} \times \mathbf{E} = -\omega \mathbf{B} \qquad (14\text{-}14)$$

Recall that $\mathbf{E}$ and $\mathbf{D}$ need not be parallel. Equation (14-14) indicates that $\mathbf{B}$ and hence $\mathbf{H}$ is perpendicular to $\mathbf{k}$ and $\mathbf{E}$. Equation (14-13) indicates that $\mathbf{D}$ is perpendicular to $\mathbf{k}$ and $\mathbf{H}$. Thus, $\mathbf{D}$, $\mathbf{H}$, and $\mathbf{k}$ are an orthogonal set. (See Fig. 14-3.)

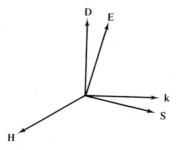

Figure 14-3 D, H, k and E, H, S are orthogonal sets.

The energy flow is given by the Poynting vector

$$\mathbf{S} = \mathbf{E} \times \mathbf{H} \qquad (14\text{-}15)$$

so that $\mathbf{S}$, $\mathbf{E}$, and $\mathbf{H}$ also form an orthogonal set. Note that the energy flows in the direction of $\mathbf{S}$ which can be different from $\mathbf{k}$, the direction of propagation of the phase front.

Substituting Eq. (14-14) into Eq. (14-13) yields

$$\frac{1}{\mu_0 \omega^2} \mathbf{k} \times (\mathbf{k} \times \mathbf{E}) = -\mathbf{D} \qquad (14\text{-}16)$$

We define the vector $\tilde{\mathbf{k}}$ by

$$\tilde{\mathbf{k}} = \frac{c}{\omega} \mathbf{k} = n\hat{\mathbf{k}} \qquad (14\text{-}17)$$

where $\hat{\mathbf{k}}$ is the unit vector in the direction of $\mathbf{k}$. Equation (14-16) can be written as

$$\frac{1}{\mu_0 c^2} \tilde{\mathbf{k}} \times (\tilde{\mathbf{k}} \times \mathbf{E}) = -\mathbf{D} \qquad (14\text{-}18)$$

Using the vector identity

$$\mathbf{A} \times (\mathbf{B} \times \mathbf{C}) = (\mathbf{A} \cdot \mathbf{C})\mathbf{B} - (\mathbf{A} \cdot \mathbf{B})\mathbf{C} \qquad (14\text{-}19)$$

we can rewrite Eq. (14-18) as

$$-(\tilde{\mathbf{k}} \cdot \tilde{\mathbf{k}})\mathbf{E} + (\tilde{\mathbf{k}} \cdot \mathbf{E})\tilde{\mathbf{k}} = -\mu_0 c^2 \mathbf{D} \qquad (14\text{-}20)$$

The quantity $\tilde{\mathbf{k}} \cdot \tilde{\mathbf{k}}$ is simply n^2, and the quantity $\tilde{\mathbf{k}} \cdot \mathbf{E}$ can be written as $\sum_j \tilde{k}_j E_j$. With these substitutions and the use of Eq. (14-6), Eq. (14-20) becomes

$$\mu_0 c^2 \sum_j \epsilon_0 \epsilon_{ij} E_j + \tilde{k}_i \sum_j \tilde{k}_j E_j = n^2 E_i$$

or

$$\boxed{\sum_j (\epsilon_{ij} + \tilde{k}_i \tilde{k}_j) E_j = n^2 E_i} \qquad (14\text{-}21)$$

This is one of the most important equations in crystal optics, describing the propagation of light in anisotropic media. It is an eigenvalue equation: the E_i are the eigenvectors, and n^2 the eigenvalues. In "solving" Eq. (14-21) in a given crystal, we start with the dielectric tensor of Table 14-1. We then select a direction of propagation $\tilde{\mathbf{k}}$ and we diagonalize the matrix.

A very remarkable historical fact is that Fresnel derived the essence of this equation 40 years prior to the development of Maxwell's wave equation, using an analogy with elastic waves in solids.

As an illustration of the procedure of finding eigenvalues and eigenvectors, we will consider a uniaxial crystal. From Table 14-1, ϵ_R is given by

$$\epsilon_{ij} = \begin{bmatrix} \epsilon_{xx} & 0 & 0 \\ 0 & \epsilon_{xx} & 0 \\ 0 & 0 & \epsilon_{zz} \end{bmatrix}$$

For $\mathbf{k}$ along the x axis, we have $\tilde{k}_i = \tilde{k}_j = \tilde{k}_x = n\hat{k}_x$ and $\tilde{k}_i \tilde{k}_j = n^2 \hat{k}_x \hat{k}_x$. Thus

$$\tilde{k}_i \tilde{k}_j = \begin{bmatrix} n^2 & 0 & 0 \\ 0 & 0 & 0 \\ 0 & 0 & 0 \end{bmatrix} \qquad (14\text{-}22)$$

In matrix notation n^2 can be written

$$n^2 = \begin{bmatrix} n^2 & 0 & 0 \\ 0 & n^2 & 0 \\ 0 & 0 & n^2 \end{bmatrix} \tag{14-23}$$

and Eq. (14-20) becomes

$$\begin{vmatrix} \epsilon_{xx} + n^2 & 0 & 0 \\ 0 & \epsilon_{xx} & 0 \\ 0 & 0 & \epsilon_{zz} \end{vmatrix} \begin{vmatrix} E_x \\ E_y \\ E_x \end{vmatrix} = \begin{vmatrix} n^2 & 0 & 0 \\ 0 & n^2 & 0 \\ 0 & 0 & n^2 \end{vmatrix} \begin{vmatrix} E_x \\ E_y \\ E_z \end{vmatrix}$$

or

$$\begin{vmatrix} \epsilon_{xx} & 0 & 0 \\ 0 & \epsilon_{xx} - n^2 & 0 \\ 0 & 0 & \epsilon_{zz} - n^2 \end{vmatrix} \begin{vmatrix} E_x \\ E_y \\ E_z \end{vmatrix} = 0 \tag{14-24}$$

Writing the three equations separately gives

$$\epsilon_{xx} E_x = 0 \tag{14-25a}$$

$$(\epsilon_{xx} - n^2) E_y = 0 \tag{14-25b}$$

$$(\epsilon_{zz} - n^2) E_z = 0 \tag{14-25c}$$

The first equation requires that either $\epsilon_{xx} = 0$ or $E_x = 0$. In crystal optics we deal with transverse waves, so that $E_x = 0$ is the appropriate solution. (In plasmas, longitudinal waves exist and $\mathbf{E}$ is parallel to $\mathbf{k}$. Such waves require that $\epsilon_{xx} = 0$).

The other two equations yield

$$n_1^2 = \epsilon_{xx}$$

$$n_2^2 = \epsilon_{zz}$$

Therefore in our example of uniaxial crystals for $\mathbf{k}$ along the x axis we find eigenvectors E_x (zero), E_y, and E_z, and eigenvalues ϵ_{xx} (undetermined), ϵ_{xx}, and ϵ_{zz}.

As a more complicated example, we will now solve Eq. (14-21) for a uniaxial crystal with $\tilde{\mathbf{k}}$ in the y-z plane. (See Fig. 14-4). Under this condition

$$\tilde{k}_x = 0$$

$$\tilde{k}_y = n \sin \theta$$

$$\tilde{k}_z = n \cos \theta$$

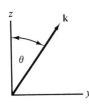

Figure 14-4 The vector **k** lies in the y-z plane and is inclined at an angle θ with respect to the z axis.

The matrix $\tilde{k}_i \tilde{k}_j$ is

$$\tilde{k}_i \tilde{k}_j = \begin{vmatrix} 0 & 0 & 0 \\ 0 & n^2 \sin^2 \theta & n^2 \sin \theta \cos \theta \\ 0 & n^2 \sin \theta \cos \theta & n^2 \cos^2 \theta \end{vmatrix} \qquad (14\text{-}26)$$

and the matrix ϵ_{ij} is given in Table 4-1 for a uniaxial crystal. Equation (14-21), under these conditions, becomes

$$\begin{vmatrix} \epsilon_{xx} & 0 & 0 \\ 0 & \epsilon_{xx} + n^2 \sin^2 \theta & n^2 \sin \theta \cos \theta \\ 0 & n^2 \sin \theta \cos \theta & \epsilon_{zz} + n^2 \cos^2 \theta \end{vmatrix} \begin{vmatrix} E_x \\ E_y \\ E_z \end{vmatrix} = \begin{vmatrix} n^2 & 0 & 0 \\ 0 & n^2 & 0 \\ 0 & 0 & n^2 \end{vmatrix} \begin{vmatrix} E_x \\ E_y \\ E_z \end{vmatrix}$$

The eigenvalues n^2 are obtained by setting the determinant of the coefficients equal to zero and solving the equation

$$\begin{vmatrix} \epsilon_{xx} - n^2 & 0 & 0 \\ 0 & \epsilon_{xx} + n^2(\sin^2 \theta - 1) & n^2 \sin \theta \cos \theta \\ 0 & n^2 \sin \theta \cos \theta & \epsilon_{zz} + n^2(\cos^2 \theta - 1) \end{vmatrix} = 0 \qquad (14\text{-}27)$$

or

$$\{\epsilon_{xx} - n^2\}\{(\epsilon_{xx} - n^2 \cos^2 \theta)(\epsilon_{zz} - n^2 \sin^2 \theta) - n^4 \sin^2 \theta \cos^2 \theta\} = 0$$

One solution is

$$n_1^2 = \epsilon_{xx} \qquad (14\text{-}28a)$$

and the other is

$$\frac{1}{n_2^2} = \frac{\cos^2 \theta}{\epsilon_{xx}} + \frac{\sin^2 \theta}{\epsilon_{zz}}$$

or

$$\frac{1}{n_2^2} = \frac{\cos^2 \theta}{n_0^2} + \frac{\sin^2 \theta}{n_e^2} \qquad (14\text{-}28b)$$

Thus the second index of refraction varies between n_o and n_e as the propagation vector varies from $\mathbf{k}_y$ to $\mathbf{k}_z$. This equation was encountered in Section 13–15 and discussed in terms of the index ellipsoid.

The eigenvectors are obtained by substituting each eigenvalue separately into Eq. (14-21) and solving for the corresponding eigenvector. Starting with n_1^2, we obtain

$$\epsilon_{xx} E_x = \epsilon_{xx} E_x$$
$$(\epsilon_{xx} - \epsilon_{xx} \cos^2 \theta)E_y + \epsilon_{xx} \sin \theta \cos \theta \, E_z = \epsilon_{xx} E_y$$
$$\epsilon_{xx} \sin \theta \cos \theta \, E_y + (\epsilon_{zz} - \epsilon_{xx} \sin^2 \theta)E_z = \epsilon_{xx} E_z \qquad (14\text{-}29a)$$

or

$$\epsilon_{xx} E_x = \epsilon_{xx} E_x$$
$$\sin \theta E_z - \cos \theta E_y = 0$$
$$\sin \theta \cos \theta \, E_y + \left[\left(\frac{\epsilon_{zz} - \epsilon_{xx}}{\epsilon_{xx}} \right) - \sin^2 \theta \right] E_z = 0 \qquad (14\text{-}29b)$$

The solution is

$$E_y = E_z = 0 \tag{14-30}$$

while E_x can take any value. Thus the eigenvalue n_1^2 corresponds to the eigen-

vector $\mathbf{E} = E_0 \begin{bmatrix} 1 \\ 0 \\ 0 \end{bmatrix}$. The other eigenvector is obtained by substituting n_2^2 into

Eq. (14-21) and solving for $\mathbf{E}$. The first equation is

$$\epsilon_{xx} E_x = n_2^2 E_x$$

Since n_2^2 can not be equal to ϵ_{xx}, we conclude that $E_x = 0$.

Thus the electric field lies in the y-z plane for this wave. To find the eigenvector, the algebra becomes tedious. The result is that $\mathbf{E}$ is not exactly perpendicular to $\tilde{\mathbf{k}}$, but has the form

$$\mathbf{E}_2 = E_0 \begin{bmatrix} 0 \\ \cos \psi \\ -\sin \psi \end{bmatrix}$$

where ψ is slightly different from θ.

To summarize the calculations for a uniaxial crystal, there are always three characteristic waves. For $\mathbf{k}$ along a principal axis, say x, one wave is longitudinal with eigenvalue $\epsilon_{xx} = 0$ and eigenvector E_x. The two transverse

waves have $n_1^2 = \epsilon_{xx}$, $\mathbf{E}_1 = E_0 \begin{bmatrix} 0 \\ 1 \\ 0 \end{bmatrix}$ and $n_2^2 = \epsilon_{zz}$, $\mathbf{E}_2 = E_0 \begin{bmatrix} 0 \\ 0 \\ 1 \end{bmatrix}$. For propaga-

tion at an arbitrary angle θ with respect to z, the solutions yield a transverse wave with $n_1^2 = \epsilon_{xx}$ and $\mathbf{E}$ perpendicular to the plane containing $\tilde{\mathbf{k}}$ and the z axis, and a quasitransverse wave with $\mathbf{E}_2$ in the k-z plane and n_2^2 given by Eq. (14-28b).

Problem 14-1

Derive Eq. 13-46. ∎

14-4 Nonlinear Optics

In the preceding section, the optical effects that were treated are adequately described by a linear relationship between an applied electric field $\mathbf{E}$ and the material response $\mathbf{P}$. Anisotropy was taken into account by representing χ as a tensor. There are many optical effects that can only be described by assuming nonlinear or other more general relationships between $\mathbf{P}$ and $\mathbf{E}$. Examples of these are electro- and magneto-optical effects, circular birefringence and higher-order birefringence, harmonic generation, and stimulated effects.

These are important because they are the keys to the control and manipulation of light waves. One of the beauties of optics is that all of these nonlinear effects can be described in a unified treatment that brings out their close interrelationship.

The theory starts by expanding the polarization **P** in a general power series in the electric and magnetic fields and the spatial derivatives. (The time derivative only multiplies the wave by $i\omega$, so we do not obtain anything new by taking it into consideration.) The Taylor series has the form

$$P_i^\omega = P_i^0 + \chi_{ij}E_j^\omega + \chi_{ijl}\nabla_l E_j^\omega + \chi_{ijl}E_j^{\omega_1}E_l^{\omega_2} + \chi_{ijlm}E_j^{\omega_1}E_l^{\omega_2}E_m^{\omega_3}$$
$$+ \chi_{ijl}E_j^{\omega_1}B_l^{\omega_2} + \chi_{ijlm}E_j^{\omega_1}B_l^{\omega_2}B_m^{\omega_3} \qquad (14\text{-}31)$$

where the superscripts denote the frequency of the field and the subscripts denote cartesian components. Each term in this series describes one or more effects; for example, the fourth term describes the linear electro-optic effect when $\omega_1 = \omega$ and $\omega_2 = 0$, giving $P_i^\omega = X_{ijl}E_j^\omega E_l^0$. This describes the effect of a constant electric field on the optical properties of crystals, and it can be used to modulate light beams, leading to optical communication. The same term describes second harmonic generation when $\omega_1 = \omega_2 = \omega/2$, so that $P_i^\omega = X_{ijl}E_j^{\omega/2}E_l^{\omega/2}$. This is the generation of a light wave at twice the frequency of an incident wave, and it is analogous to the generation of harmonics in electronic circuits due to the nonlinear response of circuit elements.

14-5 Polar Crystals: Pyroelectricity and Ferroelectricity

The first term in Eq. (14-31) ($P_i^\omega = P_i^0$) simply describes a spontaneous dc ($\omega = 0$) polarization of a material. Such a polarization can arise if the unit cell of a crystal is polar, i.e., has a net dipole moment. In such unit cells, the centers of positive and negative charge do not coincide. One outer surface has a positive charge and the other has a negative charge, giving an external electric field. These materials are analogous to ferromagnets which possess a permanent magnetic field. With time, however, compensating charges accumulate on the surface to neutralize the polarization charges.

Pyroelectric materials are those for which the polarization changes with temperature. It was known for a long time that tourmaline, when placed in hot ashes, attracted the ashes much like a charged comb attracts bits of paper. This was first recognized in Ceylon and India. Tourmaline was sometimes called the "Ceylon Magnet" because of this behavior. Since **P** eventually decays, only changes in **P**, caused by temperature changes, are detectable. Like permanent magnets, the spontaneous fields disappear when the material is heated to a transition temperature, the *Curie temperature*.

Electrets are made of paraffins which have long polar molecules. Wax is melted and allowed to cool between two parallel plates which have a high voltage across them. The polar molecules line up parallel to the field, and as the wax solidifies, the molecules are "frozen" into this position. When the plates are removed, the electret remains polarized and can generate a high electric field in its vicinity for many months.[14-4]

Ferroelectrics are pyroelectric materials in which the spontaneous polarization can be reversed by the application of a sufficiently strong field contrary to **P**.

Both pyroelectric and ferroelectric crystals have a domain structure; a discussion of this subject will be found in Ref. 14-5.

Although these effects, strictly speaking, are not optical effects (since the polarization is at zero frequency rather than at optical frequencies), they nevertheless affect the optical properties of crystals, since a multiple-domain crystal behaves quite differently than a single-domain crystal. See Ref. 14-7.

14-5a Crystal Symmetry and Spontaneous Polarization

Restrictions on the possible existence of spontaneous polarization are imposed by crystal symmetry. Since this effect is described by a vector, it can not exist in a direction perpendicular to a mirror plane in a crystal. A mirror plane normal to the ith axis reverses the direction of the ith component of **P**. By definition, a symmetry operation is one that leaves the material in a state that is indistinguishable from its original state. The only way P_i can be identical to $-P_i$ is for it to vanish. Nor can **P** exist in a crystal having a center of inversion. Inversion of a coordinate system means that the direction of each axis x, y, z changes sign, and any vector which is converted into its own negative must vanish by the argument just given.

Mathematically, symmetry operations can be represented by matrices.[14-2] A mirror plane, m_x perpendicular to the x axis changes the sign of the x component of a vector but leaves the y and z components unchanged. It is represented as

$$m_x = \begin{bmatrix} -1 & 0 & 0 \\ 0 & 1 & 0 \\ 0 & 0 & 1 \end{bmatrix} \tag{14-32}$$

A center of inversion $\bar{1}$ changes the sign of each component

$$\bar{1} = \begin{bmatrix} -1 & 0 & 0 \\ 0 & -1 & 0 \\ 0 & 0 & -1 \end{bmatrix} \tag{14-33}$$

A twofold rotation 2_x $(360°/2)$ about the x axis changes the sign of y and z

$$2_x = \begin{bmatrix} 1 & 0 & 0 \\ 0 & -1 & 0 \\ 0 & 0 & -1 \end{bmatrix} \tag{14-34}$$

and a fourfold rotation $(360°/4)$ about y is described as

$$4_y = \begin{bmatrix} 0 & 0 & 1 \\ 0 & 1 & 0 \\ -1 & 0 & 0 \end{bmatrix} \qquad (14\text{-}35)$$

By the definition of symmetry, any physical property of the crystal must be unchanged by the application of the symmetry operator, S. This can be described, for a vector $\mathbf{P}$, as

$$SP = P \qquad (14\text{-}36)$$

Example 14-1

Let us determine the allowed direction for $\mathbf{P}$ in a crystal of sodium nitrite ($NaNO_2$). This substance is a member crystal class* $mm2$ which has the following symmetry elements: a mirror plane perpendicular to x, one perpendicular to z, and a twofold rotation about y. Using Eq. (14-32) in Eq. (14-36) gives

$$\begin{bmatrix} -1 & 0 & 0 \\ 0 & 1 & 0 \\ 0 & 0 & 1 \end{bmatrix} \begin{bmatrix} P_x \\ P_y \\ P_z \end{bmatrix} = \begin{bmatrix} P_x \\ P_y \\ P_z \end{bmatrix}$$

or

$$\begin{bmatrix} -P_x \\ P_y \\ P_z \end{bmatrix} = \begin{bmatrix} P_x \\ P_y \\ P_z \end{bmatrix}$$

From this we conclude that $P_x = 0$. Using m_z in a similar way yields $P_z = 0$, but the 2_y operation does not yield any new information. Therefore the spontaneous polarization can exist only along the y axis for a crystal in this class.

It turns out that spontaneous polarization is allowed in 10 of the 32 crystal classes (Table 14-2) but not every crystal in these classes exhibits this effect. ▄

Table 14-2
Crystal Classes in which a Spontaneous
Polarization is allowed by Symmetry

1	m	$mm2$	4	3	6
	2		$4mm$	$3m$	$6mm$

*The concept of *crystal class* describes the allowed combinations of symmetry in a given material. See Ref. (14-2) for a full discussion, including notation.

14-6 Optical Activity

When a plane-polarized wave propagates in some materials, the plane of polarization is gradually rotated, and the amount of rotation is proportional to the thickness of the material. The effect known as *optical activity* was discovered by Arago in 1811. See Fig. 14-5. It must be distinguished from

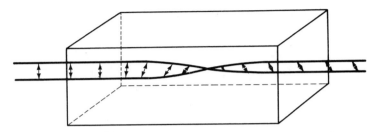

Figure 14-5 Optical activity is the gradual rotation of the plane of polarization as a wave passes through a material.

double refraction, which involves the conversion of a plane-polarized wave to an elliptically polarized one.

Many substances exhibit optical activity, among them are sugar solutions, turpentine, sodium chlorate (cubic), quartz (uniaxial), and sodium nitrite (biaxial). In uniaxial and biaxial crystals, the effect is most easily observed along the optic axis.

The explanation of the effect was given by Fresnel, who assumed that circularly polarized light is propagated without change in these materials, but that the right circularly polarized wave travels with a slightly different velocity than the left circularly polarized one. Fresnel proposed that the incident plane-polarized wave could be represented by the sum of two circularly polarized waves with opposite senses of rotation (Fig. 14-6). After traveling through the material, one wave is shifted in phase with respect to

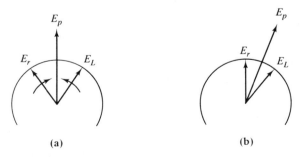

Figure 14-6 A plane-polarized wave, E_p, can be represented by two oppositely rotating circularly polarized waves, E_r and E_l.

the other. When they are recombined at the exit face, the plane of polarization is rotated.

After proposing his explanation of optical activity, Fresnel experimentally demonstrated the existence of circular birefringence in quartz. He reasoned that a prism made of quartz, with the optic axis parallel to its base, would refract the two circularly polarized waves into different directions. To increase the effect, he made a multiple prism (Fig. 14-7) of alternating

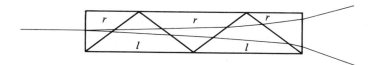

Figure 14-7 Fresnel's prism for demonstrating circular birefringence in quartz.

sections of left- and right-handed quartz (Section 14-6c) and found two emerging waves for each incident plane-polarized wave, as predicted.

Fresnel's explanation can be demonstrated with Jones' calculus. The incident wave, being plane-polarized, is represented as the sum of two circularly polarized waves, giving

$$\mathbf{E}_{in}\begin{bmatrix}1\\0\end{bmatrix} = \frac{E_0}{2}\begin{bmatrix}1\\i\end{bmatrix} + \frac{E_0}{2}\begin{bmatrix}1\\-i\end{bmatrix} \qquad (14\text{-}37)$$

At a distance l, the phase of a wave is shifted by an amount $\delta = 2\pi n l/\lambda$ with respect to the incident wave, and the emerging wave is then

$$\mathbf{E}_{out} = \frac{E_0}{2}\begin{vmatrix}1\\i\end{vmatrix}e^{i2\pi n_r l/\lambda} + \frac{E_0}{2}\begin{vmatrix}1\\-i\end{vmatrix}e^{i2\pi n_l l/\lambda} \qquad (14\text{-}38)$$

where n_r is the index for the right circularly polarized wave and n_l the index for the left circularly polarized one. Equation (14-38) can be written in the form

$$\mathbf{E}_{out} = \frac{E_0}{2}e^{i2\pi(n_r+n_l)l/2\lambda}\left\{\begin{bmatrix}1\\i\end{bmatrix}e^{i2\pi(n_r-n_l)l/2\lambda} + \begin{vmatrix}1\\-i\end{vmatrix}e^{-i2\pi(n_r-n_l)l/2\lambda}\right\} \qquad (14\text{-}39)$$

Letting

$$\frac{2\pi(n_r + n_l)l}{2\lambda} = \psi \qquad (14\text{-}40)$$

and

$$\frac{2\pi(n_r - n_l)l}{2\lambda} = \delta \qquad (14\text{-}41)$$

gives

$$\mathbf{E}_{out} = E_0 e^{i\psi}\left\{\begin{bmatrix}1\\0\end{bmatrix}\left(\frac{e^{i\delta} + e^{-i\delta}}{2}\right) + \begin{bmatrix}0\\1\end{bmatrix}\left(\frac{e^{i\delta} - e^{-i\delta}}{2i}\right)\right\}$$

$$= E_0 e^{i\psi}\left\{\begin{bmatrix}1\\0\end{bmatrix}\cos\delta + \begin{bmatrix}0\\1\end{bmatrix}\sin\delta\right\} \qquad (14\text{-}42)$$

This represents a plane-polarized wave whose plane of polarization makes an angle δ with respect to the incoming wave.

Optical activity then is really circular birefringence and is related to the helical structure of a substance. For propagation down a helix, right-handed waves interact with the helix differently than left-handed waves. In a similar manner, the behavior of plane-polarized waves which interact with a set of parallel wires depends on whether the wires are parallel or perpendicular to the polarization.

Most optically-active materials exist in two forms: The pitch of the helix can be right-handed or left-handed; the first rotates the plane of polarization clockwise, and the second, counterclockwise. Such substances are said to be *enantiomorphic*.

Problem 14-2

In quartz, optical rotation is 22.0 degrees/mm at 632.8 nm. Find $n_r\text{-}n_l$. What is the angular separation of the two beams emerging from the Fresnel prism in Fig. 14.8? ◼

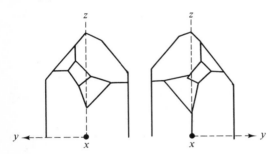

Figure 14-8 The shape of right- and left-handed quartz crystals.

14-6a Optical Activity in Liquids

Many liquids, particularly complex organic compounds, exhibit optical activity. The rotation in liquids is due to the helical structure of the molecule, each one rotating the polarization by a small amount. The random orientation in a liquid does not result in a zero average, as was the case for cigar-shaped molecules. This is because a helix viewed from either end has the same pitch and the light wave will experience the sense of the helix, independent of direction.

Sugar solutions exhibit optical activity, the amount of rotation depending on the concentration. That it is not strictly proportional to the concentration indicates that solvent effects are important. Some sugars exist in right- and left-handed forms, such as dextro-glucose (dextrose) and fructose, and a mixture can be made such that no net rotation is produced. However, bacteria prefer dextrose, since they can metabolize it easier (it must taste sweeter to them!). If bacteria are placed in a solution with equal rotations by right- and left-handed sugars, the plane of polarization of a light wave passing through the solution will be rotated to the left.

An industrial application of optical activity is *polarimetry*: the concentration of sugar solutions is rapidly and accurately determined by measuring the rotation that they produce.

14-6b   Optical Activity in Cubic Crystals

A number of cubic crystals such as sodium chlorate ($NaClO_3$) can exhibit optical activity for any direction of propagation. The importance of this observation is that it shows cubic crystals can have birefringence. This is a consequence of their helical or screw-axis structure. Half of the $NaClO_3$ crystals measured are found to be right-handed, and the other half are left-handed.

It may seem puzzling that a structure as symmetric as a cube can have a helical arrangement associated with it. This is because the helical nature comes from the arrangement of individual atoms or ions at the periodically repeated lattice points. A complete explanation of these aspects of crystallography will be found in Ref. 14-3.

14-6c   Optical Activity in Quartz

In uniaxial crystals, ordinary linear birefringence vanishes for propagation down the optic axis, but some uniaxial crystals such as quartz exhibit circular birefringence in this direction. Quartz (SiO_2) exists in right- and left-handed forms which are mirror images of each other (Fig. 14-8).

The atomic structure of quartz can be determined by X-ray diffraction, which shows the silicon and oxygen atoms arranged in a spiral structure along the z axis. (See Fig. 14-9.)

Fused silica, on the other hand, consists of SiO_2 molecules in a random array—it is actually a supercooled liquid, and does not rotate the plane of polarization.

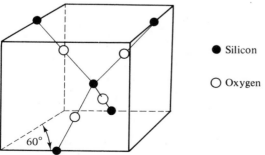

● Silicon

○ Oxygen

Figure 14-9 The crystal structure of quartz.

14-6d   Optical Activity in Sodium Nitrite

Sodium nitrite ($NaNO_2$) is a biaxial crystal with two mirror planes at right angles and a twofold axis (class mm2). The optic axes lie in the *x-z* plane, and the optic angle is approximately 66°.

Measurements in these crystals show that light at 632.8 nm is rotated 15.7°/mm clockwise for propagation down one axis and the same amount counterclockwise down the other axis; from these observations, the structure can be deduced. It must consist of a right-handed helix oriented along one axis and a left-handed helix oriented along the other, with the angles between the axes of the helices bisected by mirror planes (Fig. 14-10). A mirror plane

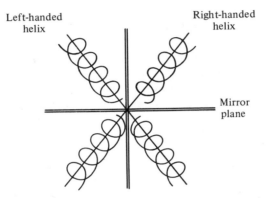

Figure 14-10 The structure of $NaNO_2$ as deduced from the observation of optical rotation.

converts a right-handed helix into a left-handed one in the same way that a right-handed screw looks left-handed in a mirror.

X-ray diffraction reveals the atomic structure of this interesting crystal (Fig. 14-11). The negative charge is shared by the three atoms in the NO_2^- ion.

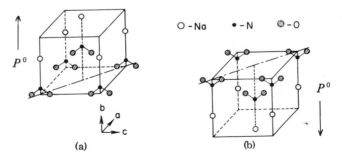

Figure 14-11 The structure of $NaNO_2$ as deduced from *X*-ray diffraction data.

An optic axis corresponds closely to the direction of a diagonal on the bottom face of the unit cell. Along one diagonal, a wave would encounter a clockwise arrangement of these ions, but along the direction of the other diagonal, the same ions would appear in a counterclockwise arrangement. This crystal is particularly interesting because it is ferroelectric.* The direction of the spontaneous polarization is upward for the cell in Fig. 14-11(a).

14-6e Theory of Optical Activity

Optical activity is described by the polarization produced by the third term on the right-hand side of Eq. (14-31), which gives

$$P_i^\omega = \chi_{ijl}\nabla_l E_j^\omega \tag{14-43}$$

An expression involving the gradient is equivalent to considering the variation of the electric field across a molecule or unit cell of a crystal, rather than considering them to be so small that $\mathbf{E}$ is everywhere constant. This variation is also known as *spatial dispersion*. To treat these effects, we expand the electric field itself in a power series in ρ, the molecular coordinate, obtaining

$$\mathbf{E}(\rho) = \mathbf{E}|_{\rho=0} + (\rho\cdot\nabla)\mathbf{E}|_{\rho=0} + (\rho\cdot\nabla)^2\mathbf{E}|_{\rho=0} + \cdots \tag{14-44}$$

Since the electric field of a plane wave can be represented by the expression $\mathbf{E} = \mathbf{E}_0 e^{i\mathbf{k}\cdot\mathbf{r}}$, where $\mathbf{E}_0$ is a constant, the gradient operation is equivalent to multiplying the original expression for $\mathbf{E}$ by the quantity $i\mathbf{k}$. Each time the operation is performed, the rank of the corresponding susceptibility tensor is increased by one. The quantity ρ is incorporated in χ and does not appear explicitly in Eq. (14-43). Each successive term in the above expansion is reduced by a factor $2\pi na/\lambda$, where a is the characteristic length for the media, such as a molecular radius or lattice constant. In the optical region, the value of this quantity is approximately 10^{-3}. Therefore, we expect that optical activity or circular birefringence will be three orders of magnitude weaker than linear birefringence, which is a reasonably close estimate. In quartz, for example, linear birefringence is small (0.009), but circular birefringence has a value of 0.00007 (in the visible region).

After performing the gradient operation, we can write Eq. (14-43) as

$$P_i^\omega = i\frac{\omega}{c}\chi_{ijl}E_j^\omega\tilde{k}_l \tag{14-45}$$

In the treatment of spatial dispersion, electro- and magneto-optical effects, and the intensity-dependent index of refraction, the higher-order terms in the expansion of $\mathbf{P}$ given by Eq. (14-31) are treated as contributions or perturbations to the dielectric tensor ϵ_{ij}. Considering optical activity, the effective value of ϵ_{ij} is the sum of the ordinary value ϵ_{ij} and the term

*See Ref. 14-7.

$i(\omega/c)\chi_{ijl}\tilde{k}_l$ or

$$\text{eff}\,(\epsilon_{ij}) = \epsilon_{ij} + i\frac{\omega}{c}\chi_{ijl}\tilde{k}_l \tag{14-46a}$$

In the absence of absorption, ϵ_{ij} must be Hermitian, that is, it obeys the relation*

$$\epsilon_{ij} = \epsilon_{ji}^* \tag{14-46b}$$

Both the dielectric tensor and the susceptibility tensor have real and imaginary parts which we write as

$$\epsilon_{ij} = \epsilon_{ij}' + i\epsilon_{ij}'' \tag{14-47a}$$

$$\chi_{ijl} = \chi_{ijl}' + i\chi_{ijl}'' \tag{14-47b}$$

The dielectric tensor ϵ_{ij} must be real and symmetric in the cartesian coordinate system; this is indicated as

$$\epsilon_{ij} = \epsilon_{ji} \quad \text{(i.e., symmetric)} \tag{14-48a}$$

and

$$\epsilon_{ji}^* = \epsilon_{ji} \quad \text{(i.e., real)} \tag{14-48b}$$

where the imaginary part describes absorption.†

The restrictions on χ_{ijl} are not so simple. It also must be Hermitian, so there are two classes of solutions, as follows:

(a) Imaginary coefficients, antisymmetric upon an interchange of i and j.
(b) Real coefficients, symmetric upon an interchange of i and j.

Optical activity is described by the first of these conditions. In order for a crystal to be optically active, its symmetry must be such that at least two coefficients satisfy the relation

$$\chi_{ijl} = -\chi_{jil} \tag{14-49}$$

and in quartz $\chi_{xyz} = -\chi_{yxz}$. Thus, for propagation along the z axis, Eq. (14-45) becomes

$$P_x^\omega = i\chi_{xyz}E_y\tilde{k}_z \tag{14-50a}$$

$$P_y^\omega = i\chi_{yxz}E_x\tilde{k}_z = -i\chi_{xyz}E_x\tilde{k}_z \tag{14-50b}$$

With reference to Eq. (14-46), the effective dielectric tensor for quartz written in matrix form is

$$\text{eff}\,(\epsilon_{ij}) = \begin{vmatrix} \epsilon_{xx} & i\chi_{xyz}\tilde{k}_z & 0 \\ -i\chi_{xyz}\tilde{k}_z & \epsilon_{xx} & 0 \\ 0 & 0 & \epsilon_{zz} \end{vmatrix} \tag{14-51}$$

Equation (14-21), which describes the propagation of waves, gives

$$\begin{vmatrix} \epsilon_{xx} & i\chi_{xyz}\tilde{k}_z & 0 \\ -i\chi_{xyz}\tilde{k}_z & \epsilon_{xx} & 0 \\ 0 & 0 & \epsilon_{zz} + \tilde{k}_z\tilde{k}_z \end{vmatrix} \begin{vmatrix} E_x \\ E_y \\ E_z \end{vmatrix} = \begin{vmatrix} n^2 & 0 & 0 \\ 0 & n^2 & 0 \\ 0 & 0 & n^2 \end{vmatrix} \begin{vmatrix} E_x \\ E_y \\ E_z \end{vmatrix} \tag{14-52}$$

*See Landau and Lifshitz, *Electrodynamics of Continuous Media*, Addison-Wesley Publishing Co., Mass., p. 314.
†The reader should verify that Eq. (14-46b) applied to Eq. (14-47a) leads to Eq. (14-48a, b).

This gives the z component of **E** as the solution of

$$(\epsilon_{zz} + \tilde{k}_z\tilde{k}_z - n)^2 E_z = 0$$

but $\tilde{k}_z\tilde{k}_z = n^2$, so that

$$\epsilon_{zz}E_z = 0$$

Neglecting the plasma waves part of the solution, $E_z = 0$. The remainder of Eq. (14-52) can be written as

$$\begin{vmatrix} \epsilon_{xx} - n^2 & i\chi_{xyz}\tilde{k}_z \\ -i\chi_{xyz}\tilde{k}_z & \epsilon_{xx} - n^2 \end{vmatrix}\begin{vmatrix} E_x \\ E_y \end{vmatrix} = 0 \tag{14-53}$$

and the eigenvalues are

$$n_1^2 = \epsilon_{xx} + \chi_{xyz}\hat{k}_z \tag{14-54a}$$

$$n_2^2 = \epsilon_{xx} - \chi_{xyz}\hat{k}_z \tag{14-54b}$$

where $\hat{k}$ is a unit vector, $\hat{k}_z = \omega/c$, and the eigenvectors are

$$\mathbf{E}_1 = \begin{bmatrix} 1 \\ i \end{bmatrix}, \qquad \mathbf{E}_2 = \begin{bmatrix} 1 \\ -i \end{bmatrix} \tag{14-54c, d}$$

These eigenvectors represent circularly polarized light. Since the index of refraction is different for the two waves, the effect of the gradient of the electric field in Eq. (14-43) for a material which is antisymmetric to an interchange of i and j is to induce circular birefringence. From Section 14-6, we know that this rotates the plane of polarization of the wave as it propagates through the material.

In right-handed quartz, χ_{xyz} is positive, and in left-handed quartz, it is negative. Changing the sign of χ_{xyz} interchanges the eigenvalues of Eq. (14-54a) and Eq. (14-54b) that correspond to the eigenvectors of Eq. (14-54c) and Eq. (14-54d).

Approximate expressions for $(n_r - n_l)$ can be obtained by assuming that χ_{xyz} is small compared to ϵ_{xx} and expanding the right-hand side of the expression for n_1 and n_2. We note that $E_1 = E_r$ and $E_2 = E_l$. Then

$$n_r - n_l = \frac{\omega}{nc}\chi_{xyz} = \chi_{xyz}\frac{k_z}{n^2} \tag{14-55}$$

An approximate expression for θ, the angle of rotation, can be obtained from Eq. (12-41). We have

$$\theta = \frac{\pi}{\lambda}\frac{l}{n^2}\chi_{xyz}k_z$$

Since $k_z = 2\pi n/\lambda$, this can be written as

$$\theta = \frac{2\pi^2 l}{n\lambda^2}\chi_{xyz} \tag{14-56}$$

Thus, the angle of rotation varies as λ^{-2}, with one factor of λ^{-1} coming from ∇ in Eq. (14-45) and the other from $\delta = (2\pi/\lambda)\Delta$.

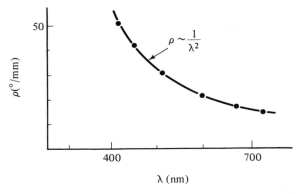

Figure 14-12a The optical rotation produced by a 1-mm thick plate of quartz. The curve plotted is for $\rho = K/\lambda^2$.

14-6f Eigenvectors in Quartz

We will now summarize our findings about the eigenvectors in quartz. First, we will consider the effects produced by the anisotropy in ϵ_{ij}. For propagation down the optic axis, there is no linear birefringence. For propagation perpendicular to the optic axis, there is an ordinary wave polarized perpendicular to the optic axis and an extraordinary wave parallel to it. For propagation at other angles with respect to the axis, there is an ordinary and extraordinary wave. The index of refraction for the extraordinary wave varies from n_o to n_e. Next, we will consider the effects produced by the term $\chi_{ijl}k_l$. For propagation down the optic axis, we now have circular birefringence. At other directions of propagation, both linear birefringence and circular birefringence occur in combination; the eigenvectors represent

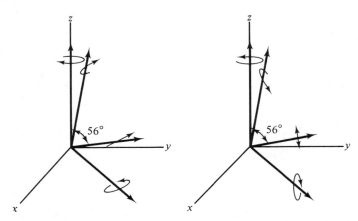

Figure 14-12b The eigenvectors for several directions of propagation in quartz.

elliptically polarized waves, but the degree of ellipticity is small and difficult to detect. Because the coefficient for propagation normal to the optic axis changes sign, it must be zero at some orientation between 0° and 90°. It turns out that circular birefringence is zero at 56°, and a pure plane-polarized wave is the eigenvector at this angle.

Problem 14-3

In quartz $\chi_{xyz}/\chi_{yzx} = -2.25$ at 510.0 nm. Show that the eigenvectors are linearly polarized at $\theta = 56°$. ◼

14-6g Circular Birefringence and Circular Dichroism

We have just seen that the imaginary part of the susceptibility in a material antisymmetric under an interchange of i and j leads to circular birefringence. The real part in the same material is not Hermitian and leads to circular dichroism, which represents the preferential absorption of one sense of circularly polarized light. The absorption frequencies that produce the strongest rotations in the visible region usually lie in the ultraviolet region. Nickel sulfate ($NiSO_4 \cdot 6H_2O$), a green crystal with a fourfold and two twofold mutually perpendicular, symmetry axes (class 422) is unusual in that the rotation changes sign at 500 nm in the visible region. The crystal has a strong absorption band from 580–700 nm, which makes a contribution to the rotation. A second band from 1.0 μm–1.5 μm produces stronger rotations in the near infrared region. At 2.0 μm, another band strongly absorbs linearly polarized light but produces no rotation or circular birefringence (Fig. 14-13).

14-6h Higher-Order Terms in ∇E

When $\mathbf{E}$ is expanded in a power series involving the gradient of $\mathbf{E}$ (Eq. (14-44)), there are higher-order terms in the series beyond those appearing in Eq. (14-45). The second-order term can be written

$$P_i^\omega = \chi_{ijlm} E_j^\omega k_l k_m \qquad (14\text{-}57)$$

where χ_{ijlm} is a fourth-rank tensor. The restrictions imposed by crystal symmetry on the existence of this tensor have different results than for third-rank tensors. In particular, this coefficient is allowed in materials that have a center of inversion. The effects of Eq. (14-57) have been observed in cuprous oxide (Cu_2O), a red crystal with threefold symmetry plus mirror planes (class m3m). The contribution from this equation means that birefrinence occurs for propagation across a diagonal of the face of a cube. The crystal has seven optic axes, three in the principal directions x, y, z and four along the main diagonals of the cube; absorption occurs in some directions and not in others. These observations point out that materials of cubic symmetry are not the same as isotropic materials. For second-rank tensors, such as ϵ_{ij}, the three principal values are equal in both isotropic and cubic substances.

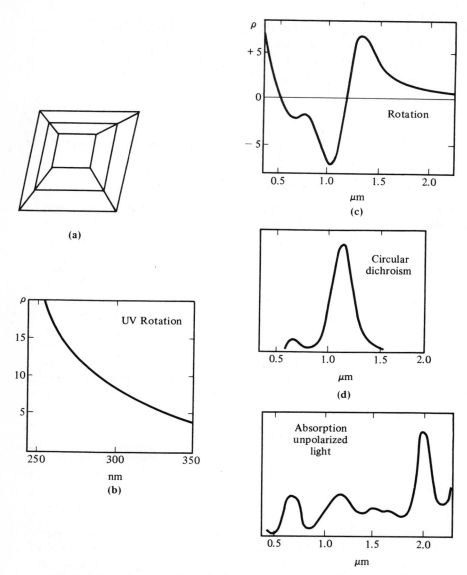

Figure 14-13 A nickel sulfate ($NiSO_4 \cdot 6H_2O$) crystal (a) as grown from water solution. Plates perpendicular to the optic axis can be easily cleaved. Rotation is strongest in the ultraviolet (b). The optical rotation in the visible and infrared (c) is related to circular dichroism (d). Not every absorption band contributes to circular dichroism (e).

For effects described by higher-order tensors, however, cubic substances are anisotropic.

In 1879, Lorentz predicted that cubic crystals should exhibit the above described anisotropy and conducted an unsuccessful search for the effect in rock salt. He found birefringence but could not separate it from stress-induced birefringence.

14-7 The Linear Electro-Optic Effect

In 1883, Kundt and Röntgen working independently and in different countries simultaneously discovered that an applied electric field induced birefringence in quartz. The induced birefringence was linearly proportional to the strength of the electric field and occurred in addition to natural birefringence. In 1893, Pockels published a general theory of this effect, invoking crystal symmetry to determine in which crystal classes it could take place. The linear electro-optic effect is frequently called the *Pockels' effect*, even though it was discovered 10 years earlier by Kundt and Röntgen.

For many years the effect was regarded as a curiosity, but more recently, its practical applications have been widely appreciated, for it can be used to modulate light waves. It is especially useful for modulating laser beams which have a small angular divergence, and a considerable amount of research effort is currently being expended to find better electro-optic materials. Crystals with large electro-optic coefficients and those that can be used throughout the infrared region are of particular interest.

The linear electro-optic effect is described by the fourth term on the right-hand side of Eq. (14-31) with $\omega_l = 0$, giving

$$P_i^\omega = \chi_{ijl} E_j^\omega E_l^0 \qquad (14\text{-}58)$$

In potassium dihydrogen phosphate (KH_2PO_4) abbreviated as KDP, the effect can be used in the longitudinal configuration (applied electric field parallel to **k**, Fig. 14-14). Electrodes are deposited on the two polished faces perpendicular to the optic axis. Vacuum deposition is often used, but conducting pastes are becoming increasingly popular. A small spot is left in the center of the faces for light to enter and leave the crystal. The applied field introduces linear birefringence, with the index for the wave polarized at $+45°$ with respect to a crystallographic axis being greater than n_o, and the index for the wave polarized at $-45°$ less than n_o.

As will be shown later, the phase shift introduced by the field is given by

$$\delta = \frac{2\pi}{\lambda} l \Delta n = \frac{2\pi}{\lambda} \frac{l \chi E^0}{2} \qquad (14\text{-}59)$$

where χ is the electro-optic coefficient. Note that δ is proportional to $l\, E^0$. Since $E^0 = V/l$, where V is the applied voltage in the longitudinal configuration, the phase shift depends only on the total applied voltage and is inde-

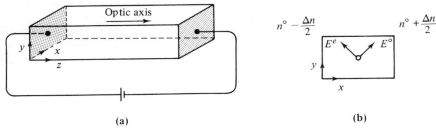

Figure 14-14 The longitudinal configuration for the linear electro-optic effect in KDP. In (a), the field is shown parallel to the optic axis. In (b), the direction of induced birefringence is shown. The wave polarized across one diagonal has a different index than the one polarized across the other diagonal.

pendent of the length of the crystal. The crystal must be long enough so that electrical breakdown does not take place either in the crystal or through the air outside. The disadvantage of this arrangement is that increasing the length of the crystal does not lower the required voltage. (In present day applications, KDP is cut in a different direction and used in the transverse configuration.)

The incoming wave is polarized parallel to x and $\mathbf{E}_{\text{in}} = E_0 \begin{bmatrix} 1 \\ 0 \end{bmatrix}$. With no voltage, the wave propagates unaltered through the crystal. When the voltage is turned on, $\mathbf{E}_{\text{in}}$ must be represented by the sum of the two characteristic waves $\mathbf{E}_1 = \dfrac{E_0}{2} \begin{bmatrix} 1 \\ +1 \end{bmatrix}$ and $\mathbf{E}_2 = \dfrac{E_0}{2} \begin{bmatrix} 1 \\ -1 \end{bmatrix}$. For the *half-wave voltage*, the phase shift between these two components is π; at the exit face they combine in such a manner that the polarization is rotated 90° and $\mathbf{E}_{\text{out}} = E_0 \begin{bmatrix} 0 \\ 1 \end{bmatrix}$. The half-wave voltage in KDP at 632.8 nm is 6600 volts.

The importance of the effect is that it occurs almost instantaneously; that is, the time it takes electrons to respond to the applied voltage. Modulation at 9 MHz has been obtained. A block diagram of a light modulator is shown in Fig. 14-15.

In some applications, a quarter-wave plate is inserted between the crystal and the analyzer, reducing the voltage required to obtain modulation to that for which $\delta = \pi/2$ (or one half of that for $\delta = \pi$).

We will now calculate the transmitted intensity for the arrangement shown in Fig. 14-15. At the entrance face of the crystal

$$\mathbf{E}_{\text{in}} = E_0 \begin{bmatrix} 1 \\ 0 \end{bmatrix} = \frac{E_0}{2} \begin{bmatrix} 1 \\ 1 \end{bmatrix} + \frac{E_0}{2} \begin{bmatrix} 1 \\ -1 \end{bmatrix} \qquad (14\text{-}60)$$

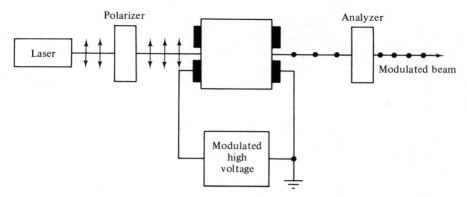

Figure 14-15 An electro-optic crystal is used to modulate a light beam.

and at the exit face

$$\mathbf{E}_{\text{out}} = \frac{E_0}{2} e^{i(2\pi n_0 l/\lambda)} \left\{ \begin{bmatrix} 1 \\ 1 \end{bmatrix} e^{i\delta/2} + \begin{bmatrix} 1 \\ -1 \end{bmatrix} e^{-i\delta/2} \right\} \qquad (14\text{-}61)$$

where δ is given by Eq. (14-59). This can be written as

$$\mathbf{E}_{\text{out}} = \frac{E_0}{2} e^{i(2\pi n_0 l/\lambda)} \left\{ \begin{bmatrix} 1 \\ 0 \end{bmatrix} (e^{i\delta/2} + e^{-i\delta/2}) + \begin{bmatrix} 0 \\ 1 \end{bmatrix} (e^{i\delta/2} - e^{-i\delta/2}) \right\}$$

$$= E_0 e^{i(2\pi n_0 l/\lambda)} \left\{ \begin{bmatrix} 1 \\ 0 \end{bmatrix} \cos \frac{\delta}{2} + i \begin{bmatrix} 0 \\ 1 \end{bmatrix} \sin \frac{\delta}{2} \right\} \qquad (14\text{-}62)$$

The analyzer is oriented to transmit the y component of $\mathbf{E}_{\text{out}}$. Then the intensity transmitted, I_{trans}, is $E_y E_y^*$ and is given by

$$\frac{I_{\text{trans}}}{I_0} = \sin^2 \left(\frac{\delta}{2} \right) \qquad (14\text{-}63)$$

If the induced phase shift is of the form

$$\delta = \delta_0 + \delta_m \cos \omega_m t \qquad (14\text{-}64)$$

the output will have a component at ω_m which will be a maximum when $\delta_0 = \pi/2$. This is accomplished either with a dc bias or by inserting the quarter-wave plate between the analyzer and crystal.

In cubic crystals, such as ZnS, a member of class $\bar{4}3m$,[14-2] the transverse configuration is used. Electrodes are deposited on the surfaces normal to one crystallographic axis, and the wave propagates at right angles to the dc field E_0. One device that uses this configuration is a beam deflector (Fig. 14-16). A prism is made of a suitable material, and the angle of deviation depends on its index of refraction. Changing the dc field changes n and hence changes the position of the output beam. In this manner, the output beam can be made to scan over a region.

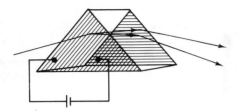

Figure 14-16 A beam deflector using the electro-optic effect.

14-7a   Theory of the Electro-optic Effect—Class $\overline{4}2m$

The starting point in the theory of the electro-optic effect is Eq. (14-58):

$$P_i^\omega = \chi_{ijl} E_j^\omega E_l^0 \tag{14-58}$$

Crystal symmetry determines which of the 27 coefficients χ_{ijl} $(i, j, l = x, y, z)$ are allowed and often gives relations between these allowed coefficients. In class $\overline{4}2m$ (KDP is a member of this class), six coefficients are nonzero and are related as follows:

$$\chi_{xyz} = \chi_{yxz}$$
$$\chi_{yzx} = \chi_{xzy}$$
$$\chi_{zxy} = \chi_{zyx}$$

We will consider an applied electric field in the z direction. For this case, only two coefficients apply and Eq. (14-58) becomes

$$P_x^\omega = \chi_{xyz} E_y^\omega E_z^0 \tag{14-65a}$$

$$P_y^\omega = \chi_{yxz} E_x^\omega E_z^0 \tag{14-65b}$$

Since χ_{ijl} must be Hermitian for an interchange of i and j, the coefficients are real. The effective dielectric tensor is

$$\text{eff}\,(\epsilon_{ij}) = \begin{bmatrix} \epsilon_{xx} & \chi_{xyz} E_z^0 & 0 \\ \chi_{xyz} E_z^0 & \epsilon_{xx} & 0 \\ 0 & 0 & \epsilon_{zz} \end{bmatrix} \tag{14-66}$$

where use has been made of the fact that $\chi_{xyz} = \chi_{yxz}$. We will pick the direction of propagation as the z axis $(\mathbf{k} = \mathbf{k}_z)$ and solve for the eigenvectors and eigenvalues. Equation (14-21), which describes the propagation of electromagnetic waves in anisotropic materials, becomes

$$\begin{bmatrix} \epsilon_{xx} & \chi_{xyz} E_z^0 & 0 \\ \chi_{xyz} E_z^0 & \epsilon_{xx} & 0 \\ 0 & 0 & \epsilon_{zz} + n^2 \end{bmatrix} \begin{bmatrix} E_x \\ E_y \\ E_z \end{bmatrix} = \begin{bmatrix} n^2 & 0 & 0 \\ 0 & n^2 & 0 \\ 0 & 0 & n^2 \end{bmatrix} \begin{bmatrix} E_x \\ E_y \\ E_z \end{bmatrix} \tag{14-67}$$

The equation for E_z is then

$$(\epsilon_{zz} + n^2) E_z = n^2 E_z \tag{14-68}$$

from which $E_z = 0$, i.e., transverse waves. The solution to Eq. (14-67) can be obtained if we equate the determinant of the coefficients of $\bar{E}$ to zero, or

$$\det \begin{vmatrix} \epsilon_{xx} - n^2 & X_{xyz}E_z^0 \\ X_{xyz}E_z^0 & \epsilon_{xx} - n^2 \end{vmatrix} = 0 \tag{14-69}$$

which yields

$$n_1^2 = \epsilon_{xx} + X_{xyz}E_z^0 \tag{14-70a}$$

$$n_2^2 = \epsilon_{xx} - X_{xyz}E_z^0 \tag{14-70b}$$

Substituting each value of n^2 into Eq. (14-67) and solving for the eigenvectors yields

$$E_x = E_y \quad \text{or} \quad \mathbf{E}_1 = E_0 \begin{bmatrix} 1 \\ 1 \end{bmatrix} \tag{14-70c}$$

$$E_x = -E_y \quad \text{or} \quad \mathbf{E}_2 = E_0 \begin{bmatrix} 1 \\ -1 \end{bmatrix} \tag{14-70d}$$

Thus the application of the dc electric field along the z direction of a crystal of class $\bar{4}$2m induces linear birefringence for a wave propagating along the optic axis. The relative phase shift is approximately given by

$$\delta \sim \frac{2\pi l X_{xyz}E_z^0}{\lambda n} \tag{14-71}$$

and the eigenmodes are $\begin{bmatrix} 1 \\ 1 \end{bmatrix}$ and $\begin{bmatrix} 1 \\ -1 \end{bmatrix}$.

14-7b   Theory of the Electro-optic Effect in Class $\bar{4}$3m

Let us now consider the effect in cubic crystals of class $\bar{4}$3m. Electro-optic crystals that belong to this class are CuCl, ZnS, GaAs, and ZnTe. The following coefficients are allowed by symmetry:

$$X_{xyz} = X_{yzx} = X_{zxy} = X_{yxz} = X_{xzy} = X_{zyx} \tag{14-72}$$

In this case we will apply the field in the z direction and propagate on a diagonal of an xy face (transverse configuration). Then

$$\tilde{k}_x = \tilde{k} \cos 45° = n\frac{\sqrt{2}}{2} \tag{14-73a}$$

$$\tilde{k}_y = \tilde{k} \sin 45° = n\frac{\sqrt{2}}{2} \tag{14-73b}$$

and

$$\tilde{k}_x\tilde{k}_x = \tilde{k}_x\tilde{k}_y = \tilde{k}_y\tilde{k}_x = \tilde{k}_y\tilde{k}_y = \frac{n^2}{2} \tag{14-73c}$$

Substituting these expressions into Eq. (14-21) and noting that $\epsilon_{xx} = \epsilon_{yy} = \epsilon_{zz}$

for a cubic crystal, we obtain

$$
\begin{bmatrix}
\epsilon + \dfrac{n^2}{2} & \dfrac{n^2}{2} + \chi_{xyz}E_z^0 & 0 \\[2mm]
\dfrac{n^2}{2} + \chi_{xyz}E_z^0 & \epsilon + \dfrac{n^2}{2} & 0 \\[2mm]
0 & 0 & \epsilon
\end{bmatrix}
\begin{bmatrix} E_x \\ E_y \\ E_z \end{bmatrix}
=
\begin{bmatrix}
n^2 & 0 & 0 \\
0 & n^2 & 0 \\
0 & 0 & n^2
\end{bmatrix}
\begin{bmatrix} E_x \\ E_y \\ E_z \end{bmatrix}
\tag{14-74}
$$

The eigenvalues are

$$ n_1^2 = \epsilon \tag{14-75a} $$

$$ n_2^2 = \epsilon - \chi_{xyz}E_z^0 \tag{14-75b} $$

(The third eigenvalue is $n^2 = 0$ or $\epsilon + \chi_{xyz}E_z^0 = 0$. This is the plasma wave which we shall ignore.)

The eigenvectors are

$$ \mathbf{E}_1 = E_0 \begin{bmatrix} 0 \\ 0 \\ 1 \end{bmatrix} \tag{14-75c} $$

$$ E_y = -E_x \quad \text{or} \quad \mathbf{E}_2 = E_0 \begin{bmatrix} 1 \\ -1 \\ 0 \end{bmatrix} \tag{14-75d} $$

$\mathbf{E}_1$ is along the direction of the applied field and is unchanged by the field. $\mathbf{E}_2$ is perpendicular to $\mathbf{k}$ and the applied field and is modified by the field strength.

Problem 14-4

The symmetry elements of class 422 consist of a 4 fold rotation about z, a two fold rotation about x, and a two fold rotation about y. Find the non-zero elements for a third rank tensor χ_{ijl}. If a steady electric field is applied along the z axis, what effect does it have on light propagating along the z axis? ■

14-7c Dichroism and Birefringence

In a previous section, we found that absorption and dispersion were related. When an electric field induces linear birefringence, it must therefore induce linear dichroism at the same time. The splitting of an atomic energy level by an applied electric field is known as the *Stark effect* and gives rise to linear dichroism in electro-optic materials.

14-8 Quadratic Electro-Optic Effects

Faraday believed that light and electromagnetism were intimately related and for many years tried unsuccessfully to show that an electric field would modify the passage of light in a material. (He later demonstrated that a magnetic field would do this as described in the next section.) In 1875, Kerr discovered the effect in glass. He drilled two holes separated by $\frac{1}{4}''$ in a block of glass. Electrodes were inserted in the holes and the apparatus placed between crossed Nicol polarizers in such a manner that light passed between the electrodes. When Kerr applied the field, light was transmitted through the analyzer, and it could not be extinguished by rotating the analyzer. This indicated that the light had become elliptically polarized and that the field induced linear birefringence. The effect was independent of the polarity of the field and proportional to E^2. It did not occur instantaneously but took 30 seconds to develop or disappear after the field was turned on or off. On a microscopic basis, glass is isotropic, with a center of inversion, so that all elements of the odd-rank tensor χ_{ijl} which describes the linear electro-optic effect are zero. The lowest-order term allowed in glass is the one quadratic in E^0; it is

$$P_i^\omega = \chi_{ijlm} E_j^\omega E_l^0 E_m^0 \tag{14-76}$$

The coefficient χ_{ijlm} for most solids is many orders of magnitude weaker than the linear coefficient, and the effect is not very useful. An exception occurs in ferroelectric materials. At temperatures near the Curie temperature, the coefficient χ_{ijlm} becomes so large that the quadratic effect is as readily observable as the linear effect. Some crystals which exhibit this behavior are perovskites, members of class m3m. Barium titanate ($BaTiO_3$) is one example. By altering the composition of mixed crystals (for example, in $KTa_{(1-x)}Nb_xO_3$ the parameter x can be adjusted), the Curie temperature can be set near room temperature. Under these conditions, the effect has practical applications.

The time lag observed in glass suggested that molecules were lining up along the direction of the field. Kerr also discovered that some liquids exhibit double refraction upon application of an electric field, but without any measurable time lag. In liquids, the molecules are free to rotate and respond in 10^{-9} second or less. Anisotropic molecules tend to line up with their axis of greatest polarizability parallel to $\mathbf{E}^0$, and the liquid in this state resembles a uniaxial crystal. Thermal agitation works to misalign the molecules so that only a small fraction are aligned for fields normally employed.

The polarization $\mathbf{p}$ of a molecule equals simply its induced dipole moment and has the dimensions of charge e times distance r. This induced polarization is proportional to the applied electric field, and for anisotropic molecules the relation is[5-2]

$$p_i = \alpha_{ij} E_j \tag{14-77}$$

where α_{ij} is the polarizability tensor. The energy U of a dipole in an electric field is

$$U = \mathbf{p} \cdot \mathbf{E} \tag{14-78}$$

or

$$U = \alpha_{ij} E_i E_j \tag{14-79}$$

A torque **L** is exerted on the dipole, causing it to line up parallel to **E**, where

$$\mathbf{L} = \mathbf{p} \times \mathbf{E} \qquad (14\text{-}80)$$

The Kerr effect has been used in electro-optic shutters to measure the speed of light. Nitrobenzene is frequently used as the liquid. Extreme care must be taken to maintain cleanliness in assembling a Kerr cell, since dirt or impurities lead to electrical arcing inside the cell, which will damage it.

The phase shift introduced by the Kerr effect is given by

$$\delta = \frac{2\pi l \chi E^2}{\lambda} \qquad (14\text{-}81)$$

where χ is a scalar for the liquid. It is approximately constant for optical frequencies* and is proportional to E^2.

When a Kerr cell is placed between crossed polarizers, the transmission T varies as

$$T = \sin^2 \delta \qquad (14\text{-}82)$$

as is shown in Fig. 14-17.

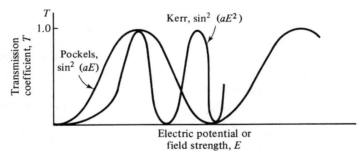

Figure 14-17 The transmission for Kerr and Pockels cells placed between crossed polarizers.

Table 14-3
Half- and Quarter-Wave Voltages for a Kerr Cell†

$\lambda(\mu m)$	$V_{\lambda/4}(kV)$	$\lambda^{-1}V_{\lambda/4}^2$	$V_{\lambda/2}(kV)$	$\lambda^{-1}V_{\lambda/2}^2$
0.5	26.2	1370	37.0	2750
0.7	32.0	1470	45.2	2930
1.06	39.4	1460	55.6	2910

†l = 1 inch, electrode spacing = 0.8 inch, nitrobenzene.

Problem 14-5

What fraction of molecules are aligned parallel to the field in a Kerr cell when the applied field is 15,000 volts/inch? (Hint: consider two energy levels, one

*In many references the induced birefringence is given by the expression $\Delta n = B\lambda l(E^2d^2)$, where B is the Kerr constant. This leads to a phase shift of $\delta = 2\pi Bl(E^2/d^2)$. The problem is that B then is not a constant; it varies as λ^{-1} and must be tabulated for each wavelength.

with the molecule parallel and one perpendicular to E. The order of magnitude of an induced dipole moment is the electron charge e times the molecular diameter d.)

14-9 The Faraday Effect

Michael Faraday was one of the greatest experimentalists in the history of physics. He was self-educated and did not develop sophisticated mathematical skills. Instead, he developed his own unorthodox lines of reasoning. For many years he believed that there was a connection between magnetism and light and unsuccessfully tried to discover the Zeeman effect and the Kerr effect. In 1845, culminating a ten-year search, he observed that a magnetic field applied to glass caused the plane of polarization of a light wave to rotate as it propagated in the direction of the applied field. This experiment was the first to demonstrate the connection between light and electromagnetism.

Maxwell showed in 1857 that the experimental foundations of electricity and magnetism could be expressed by four compact equations, and in 1864, partly basing his reasoning on Faraday's work, he predicted electromagnetic waves. It is interesting to note that while Maxwell was working on solving his equations the very light with which he was viewing them was the solution.

The amount of rotation of the plane of polarization in the Faraday effect was found to be proportional to the applied magnetic field B and the length of the medium l. Therefore,

$$\theta = VBl \tag{14-83}$$

where V is the *Verdet constant*.

One of the most important features of the Faraday effect is that the direction of rotation is the same whether **k** is parallel or antiparallel to **B**. In other words, if a wave passes through a Faraday cell, is reflected, and retraces its path, the rotation is 2θ. (See Fig. 14-18.)

This is not the case for optical activity, which is produced by a screw-like structure. When a nut advances along a screw, it turns one way; when it

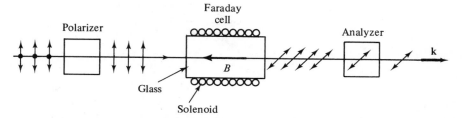

Figure 14-18 Apparatus for measuring the rotation of the plane of polarization by the Faraday effect. The analyzer is mounted on a divided circle, graduated in minutes of arc, which can be rotated.

comes back, it turns the other way, ending up in the same position at which it started. Equation (14-56) shows that the rotation is proportional to **k**, and changing the sign of **k** changes the sense of rotation. When a wave passes through an optically-active material, and retraces its path, the net rotation is zero.

One of the most important uses of the Faraday effect is in the *Faraday isolator*. Sometimes it is desirable to use a laser beam in an interferometer or other device that reflects a large portion of the beam back upon itself. The accurate alignment of the interferometer usually requires that the reflected beam return *exactly* on itself. This returning beam interacts with the laser and greatly modifies its operation, often producing large intensity fluctuations. This is a case where the process of measurement greatly modifies the tool performing the measurement.

The solution to this problem is to pass the output of the laser through the polarizer and then a Faraday cell, which rotates its plane of polarization through 45°. When the wave is reflected from the interferometer and attempts to reenter the laser cavity, its plane of polarization is rotated another 45° by the Faraday cell, and it is then rejected by the polarizer, which decouples the measurement apparatus from the source. An absorber would do this, but the amount of useful light would be seriously reduced.

A microwave isolator was designed by Hogan in 1952 using the Faraday effect. A ferrite rod was incorporated in a twisted wave guide to allow microwaves to travel through the device in only one direction. Ferrite is ferromagnetic—it has a permanent magnetization which gives rise to an internal field, so that an external field need not be applied. The length of ferrite is selected so that the plane of polarization of the microwave signal is rotated 45°, and the output wave guide is twisted 45° with respect to the input wave guide. Only microwaves of one polarization can propagate in the wave guide, and when they pass through the ferrite from one direction, the plane of polarization is rotated so that it is aligned with the output wave guide. When the microwaves come from the other direction, the polarization is rotated so that it is wrong for the other wave guide (Fig. 14-19).

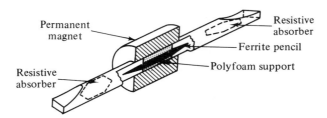

Figure 14-19 The microwave Faraday isolator. The rightward traveling wave is rotated 45° by the ferrite and is matched to the second wave guide. The leftward traveling wave also rotates 45° and does not propagate in the first wave guide.

Kundt discovered that ferrous metal films of thickness 5.5×10^{-6} cm rotated the plane of polarization of visible light through an angle of $1° 48'$. This is 3×10^4 times as large as typical values observed in liquids. The metal films have a very large permanent magnetic field which gives rise to the effect.

14-9a Theory of the Faraday Effect

The Faraday effect is described by the equation

$$P_i^\omega = \chi_{ijl} E_j^\omega B_l^0 \tag{14-84a}$$

The tensor χ_{ijl} is a pseudotensor of rank three.* For **k** and **B** along the z axis, the two significant components of Eq. (14-84a) are

$$P_x^\omega = i\chi_{xyz} E_y^\omega B_z^0 \tag{14-85a}$$

and

$$P_y^\omega = i\chi_{yxz} E_x^\omega B_z^0 \tag{14-85b}$$

The Faraday effect is described by the part of χ_{ijl} that is antisymmetric for the interchange of i and j. To be Hermitian, it must therefore be imaginary, and

$$\chi_{xyz} = -\chi_{yxz} \tag{14-86}$$

The effective dielectric constant in isotropic glass is

$$\text{eff}(\epsilon_{ij}) = \begin{vmatrix} \epsilon & i\chi_{xyz}B_z^0 & 0 \\ -i\chi_{xyz}B_z^0 & \epsilon & 0 \\ 0 & 0 & \epsilon \end{vmatrix} \tag{14-87}$$

The calculation of the eigenvectors and eigenvalues is similar to that for optical activity and yields

$$n_1^2 = \epsilon + \chi_{xyz}B_z^0 \tag{14-88a}$$

$$n_2^2 = \epsilon - \chi_{xyz}B_z^0 \tag{14-88b}$$

$$\mathbf{E}_1 = \frac{E_0}{\sqrt{2}} \begin{bmatrix} 1 \\ i \\ 0 \end{bmatrix} \tag{14-88c}$$

$$\mathbf{E}_2 = \frac{E_0}{\sqrt{2}} \begin{bmatrix} 1 \\ -i \\ 0 \end{bmatrix} \tag{14-88d}$$

---

*A more accurate way to write Eq. (14-84a) is

$$P\omega_i = \chi_{ijlm}\epsilon_{lmn}E^\omega_j B^0_n \tag{14-84b}$$

in which the permutation tensor ϵ_{lmn} is employed to insure proper behavior of the expression under transformations; $\epsilon_{lmn} = +1$ for xyz, yzx, and zxy; $\epsilon_{lmn} = -1$ for xzy, zyx, and yxz, $\epsilon_{lmn} = 0$ for all other permutations, such as xxy. We see that the susceptibility χ_{ijlm} is actually a fourth-rank tensor. We will employ the notation of Eq. (14-84a), remembering that, χ_{ijl} is a pseudotensor.

The difference in refractive index is $n_1 - n_2 \simeq \chi_{xyz} B_z^0$. The phase difference for a path of length l is $\delta = 2\pi l \chi_{xyz} B_z^0 / \lambda$ and the angle of rotation θ is

$$\theta = \frac{\pi l \chi_{xyz} B_z^0}{\lambda} \tag{14-89}$$

The real part of χ_{xyz} produces circular dichroism. This effect is related to the Zeeman effect in which the energy levels are split by the magnetic field. For the Zeeman effect, the light emitted parallel to the field is circularly polarized, and right-handed circularly polarized waves have a different frequency for their peak absorption than do the left-handed waves.

14-10 Cotton-Mouton Effect

In liquids, an applied magnetic field causes anisotropic molecules with a magnetic moment to align, introducing birefringence. This is the *Cotton-Mouton effect*, which is the magnetic analogue of the Kerr effect in liquids. It is a quadratic magneto-optic effect.

14-11 Second-Harmonic Generation

Most of the nonlinear optical effects described in the previous sections were discovered between 1810 and 1910. During this period, it became possible to produce large and controlled electric and magnetic fields in the laboratory. Investigators turned their attention to studying the effects of these fields on the optical properties of materials, and the classical phenomena of optical activity, linear and quadratic electro-optic effects, the Stark effect, the Faraday and Zeeman effects, and the Cotton-Mouton effect were discovered one by one. These small effects could be observed by virtue of their influence on the index of refraction of the primary light beam, since interference and polarization phenomena provide a very sensitive test of changes in this property. By 1910, the effects that were easy to observe at the then-existing level of technology had been discovered and there was quiescence in the field of nonlinear optics until 1960. During the interim, the major developments were the use of the Kerr cell to measure the speed of light and the Carpenter and Billings study of the electro-optic effect in KDP and its application to the modulation of light beams. The system worked only for light almost parallel to the optic axis and required a bright source with narrow beam divergence to reach its full utility.

With the discovery of the laser in 1959, it became possible to produce intense beams of coherent monochromatic light; field strengths of 10^5 V/cm could be produced at the focus of a 10-cm focal-length lens. Investigators then studied the effects of the focused beams on the optical properties of materials, and the decade after 1959 was profuse in the discovery of a new generation of effects. Harmonic generation, continuously tunable sources which operate on the principle of parametric amplification, and stimulated Raman and Brillouin scattering have been discovered. Today, commercial devices are available which employ these effects.

14-11a Nonlinear Electronics

Perhaps the best way to introduce the subject of harmonic generation is by analogy with electronic circuits. Consider a circuit element such as the transistor of Fig. 14-20. An input voltage V is applied to the circuit and

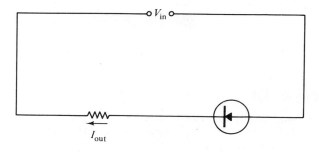

Figure 14-20 A transistor circuit.

a current I appears at the output. The relationship between the input voltage and output current (Fig. 14-21) can be approximately described by the equation

$$I = aV + bV^2 + \cdots \qquad (14\text{-}90)$$

The first term on the right is the linear one, represented by the tangent at the origin in the figure. For small input voltages, the curve follows closely the straight line, but as V increases, the second term bV^2 becomes noticeable and the curve becomes nonlinear.

The nonlinearities produce harmonics of the input frequency. If V varies

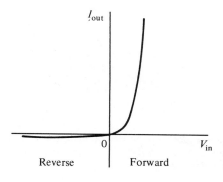

Figure 14-21 The nonlinear response of the transistor circuit.

as $\sin \omega t$, then V^2 varies as $\sin^2 \omega t$. By the trigonometric identity

$$\sin^2 \omega t = \frac{1 - \cos 2\omega t}{2} \qquad (14\text{-}91)$$

we see that I has a component with frequency 2ω, giving a harmonic contribution. In audio amplifiers, harmonics cause distortion. The upper curve of Fig. 14-22 shows the amplified portion of a sine wave. If this represents the

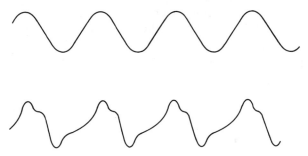

Figure 14-22 The upper curve shows an amplified sine wave from the audio amplifier when the volume control is low. The lower curve is the output when the volume control is turned up. The nonlinear response of the amplifier generates harmonics which cause distortion.

amplifier output for a small input signal, then the lower curve may be regarded as the wave shape for an input of larger amplitude. The listener may easily detect the difference in the sound of these two waves.

14-11b Nonlinear Atoms

It turns out that atoms behave in a similar way. Figure 14-23(a) shows a free atom; the positive nucleus is surrounded by a negatively charged electron cloud. On the right, an electric field **E** of a light wave is applied to the atom, causing the positive and negative charges to separate by a distance d; the

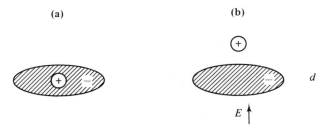

Figure 14-23 When an atom experiences an electric field, the centers of positive and negative charges separate a distance d. The polarization of the atom is $p = ed$ where e is the charge.

resulting dipole has a moment $p = ed$ or αE, where e is the charge on the electron and α is the polarizability.

The relation between the polarization and the applied field E, which we have assumed to be linear, actually has the form indicated in Fig. 14-24. If a laser generates the field E, the higher-order term bE^2 becomes noticeable

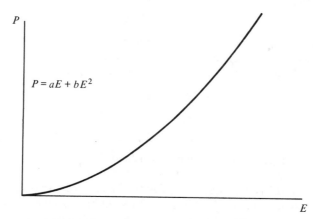

$P = aE + bE^2$

Figure 14-24 The polarization, in general, is a nonlinear function of E. For low values of E, the relation can be approximated by the line $P = aE$. For higher values of E, the nonlinear term is significant.

and produces harmonics. If the initial wave is in the infrared at 1.06 μm, the harmonic will be in the green at 0.53 μm, or twice the fundamental frequency.

This is a very surprising result, for it predicts that when a sufficiently intense beam of light of one color passes through a crystal, the emerging light will have two colors—one very different from the incoming color (Fig. 14-25).

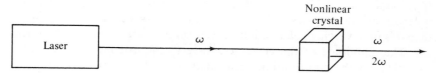

Figure 14-25 Harmonic generation.

The effect is proportional to E^2; focusing the beam to increase E facilitates the observation of the harmonic.

Second-harmonic generation was discovered in 1960 by Franken, Hill, Peters, and Weinreich. Light from a ruby laser at 694.3 nm was focused in a quartz plate, and the emerging light was analyzed with a prism spectrometer. A spectral line appeared at 347.2 nm and was photographed. When the quartz was removed, the ultraviolet signal disappeared. It is not generally realized that Blaton predicted in 1934 that two photons of frequency ω could combine to produce one photon of frequency 2ω in a suitable material; he performed the first quantum mechanical calculations for the frequency dependence of the associated coefficient. Second-harmonic generation is described by the fourth term of the right side of Eq. (14-31); the pertinent equation is then

$$P_i^{2\omega} = \chi_{ijl}E_j^\omega E_l^\omega \qquad (14\text{-}92)$$

A very important role is played by crystal symmetry restrictions in dictating for which materials and orientations the effect can be observed. In class $\bar{4}2m$ (KDP and ADP, for example) the allowed coefficients are

$$\chi_{zxy} = \chi_{zyx} \qquad (14\text{-}93a)$$

$$\chi_{xzy} = \chi_{yzx} \qquad (14\text{-}93b)$$

$$\chi_{yxz} = \chi_{xyz} \qquad (14\text{-}93c)$$

The order in which two components of a field E are written is immaterial; consequently, $E_x^\omega E_y^\omega$ cannot be distinguished from $E_y^\omega E_x^\omega$, (This is not true for components at different frequencies; for the linear electro-optic effect, $E_x^\omega E_y^0$ is distinguishable from $E_y^\omega E_x^0$). Thus, all physical measurements of harmonic generation involve pairs of coefficients. The z component of the induced polarization is written as

$$P_z^{2\omega} = (\chi_{zxy} + \chi_{zyx})E_x^\omega E_y^\omega \qquad (14\text{-}94)$$

In KDP, the two coefficients happen to be equal by symmetry (see Eq. (14-93)). Sometimes one is the negative of the other (as is the case for one pair in quartz), and the sum is zero.

14-11c Velocity Matching

If a plane wave of frequency ω enters a slab of material and induces a polarization which varies with angular frequency 2ω, as given by Eq. (14-92), then this time-varying polarization will in turn generate a wave with frequency 2ω. The amplitude of the generated wave at the exit face is obtained by summing up the contributions generated by each volume element in the material. However, interference between waves generated from different volume elements takes place, and it will be necessary to keep track of the relative phases of the components.

The phase of each wavelet at its point of generation is locked to the phase of the driving wave, as determined by $k_z^\omega z$, where the z axis is specified as the direction of propagation, and we take the zero reference for phase at $z = 0$. The phase at a distant point depends on $\mathbf{k}^{2\omega} \cdot \mathbf{r}$, where $\mathbf{r}$ is the vector from the source to the point of observation, being equal to $k^{2\omega}z$ in the forward direction.

Using Eq. (14-90) as a nonlinear-source term in the wave equation, linearizing the differential equation, and then integrating, we obtain for $E_{oi}^{2\omega}$, the amplitude of the second-harmonic electric field intensity at the output face, the equation

$$E_{oi}^{2\omega} = \frac{k^{2\omega}}{\epsilon_{ii}^{2\omega}} \int_0^l P_i^{2\omega} e^{-ik^{2\omega}z} \, dz \qquad (14\text{-}95)$$

where l is the thickness of the slab in the z direction. Both the fundamental and harmonic beams propagate along z.

Substituting Eq. (14-92) into Eq. (14-95) and noting that $E_l^\omega = E_l e^{i(k_l^\omega z - \omega t)}$ then yields

$$E_{oi} = \int_0^l \frac{k_i^{2\omega}}{\epsilon_{ii}^{2\omega}} \chi_{ijl} E_j E_l e^{i\Delta k z} \, dz \qquad (14\text{-}96)$$

where $\Delta k = k_j^\omega + k_l^\omega - k_i^{2\omega}$.

Assuming that the amplitude of the fundamental E^ω is approximately constant, Eq. (14-96) simplifies to

$$E_{oi} = \frac{k_i^{2\omega}}{\epsilon_{ii}^{2\omega}} \chi_{ijl} E_j^\omega E_l^\omega \int_0^l e^{i\Delta k z} \, dz \qquad (14\text{-}97)$$

Thus

$$E_{oi} \sim \frac{e^{i\Delta k l} - 1}{+i\Delta k}$$

$$\sim \frac{\sin\left(\dfrac{\Delta k l}{2}\right)}{\Delta k} e^{i\Delta k l/2} \qquad (14\text{-}98)$$

and the output intensity varies as

$$I_{oi} = \frac{(k_i^{2\omega})^2}{(\epsilon_{ii}^{2\omega})^2} \chi_{ijl}^2 E_j^2 E_i^2 l^2 \frac{\sin^2(\Delta k l/2)}{(\Delta k l/2)^2} \qquad (14\text{-}99)$$

The output intensity of the harmonic depends on the factor Δk and is a maximum when $\Delta k = 0$, or when

$$n_i^{2\omega} = \frac{n_j^\omega + n_i^\omega}{2} \qquad (14\text{-}100)$$

This describes the condition that the velocity of the generated wave equals the velocity of the driving wave and is called the *index-matched* or *velocity-matched* condition, (Fig. 14-26). When the condition holds, the output intensity grows as l^2, so that increasing the length of the crystal generating the harmonic, results in increasing the strength of the harmonic signal.

On the other hand, when $\Delta k \neq 0$, Eq. (14-99) indicates that the maximum harmonic is generated when

$$\frac{\Delta k l}{2} = \frac{\pi}{2} \qquad (14\text{-}101)$$

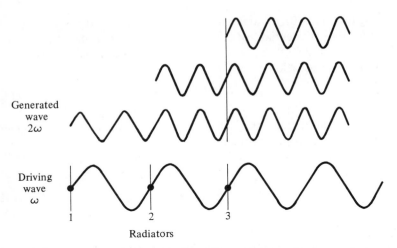

Figure 14-26 The index-matched condition. The generated wave is in phase throughout the entire region.

or odd multiples thereof, determining the maximum effective length that generates the harmonic. Equation (14-101) can be rewritten as

$$l_{max}\left[\frac{2\pi n_i^{2\omega}}{\lambda_2} - \frac{2\pi n_j^{\omega}}{\lambda_1} - \frac{2\pi n_i^{\omega}}{\lambda_1}\right] = \pi \tag{14-102}$$

Using the fact that $2\lambda_2 = \lambda_1$ yields

$$l_{max} = \frac{\lambda_1}{4\Delta n} \tag{14-103}$$

where

$$\Delta n = n_i^{2\omega} - \frac{n_j^{\omega} + n_i^{\omega}}{2} \tag{14-104}$$

and l_{max} is known as the *interaction length*. For many materials it is on the order of 10 μm. Therefore the entire harmonic is generated in this part of the crystal, and increasing the length beyond this value does not raise the output (see Fig. 14-27).

The path length of a beam through a plate can be increased by rotation of the plate. The length is given by $l = t/\cos\theta$. Figure 14-28 shows the variation in harmonic intensity generated by a quartz plate as it was rotated. The experimental points agree quite well with the curve calculated from Eq. (14-99) using known index-of-refraction data.

14-11d Velocity Matching in KDP

Because of dispersion, it is unlikely that $n^{2\omega} = n^{\omega}$ in cubic or isotropic materials. However, birefringence can be used to balance dispersion, and this is the key to the usefulness of the effect. A material in which this can be

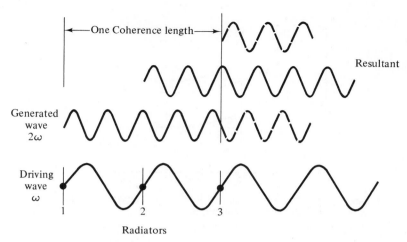

Figure 14-27 The non-index-matched condition. Phase relations set a maximum on the length that can effectively participate in the generation of the 2ω wave. The wave generated at position 3 destructively interferes with the wave generated at position 1.

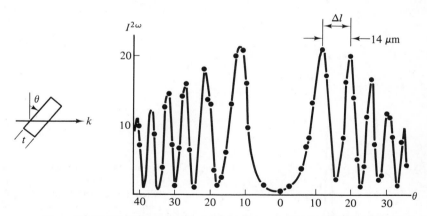

Figure 14-28 The harmonic generated in a quartz plate as its thickness is increased by rotation. The solid curve is calculated and the points are measured values.

accomplished is KDP. From Eq. (14-94), we have

$$P_z^{2\omega} = 2\chi_{zxy}E_x^\omega E_y^\omega$$

which indicates that the driving wave at ω is an ordinary wave, whereas the harmonic will be an extraordinary wave. The condition for velocity matching, Eq. (14-104), gives

$$n_e^{2\omega} = n_0^\omega$$

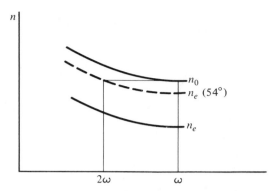

Figure 14-29 Index of refraction of KDP. ω corresponds to the output of a ruby laser. The index of refraction n_e for propagation of $54°$ to the optic axis at 2ω is matched to n_0 at ω.

Figure 14-29 shows the index of refraction vs frequency for KDP; the extraordinary index depends on the angle θ that k makes with the z axis. For a ruby laser with $\lambda = 694.3$ nm as the fundamental, the value of $n_e^{2\omega}$ ($54°$) is equal to n_0^{ω}. Therefore velocity matching takes place in this direction and the generated signal is a maximum. Figure 14-30 shows the measured harmonic

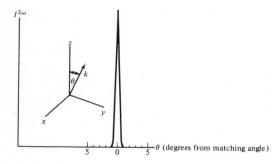

Figure 14-30 Increased production of second harmonic in KDP when the direction of propagation approaches the index-matching direction.

variation as a function of θ. These experimental points agree with Eq. (14-99) and the index-of-refraction data. The utilization of efficient index-matching techniques has enabled up to 20% of the output of a laser to be converted to an angular frequency of 2ω.

Problem 14-6

Can the velocity of the fundamental and second harmonic be matched in quartz? (Hint: consider each generating term $P_i^{2\omega} = \chi_{ijl} E_j^{\omega} E_l^{\omega}$ separately.)

If so, at what angle of propagation with respect to the optic axis? If not could quartz be as efficient a generator of second harmonic as KDP? ▬

14-12 Optical Rectification

Earlier we noted that when the polarization is proportional to E^2, the light has a component at twice the frequency of E. This follows from Eq. (14-91), where the term $\cos 2\omega t$ leads to harmonic generation, and the constant term represents the accompanying direct current or bias. This effect, known as *optical rectification*, was discovered by Bass, Franken, Ward, and Weinreich in 1962, and it can be described by

$$P_i^0 = \chi_{ijl} E_j^\omega E_l^\omega \qquad (14\text{-}105)$$

14-13 Optical Mixing

In addition to the second harmonic, sum and difference frequencies are generated by the nonlinear polarization. The general expression

$$P_i^{\omega_1} = \chi_{ijl} E_j^{\omega_2} E_l^{\omega_3}$$

describes optical mixing when $\omega_1 = \omega_2 \pm \omega_3$.

The first observation of the effect where ω_1, ω_2, and ω_3 were all in the visible or near visible region was made by Franken and co-workers. The wavelength of ruby laser radiation varies by about 1 nm (or 10 Å) as the temperature of the ruby varies from liquid nitrogen to room temperature (see Section 15-6). Using one ruby laser at room temperature and another at 77°K, the sum of the two fundamental frequencies as well as both harmonics were observed (Fig. 14-31).

Earlier, Rupp[14-6] observed mixing for ω_1 and ω_3 in the visible region and ω_2 in the megahertz region. He used an electro-optic Kerr cell to modulate a light beam at the high frequency of 10^9 Hz. He applied the well-known fact that when a carrier at ω_0 is modulated by a signal at ω_1, side bands appear at frequencies $\omega_0 \pm \omega_1$. A wave of constant amplitude can have a single

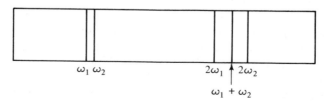

Figure 14-31 Drawing of the spectrometer photograph showing that the sum frequency was generated by ruby lasers operating at two frequencies, ω_1 and ω_2, as well as each harmonic.

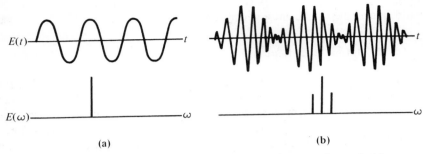

Figure 14-32 A wave of constant amplitude E can be monochromatic (a), but a modulated one has side bands (b).

frequency, but a modulated wave must have side bands, one on each side of the fundamental (Fig. 14-32).

Rupp used a thallium lamp which emitted a very narrow resonance line at 535.0 nm as his source, passing the light through an electro-optic modulator similar to the one shown in Fig. 14-15. An absorption cell with thallium vapor was placed behind the modulator, and it absorbed all of the light when there was no voltage applied to the electro-optic cell; the source appeared dark to an observer. Applying the voltage at 1 gHz produced side bands outside of the narrow absorption line of the second thallium cell, and light at the side band frequencies was transmitted.

14-14 The Tunable Source—Parametric Amplification

One of the most important applications of nonlinear optics is the *tunable source* or *parametric amplifier*. This device uses optical mixing of two waves at different frequencies to produce a third one at the difference frequency $\omega_3 = \omega_1 - \omega_2$. As in harmonic generation, a maximum in the generated signal corresponds to the condition $\Delta \mathbf{k} = 0$. The frequency ω_3 satisfying this condition can be altered by either rotating the crystal so that the direction of propagation with respect to the z axis changes or by temperature adjustment of the index of refraction.

A quantum effect—spontaneous parametric fluorescence—accounts for the initial division of a photon of frequency ω_1 into two photons of frequencies ω_2 and ω_3 which satisfy the relation $\omega_1 = \omega_2 + \omega_3$. The probability of this event taking place is very small, but is a maximum when $\Delta \mathbf{k} = 0$. As the signal at frequency ω_2 builds up, it is enhanced by mirrors with high reflectivity in the region of ω_2. Then the signal at ω_2 combines with the pump signal at ω_1 to generate ω_3.

The device takes a laser source at frequency ω_1 and produces a laser-like beam at frequency ω_3 which can be tuned over a wide range. At present, tunable parametric sources operate in the near infrared and most of the visible region. The availability of a suitable laser in the ultraviolet region combined with a nonlinear crystal having the right dispersion and nonlinear coefficients would yield a source that would be completely tunable throughout the visible. This would be as exciting and potentially as useful as the laser itself.

14-15 Intensity-Dependent Index of Refraction

The polarization involving third-order terms for the electric field describes the intensity-dependent index of refraction when $\omega_2 = -\omega_3$, giving

$$P_i^\omega = \chi_{ijlm}E_j^\omega E_l^{-\omega}E_m^\omega \qquad (14\text{-}106)$$

The susceptibility can be regarded as χE^2 and leads to an expression for the index of refraction of the form

$$n = n_1 + n_2E^2 \qquad (14\text{-}107)$$

This variation of n with intensity is closely related to a two-photon absorption process—the simultaneous absorption of two waves by an atom.

14-16 Higher-Order Effects

Higher-order terms in the expansion of **P** lead to third-harmonic generation, electric- and magnetic-field-induced harmonics, and the stimulated Raman effect.[14-7]

References

14-1 A. Nussbaum, *Electromagnetic Theory for Engineers and Scientists*, Prentice-Hall, Englewood Cliffs, N.J. (1965).

14-2 A. Nussbaum, *Electromagnetic and Quantum Properties of Materials*, Prentice-Hall, Englewood Cliffs, N.J. (1966).

14-3 A. Nussbaum, *Applied Group Theory*, Prentice-Hall, Englewood Cliffs, N.J. (1971).

14-4 G. O. Smith, "Electrets," *Sci. Am. 203*, 202-4 (Nov. 1960).

14-5 J. C. Burfoot, *Ferroelectrics*, Van Nostrand, Princeton N. J. (1967)

14-6 J. Strong, *Concepts of Classical Optics*, W. H. Freeman and Co., San Francisco (1958).

14-7 R. A. Phillips, *J.O.S.A. 57*, 1407 (1967).

General References

E. Wood, *Crystals and Light*, D. Van Nostrand, Princeton, N.J. (1964).

P. A. Franken and J. F. Ward, *Reviews of Modern Physics* **35**, 23 (1963).

W. A. Wooster, *Crystal Physics*, Oxford University Press, England (1949).

J. F. Nye, *Physical Properties of Crystals*, Oxford University Press, England (1957).

L. D. Landau and E. M. Lifshitz, *Electrodynamics of Continuous Media*, Addison-Wesley Publishing Co., Reading, Mass.

15

SOURCES

15-1 Introduction

This chapter will treat the process of emission of light. We shall study several different types of sources, among them blackbodies, emission line sources, and lasers.

Spectroscopy is the measurement and interpretation of the spectral distribution of energy in a source. The main impetus for the early development of quantum mechanics was the need to explain the spectra from atoms. The present state of our knowledge of atomic and molecular physics was achieved largely from studying the characteristics of emitted and absorbed light. Similarly, most of what we know about the universe was obtained from spectroscopy. For example, a star's temperature, composition, and pressure can be deduced from its spectrum. Furthermore, helium was detected in the sun's spectrum before it was observed on earth.

It is not the purpose of this work to present a detailed study of spectra or of quantum mechanics—these are separate subjects in themselves. It is essential, however, that a familiarity with the basic properties of sources, including lasers, be achieved because they are used so often in optics.

15-2   Blackbody Radiation

It is a common observation that a substance glows when heated to a high temperature, thus serving as a *thermal source*. As the temperature increases, the color changes progressively from a dull red to orange to yellow. An analysis of the light from a hot object indicates that all wavelengths are emitted, i.e., the spectral distribution of the light is a continuum, and as the temperature of the object increases, the total flux emitted increases and the peak of the emission curve shifts to a shorter wavelength (Fig. 15-1). Further-

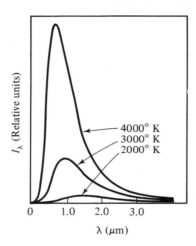

Figure 15-1 The spectral distribution of light emitted by a luminescent body. As the temperature increases, the total flux increases, and the peak of the spectral distribution curve shifts to a shorter wavelength.

more, each kind of thermal source has a characteristic energy vs wavelength curve.

In 1879, Stefan analyzed Tyndall's experimental data and concluded that the power W (the radiant emittance of Table 4-1) radiated by a hot body depends on the fourth power of the temperature according to the relation

$$W = \sigma_R \sigma T^4 \qquad (15\text{-}1)$$

where σ is the *Stefan-Boltzmann* constant and σ_R is the *emissivity*, a parameter characterizing the surface of the material.

A reflecting surface not only reflects light coming from the outside, but also light from the inside which is trying to get out. Consequently, a reflector is a poor emitter and an absorber is a good emitter. This is illustrated in a famous lecture demonstration: a section of pyrex rod is coated with lampblack, an excellent absorber, and then heated in a flame. When it is withdrawn from the flame, the region coated with lampblack glows a brighter red than does the rest of the rod.

A perfect absorber will absorb 100% of the light incident upon it at every wavelength, so that $\sigma_R = 1.0$; hence the name *blackbody*. A polished surface of a metal such as silver reflects most of the light incident upon it, and $\sigma_R \sim 0.1$.

Wien found that the wavelength λ_0 for the peak emission is inversely proportional to the temperature of the source, or

$$K = \lambda_0 T \tag{15-2}$$

which is *Wien's displacement law*. An easy mnemonic aid to remembering the constant K is to think of the sun, whose surface temperature is about 5500°K. The peak of its emission is in the yellow at approximately 550 nm or 5500 Å. This gives us K equal to about 3×10^6 nm.

It is interesting to calculate the peak wavelength emitted by room-temperature objects, such as a bottle of ink. For $T = 300°K$, Eq. (15-2) yields $\lambda_0 = 10^4$ nm or 10 μm, which is in the middle infrared region.

Problem 15-1

Find the wavelength of the peak output for the light emitted by a tungsten filament at 2800°K.   ■

Problem 15-2

At approximately what temperature does glass melt? If a plate of glass is formed by casting molten glass onto a table, at approximately what wavelength will the peak in the emission curve occur just as the glass solidifies?   ■

Problem 15-3

Does an ice cube emit radiation? If it emits energy, why doesn't it cool off?   ■

The blackbody curve for the sun (Fig. 15-2(a)) shows that 40% of the energy emitted is in the visible region, and most of the remainder is in the near infrared (see Fig. 5-3). The peak of the emission is remarkably close to that for the human eye (Fig. 15-2(b)). The similarity in the location of the peaks of these two curves is more than a coincidence. As man evolved, those individuals whose eyes were most sensitive in the spectral region where there was the most light were favored through the process of natural selection. The peak of the response curve of the general animal population thus evolved toward the peak of the blackbody curve of the sun.

The angular resolution of the human eye (about 1′ or 10^{-3} radian) is near the diffraction limit set by the size of the pupil, which is on the order of a few millimeters. The color detectors are the cones, as discussed in Chapter 16;

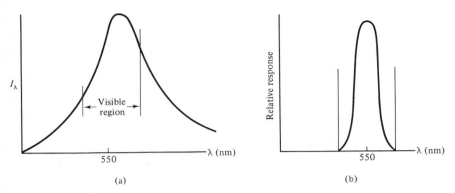

Figure 15-2 The spectral distribution of (a) light emitted from the surface of the sun, and (b) the response of the eye, are shown. The curves have maxima near 550 nm.

they are spaced very closely together at the fovea (the point of high visual acuity) to give such a high resolution.

Problem 15-4

Calculate the diffraction limit of the eye for very dim and for very bright lighting conditions. (Because of aberrations, the resolution of the eye does not increase appreciably as the pupil opens.)   ■

It is interesting to speculate what kind of eyes beings that lived on a planet orbiting a cool star would have. If the star's output were peaked at 10 μm, their "eyes" would have to be 20 times the size of ours to have the same resolution, but if they lived near a radio source, they would probably have antennas and look like reindeer. Maybe their audio sensors would be better developed than their electromagnetic sensors; high pitched sound waves have shorter wavelengths than electromagnetic waves.

A blackbody curve characterizes the output of the common incandescent light bulb. The filament is usually made of tungsten, which melts at a high temperature (3370°K). The diameter of the wire is small enough so that currents on the order of one ampere heat it to around 2800°K, about half the temperature of the sun's surface. The peak output is therefore in the near infrared. The frosted envelope of most bulbs serves to diffuse the light, making it appear softer.

In some special-purpose lamps, the filament is made in the shape of a ribbon and mounted inside unfrosted glass. These lamps are used in applications that require constant intensity illumination over a given area. With the aid of a lens, this ribbon source can be imaged onto a surface or the opening of a slit.

A recurring problem in optics is the need to illuminate a small area with a high intensity. Ordinary sources emit in all directions, but an extended source cannot be focused to a small spot (demagnified) without losing most of the light. The small image requires that it be near the focus of a lens with short focal length. For demagnification, the lens must be far from the source; even for an $f/1$ lens, the gathering efficiency is small because of the diameter of the lens with respect to the distance from the source. Lasers, which provide monochromatic light and little spread, are one answer to the problem and the concentrated arc lamp, which provides white light from a small spot, is another. The arc can be imaged with unit magnification and high gathering efficiency.

In the concentrated arc lamp, the diameter of the area emitting light is as small as 0.005 inches, and the power emitted varies between 2 and 500 watts. The arc is maintained between an anode (a sheet with a small opening) and the tip of the cathode, mounted behind the opening (Fig. 15-3). This lamp provides a close approximation to a point source. Heavy-duty arcs are capable of producing temperatures higher than 4000°K.

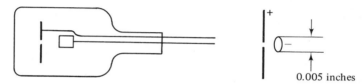

Figure 15-3 The concentrated arc lamp.

15-2a Attempts to Explain the Blackbody Curve

Near the end of the 19th century, one of the most important subjects that occupied the minds of physicists was the explanation of the blackbody radiation curve. At that time, Lord Kelvin is reputed to have stated that physics was almost a closed subject with little left to be discovered. He thought that there were only two small clouds on the horizon—the inability of the existing Rayleigh-Jeans law to predict the blackbody curve, and the negative result of the Michelson-Morely experiment to measure aether drift. (One might say that Lord Kelvin certainly knew how to pick his clouds!) The solution to the blackbody mystery was finally achieved by Planck in 1901, based on the revolutionary assumption that energy levels in matter were *quantized*. This played an important role in the development of quantum mechanics. The negative result of Michelson and Morely provided experimental basis for Einstein's special theory of relativity.

15-2b Density of Cavity Modes

An essential part of the derivation of the blackbody curve is the calculation of the number of modes of electromagnetic radiation per unit frequency allowed in a cavity. This *mode density* (or density of states) will be calculated and used in the derivation of the blackbody curve.

We make the assumption that the electromagnetic waves emitted from a source are in equilibrium with the source, providing a connection between the properties of matter (temperature) and the characteristics of the emitted light (its spectral distribution). One condition for which the assumption of equilibrium holds is when a source is placed inside a rectangular cavity with perfectly reflecting walls. Any light emitted will be trapped inside the cavity, and after a sufficiently long period of time, equilibrium will be attained.

In such a cavity, only certain wavelengths can exist under steady-state conditions. Boundary conditions[5-2] require that $\mathbf{E} = 0$ at the walls of a conductor, setting up standing waves or cavity modes. The condition for a standing wave in one dimension is $p\lambda = 2A$, where p is an integer and A is the length in that direction. This can be rearranged to yield the condition on k_x, the component of $\mathbf{k}$ along OX, as

$$k_x = \frac{p\pi}{A} \tag{15-3a}$$

Similarly, in a rectangular box in which the lengths of the other two sides are B and C, we have

$$k_y = \frac{q\pi}{B} \tag{15-3b}$$

and

$$k_z = \frac{r\pi}{C} \tag{15-3c}$$

The magnitude of $\mathbf{k}$ is then

$$k^2 = k_x^2 + k_y^2 + k_z^2 = \pi^2\left(\frac{p^2}{A^2} + \frac{q^2}{B^2} + \frac{r^2}{C^2}\right) \tag{15-4}$$

where each set of integers (p, q, r) corresponds to a cavity mode.

Now the question we pose is how many modes are there in the interval from k to $k + \Delta k$ or from ν to $\nu + \Delta\nu$? We note that Eq. (15-4) describes an ellipsoid of revolution with semiaxes kA/π, kB/π, kC/π. Each mode, which is characterized by a set of integers (p, q, r), corresponds to a corner on a cube in k-space, Fig. 15-4. The number N of modes from 0 to k is simply equal to number of cubes enclosed by the ellipsoid or to the volume enclosed, (only one octant is needed, since negative integers do not describe additional modes). Since each mode can have either of two polarizations, we must multiply our result by 2, obtaining

$$N = 2 \times \frac{1}{8} \times \frac{4\pi}{3} \times \left(k\frac{A}{\pi}\right)\left(k\frac{B}{\pi}\right)\left(k\frac{C}{\pi}\right)$$

$$= \frac{k^3}{3\pi^2}ABC = \frac{k^3 V}{3\pi^2} \tag{15-5}$$

where $V = ABC$ is the volume of the cavity. Consequently, the number n_k

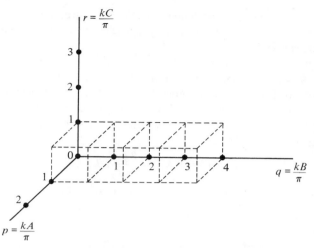

Figure 15-4 Each cavity mode is characterized by a set of integers (p, q, r). These correspond to the corners of unit cubes. The total number from 0 to k is the volume $\dfrac{1}{8} \dfrac{4\pi}{3} \dfrac{kA}{\pi} \dfrac{kB}{\pi} \dfrac{kC}{\pi}$.

of modes per unit volume having a propagation constant less than or equal to k is

$$n_k = \frac{k^3}{3\pi^2} \tag{15-6}$$

We can obtain the number g_ν *of modes per unit frequency interval per unit volume* by differentiating Eq. (15-6) as follows

$$g_\nu = \frac{dn_k}{dk} \frac{dk}{d\nu} \tag{15-7}$$

Using $k = 2\pi\nu/c$, we obtain

$$\frac{dk}{d\nu} = \frac{2\pi}{c} \tag{15-8}$$

Differentiating Eq. (15-6), substituting the result as well as Eq. (15-8) into Eq. (15-7), we end up with

$$g_\nu = \frac{8\pi\nu^2}{c^3} \tag{15-9}$$

15-2c The Rayleigh-Jeans Law

Lord Rayleigh and Sir James Jeans made early attempts to derive the formula for the spectral distribution of light emitted from a blackbody. The cornerstone of their approach was the classical concept that the temperature of a gas describes the average kinetic energy of the atoms. For each degree of freedom, an energy of $\frac{1}{2}kT$ is assigned to each atom, where k is the Boltzmann constant. A gas consisting of point particles has three degrees of freedom, i.e., translation along three mutually perpendicular axes, and each particle then has energy $\frac{3}{2}kT$.

Rayleigh and Jeans reasoned that, since electromagnetic waves carried energy away from a source, each cavity mode must have a definite number of degrees of freedom. They assigned to each mode two degrees of freedom (one for the electric field and one for the magnetic field) and an energy kT. The energy density (energy per unit volume) in the electromagnetic modes was then given by

$$u_v = g_v kT = \frac{8\pi v^2}{c^3} kT \tag{15-10}$$

and the flux, or energy flowing across a given area (Eq. 5-30), is

$$S_v = \frac{8\pi v^2}{c^2} kT \tag{15-11}$$

The law was partially successful in describing the blackbody radiation curve. It predicted that the higher the temperature of the source, the greater the flux emitted. It also predicted that the spectral intensity would increase as the square of the frequency. This turned out to be a good approximation at low frequencies, but it was obviously wrong at high frequencies. Equation (15-10) predicts that $u_v \to \infty$ as $v \to \infty$; that is, any source emits an infinite amount of energy in the ultraviolet. Since we look at sources and do not receive an infinite amount of ultraviolet light, we know this must be wrong. This result is known as the "ultraviolet catastrophy" and is implicit in classical physics. To circumvent it required the introduction of the quantum hypothesis.

15-?d   Wien's Law

In 1896, Wien proposed a radiation law which gave a good fit to the experimental data at high frequencies. Wien reasoned (incorrectly) that the different wavelengths emitted by a blackbody were due to atoms moving with different velocities. In his view, the frequency spread was essentially a Doppler shift, and the probability of an atom moving with a certain velocity was given by the Maxwell-Boltzmann distribution. He therefore incorporated a factor of $e^{-c'/\lambda T}$ in the distribution function, which gave the desired high-frequency cutoff. His law may be written as

$$S_\lambda = \frac{Cv^3}{e^{c'/\lambda T}} \tag{15-12}$$

15-2e   Planck's Law

Max Planck also became interested in the subject, attempting to fit curves of various shapes to experimental data. In 1900 he presented a paper to the

German Physical Society in which he announced that a law of the form

$$S_\nu = \frac{2\pi h\nu^3}{c^2} \frac{1}{(e^{h\nu/kT} - 1)} \tag{15-13}$$

fit the data to within experimental accuracy; six months later he presented a derivation of the law. His approach was to focus attention on the material emitting the waves. He made the revolutionary assumptions that the material consisted of harmonic oscillators which could take up or give off energy only in discrete units of $h\nu$ and, further, that the energy levels of the oscillators have an equal spacing $h\nu$ (Fig. 15-5). The energy of the αth state is $U_\alpha =$

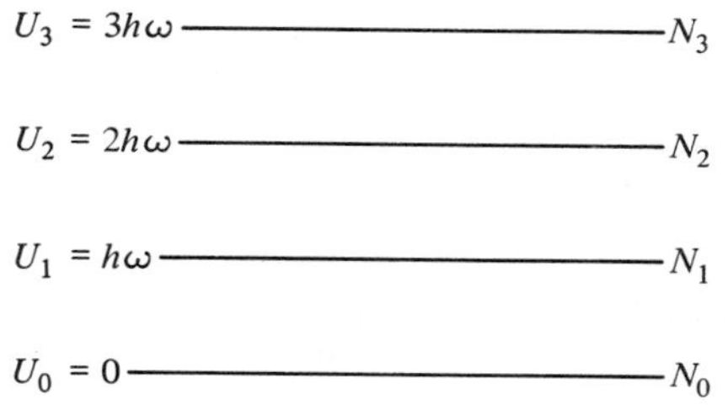

Figure 15-5 Planck assumed that the energy levels of a harmonic oscillator have equal spacing $h\nu = \hbar\omega$. The number of atoms in each level is designated by N.

$U_3 = 3\hbar\omega$ ——— N_3

$U_2 = 2\hbar\omega$ ——— N_2

$U_1 = \hbar\omega$ ——— N_1

$U_0 = 0$ ——— N_0

$\alpha h\nu = \alpha\hbar\omega$, where h is Planck's constant and $\hbar = h/2\pi$. When energy is absorbed or emitted, an oscillator made a transition from one level to another. He assumed that $\mathcal{P}(U_\alpha)$, the probability of an individual oscillator occupying a state with energy U_α is

$$\mathcal{P}(U_\alpha) = Ce^{-\alpha h\nu/kT} \tag{15-14}$$

where C is a constant.

The quantity that we will be interested in is the average energy of these oscillators. Since they are assumed to be in equilibrium with the electromagnetic waves, this will also be the average energy of the waves and will replace the factor kT in the Rayleigh-Jeans law. The average energy $\langle U \rangle$ is equal to the total energy $\sum_\alpha (N_\alpha U_\alpha)$ divided by the number of oscillators. From Eq. (15-14)

$$N_\alpha = N_0 e^{-\alpha\hbar\omega/kT} = N_0 e^{-\alpha h\nu/kT} \tag{15-15}$$

Letting $e^{-\hbar\omega/kT} = x$, we have

$$N_1 = N_0 x, \qquad N_2 = N_0 x^2, \qquad \text{etc.} \tag{15-16}$$

The total energy U_{total} is then given by the series

$$U_{total} = N_0 \times (0) + N_1 \times (\hbar\omega) + N_2 \times (2\hbar\omega) + \cdots \tag{15-17}$$

and N_{total}, the number of oscillators, is

$$N_{total} = N_0 + N_1 + N_2 + \cdots$$
$$= N_0 + N_0 x + N_0 x^2 + \cdots \tag{15-18}$$

Therefore

$$\langle U \rangle = \frac{U_{total}}{N_{total}} = \frac{N_0 \hbar\omega(0 + x + 2x^2 + \cdots)}{N_0(1 + x + x^2 + \cdots)}$$

$$= \hbar\omega \frac{x(1-x)^{-2}}{(1-x)^{-1}} = \hbar\omega \frac{x}{1-x} = \frac{\hbar\omega}{e^{\hbar\omega/kT} - 1} = \frac{h\nu}{e^{h\nu/kT} - 1} \qquad (15\text{-}19)$$

Planck's law for blackbody radiation is obtained if we multiply the number of modes g_ν per unit frequency interval from Eq. (15-9) by the average energy per mode at that frequency as given by Eq. (15-19) to obtain

$$u_\nu = \frac{8\pi h\nu^3}{c^3} \frac{1}{e^{h\nu/kT} - 1} \qquad (15\text{-}20)$$

Planck thus solved the problem which had eluded physicists for so long. His derivation included the quantum assumption which was so revolutionary that he said even he did not believe it. His idea was met with almost complete skepticism in the scientific community, as witnessed by the fact that in 1909, nine years later, Sir James Jeans presented an elaboration of Lord Rayleigh's theory. The quantum hypothesis that we now take for granted was slow to gain acceptance.

15-3   Emission Lines

When a gas is heated to a sufficiently high temperature, it emits a large amount of energy at a few discrete wavelengths in addition to the blackbody continuum; these are characteristic of the element in the gas. The patterns of a spectrum are like a fingerprint and permit the identification of the elements in the source and the measurement of their concentrations.

When light emitted from a hot gas is analyzed by a spectrometer having an entrance slit, an image of the slit appears in the output of the spectrometer at the position of each of the discrete wavelengths. The image of the entrance slit is a line; hence the discrete wavelengths are referred to as *emission lines* and are seen in the spectrum of gases carrying electric currents.

If a cool gas is illuminated from behind by a continuum source, it absorbs light at discrete wavelengths. These appear as absorption lines or dark regions in the spectrum. The sun's spectrum contains many dark lines, called Fraunhofer lines; from them the composition of the sun can be determined. In infrared spectroscopy, most measurements are made using the absorption technique. Emission is very slow and weak in this region. The lifetime of excited states is long, and the atoms collide giving up energy in the process before they can emit.

Spectral lamps which are filled with different elements are available for laboratory use. A collection of lamps permits the experimentalist to select a particular wavelength to use in a given application. Line spectra are produced by passing an electric current through a gas at low pressure, usually on the order of one torr (1 mm of Hg).* The electrons which make up the current collide with the atoms, and their kinetic energy is transferred to the atoms which simultaneously undergo transitions to excited states. As the atoms return to their ground state, they emit light. The discrete frequencies of emission arise from transitions between discrete energy levels. Often, as in the case of hydrogen, the continuum is very weak and most of the energy emitted in the visible region appears in the lines.

15-3a The Hydrogen Spectrum

The energy-level diagram for the hydrogen atom is shown in Fig. 15-6. To a first approximation, the energy of the mth level is equal to R/n^2, where R is the Rydberg constant (13.6 electron volts), which corresponds to a frequency of 3.29×10^{15} Hz. The strongest line corresponds to the transition from the first excited state $n = 2$ to the ground state $n = 1$. This "resonance" transition appears in both the emission and absorption spectra. Some of the lines in the emission spectrum are not seen in the absorption spectrum. For these lines, the emission is accompanied by the atom undergoing a transition that terminates at an excited state. The upper levels, where emission is terminated, are not populated in the absence of the electric current. The energy difference between the $n = 2$ and $n = 1$ states is $3R/4$, which corresponds to -10.2 eV or 121.5 nm in the ultraviolet. All of the bound electron states have negative energies, and the atom is ionized for positive energies.

15-3b Spectra from Other Elements

Elements such as sodium and potassium, which occupy the first column in the periodic table, have complete electron shells plus one optical or valence electron. These elements have hydrogen-like spectra. The resonance transition in each spectrum is the strongest. In these heavier elements the splitting of the lines becomes more pronounced; they are actually doublets because of the interaction of the electron spin with the magnetic field due to the relative motion of the nucleus. This splitting is present in hydrogen as well, but it is very small. The resonance transition in Na is at 589.0 and 589.6 nm; in K it is at 766.5 and 769.9 nm.

Mercury lamps are commonly used, because a large fraction of the electric current passing through the lamp is converted to light. They emit at many wavelengths in the visible. This gives them the appearance of emitting white light. At room temperature, most of the mercury condenses on the walls of the bulb. The lamp contains a buffer gas, usually neon, at a low pressure,

---

*The *SI* unit of pressure is the newton/meter², equivalent to 7.5×10^{-3} torr.

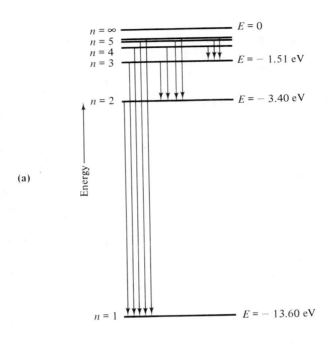

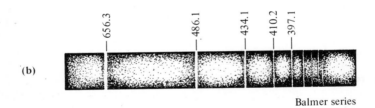

Balmer series

Figure 15-6 The energy-level diagram of a hydrogen atom. The ground-state energy is at −13.6 eV. Transitions that terminate on the ground state form the Lyman series. Those that terminate on the second level form the Balmer series. The Balmer series limit is just beyond the violet, at 3664Å.

since neon is easier to ionize. The electric field between the electrodes in the lamp actually ionizes only a small fraction of the neon atoms, which then initiate the discharge. This heats up the interior of the lamp, increasing the amount of vaporized mercury. A Tesla or spark coil is sometimes used to initiate the discharge. When the high-voltage spark hits the glass envelope of the lamp, it ionizes enough atoms inside the envelope to initiate the process. After a warm-up period of a few minutes, the mercury spectrum predominates. It consists of several bright emission lines in the visible; these lines have been used in spectroscopic work for many years.

Mercury is used in many commercial lamps because of the high intensity of light emitted from it. Fluorescent lamps have a mercury vapor discharge. The inside is coated with a powder that absorbs the strong ultraviolet lines and fluoresces white, but these tubes give off more blue light than do incandescent lamps. This produces a change in the apparent color of many objects when they are viewed under fluorescent light. Some people insist on examining clothing in sunlight before making a purchase. They want to see the "true colors" which may not be produced by the fluorescent lighting inside the salesroom.

Some individuals who wear glasses notice the stronger blues in mercury street lights. The dispersion in the glass lens focuses the blue at a different point than the other colors, so that when viewed out of the corner of the eye, the white image has a blue edge on one side (Fig. 15-7).

Some typical emission lines for H, K, Na, Hg, and Ne are listed in Table 15-1 in which wavelengths are rounded off to the nearest Angstrom and *s*, *m*, and *w* refer to strong, medium, and weak intensities. To convert to nm, multiply by 0.1.

Mercury street-lamp

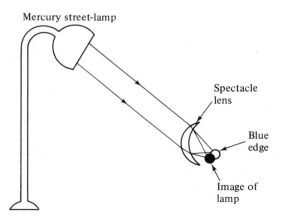

Spectacle lens

Blue edge

Image of lamp

Figure 15-7 When viewed out of the corner of the eye, a mercury street lamp appears to have a blue edge to a person wearing eyeglasses.

Table 15-1
Some Emission Lines of Elements in Å

Hydrogen	Sodium	Potassium	Mercury	Neon
6562 s	5896 s	6939 m	6908 m	6506 s
4861 s	5890 s	6911 m	6234 w	6402 s
4340 m	5688 w	5832 m	5791 s	6382 s
4102 w	5682 w	5812 m	5770 s	6334 m
		5802 m	5461 s	6328 w
		5783 m	4358 s	6266 s
			4047 s	6143 s

15-3c Light-Emitting Diodes

The light-emitting diode (LED) is a recently developed light source. When a semiconductor is doped with an impurity, energy levels appear between the valence and conduction bands. A *P*-type impurity (*P* for positive charge) accepts an electron from the valence band, creating a mobile positively charged hole; an *N*-type impurity donates an electron to the conduction band. (Further information and references will be found in Chapter 16.) A *PN*-junction is made by diffusing different impurities into a host crystal. When a current passes through the junction, the holes and electrons recombine; this process is accompanied by emission of light. In gallium arsenide phosphide, the light is centered at 660 nm and has a half-width of approximately 10 nm. Power outputs are on the order of a milliwatt from an active area with a diameter of 0.1 inch.

15-4 Lasers

The word *laser* is an acronym for *l*ight *a*mplification by *s*timulated *e*mission of *r*adiation. In 1916, Einstein showed that the existence of equilibrium between matter and electromagnetic radiation required that an undiscovered process, *stimulated emission*, take place. His discovery went unexploited until 1954, when Townes and co-workers developed a microwave amplifier (maser) using ammonia (NH_3) and operating on the principle of stimulated emission. Schawlow and Townes showed in 1958 that the maser principle could be extended into the visible region by placing the active medium inside of a Fabry-Perot cavity, and in 1960, Maiman built the first laser using ruby as the active medium. A few months later, Javan developed the helium-neon laser, which operates at 1.15 μm in the infrared.

Since 1960, the field of optics has undergone a revolution, largely due to the laser. Sales of lasers and their related components were estimated to be 400 million dollars in 1970. This is impressive testimony to their importance and widespread application.

15-4a Einstein's A and B Coefficients

Einstein made many contributions to the field of statistical mechanics. One topic he considered was the equilibrium between a blackbody and electromagnetic waves. He approached the problem from an atomic viewpoint, asking himself which processes could take place among the atoms that make up a macroscopic body. His reasoning was based on Planck's hypotheses (Section 15-2e) that the energy levels in a material were quantized and that a light quantum was emitted or absorbed whenever the material underwent a transition from one level to another.

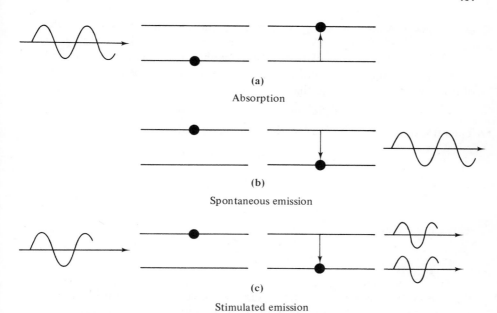

Figure 15-8 The processes of absorption, spontaneous emission, and stimulated emission are illustrated.

Einstein considered a collection of identical atoms having two energy levels each. He knew that one process that could occur was *absorption* of a photon by an atom in the ground state which simultaneously underwent a transition to the excited state 1, as shown in Fig. 15-8(a). He assumed that the probability of this event taking place was equal to $Bu_\omega(\omega)$, where B is simply a coefficient for the process and $u_\omega(\omega)$ is the energy density per frequency interval of the electromagnetic waves at the frequency of the transition. The frequency is determined by the difference $U_1 - U_0$ in energies between the excited state and ground state, using $U_1 - U_0 = \hbar\omega = h\nu$.

The second process that can occur is *spontaneous emission* of a photon from an excited atom as it returns to the ground state (Fig. 15-8(b)). The probability of this event is described by the coefficient A, equal to τ_1^{-1}, the reciprocal of the natural lifetime of the state 1.

The new idea that came out of Einstein's work was to postulate a third process, *stimulated emission* (we will shortly see why this process must be postulated). Stimulated emission (Fig. 15-8(c)) implies that an atom in the excited state is driven to the ground state by an electromagnetic wave of frequency corresponding to the transition frequency. The coefficient for this process is C, and the probability of the event taking place is $Cu_\omega(\omega)$.

In a collection of atoms, the rate of upward transitions is $N_0 Bu_\omega(\omega)$, where N_0 is the number of atoms in the ground state; the rate of downward transitions is $N_1[A + Cu_\omega(\omega)]$, where N_1 is the number of atoms in the excited state. (For simplicity, we assume a line whose width is one cycle; a more sophisticated calculation would involve an integration over $\Delta\nu$.)

For equilibrium conditions, the rate of upward transitions equals the rate of downward transitions, so that

$$N_0 Bu_\omega(\omega) = N_1[A + Cu_\omega(\omega)] \tag{15-21}$$

The ratio of N_1 to N_0 is given by the Boltzmann distribution as

$$\frac{N_1}{N_0} = e^{-(U_1 - U_0)/kT} = e^{-\hbar\omega/kT} \tag{15-22}$$

and substituting Eq. (15-22) into Eq. (15-21) yields

$$Bu_\omega(\omega) = e^{-\hbar\omega/kT}[A + Cu_\omega(\omega)] \tag{15-23}$$

This can be rearranged to obtain

$$u_\omega(\omega) = \frac{A}{Be^{\hbar\omega/kT} - C} \tag{15-24}$$

Einstein was able to evaluate the coefficients A, B, and C by comparing this expression for energy density to the one Planck obtained for blackbody radiation, since equilibrium holds in both cases. For the denominators of Eq. (15-24) and (15-20) to be equal, we first have

$$C = B \tag{15-25}$$

This means that stimulated emission and absorption are both governed by the same coefficient. Substituting B for C in Eq. (15-24) yields

$$u_\omega(\omega) = \frac{A}{B(e^{\hbar\omega/kT} - 1)} \tag{15-26}$$

By further comparison with Planck's law, Eq. (15-20) the ratio A/B can then be evaluated as

$$\frac{A}{B} = \frac{8\pi h\nu^3}{c^3} \tag{15-27}$$

From the above treatment, we now understand why Einstein had to postulate stimulated emission; if he had not done so, C would have been zero in this derivation and Eq. (15-24) would reduce to Wien's law for blackbody radiation. Planck showed that there had to be a factor of -1 in the denominator of the blackbody curve in order to fit the experimental data; this factor actually corresponds to stimulated emission!

One of the most important differences between spontaneous and stimulated emission is associated with phase. For spontaneous emission, each atom in a collection of atoms acts independently and the phase is random, but for stimulated emission the new wave is emitted with the *same phase* and in the *same direction* as the driving wave. Stimulated emission is an ordered or coherent process, whereas spontaneous emission is disordered or incoherent.

An analogy can be made with waves on the surface of a pond. If a handful of sand is thrown into the air, the individual grains will become separated in flight; when they land on the surface, each will generate a wavelet. These waves will be out of phase with each other, and they represent an incoherent source. On the other hand, if the sand is put in a bag and thrown into the air, the grains will act cooperatively when they land, and one large wave will be generated. This represents a coherent source.

Let us now use the above results to investigate the ratio of stimulated to spontaneous transitions *under equilibrium conditions*. As described earlier, the spontaneous rate is A and the stimulated rate is $Bu_\omega(\omega)$. From Eq. (15-26), this ratio is

$$\frac{Bu_\omega(\omega)}{A} = \frac{1}{e^{\hbar\omega/kT} - 1} \tag{15-28}$$

For any finite temperature, $e^{\hbar\omega/kT}$ is always greater than unity; in particular, for a blackbody at 5500°K and $v = 6 \times 10^{14}$ Hz, which is in the green part of the visible, the exponential term has a value of 188.7. This gives a ratio of about 10^{-2} for stimulated to spontaneous transitions, and at lower temperatures, this ratio would be much smaller. In the visible region, spontaneous emission is dominant.

The ammonia maser transition is in the microwave region at 24×10^9 Hz. Then, at room temperature the factor

$$\frac{\hbar\omega}{kT} = 4 \times 10^{-3}$$

and

$$\frac{1}{e^{\hbar\omega/kT} - 1} \sim \frac{1}{\hbar\omega/kT} = 250$$

which means that stimulated transitions predominate for microwave transitions.

Problem 15-5

At which wavelength does the number of stimulated transitions equal the number of spontaneous transitions for an object at room temperature? ■

Problem 15-6

At what temperature would the number of spontaneous and stimulated transitions be equal at 5500 Å? ■

15-4b Saturable Absorption

Earlier, we assumed that the ratio of the number of atoms in the excited state to the number in the ground state under equilibrium conditions was given by

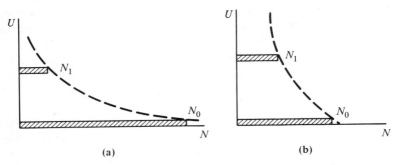

Figure 15-9 The population distribution of atoms follows the Boltzmann distribution. For equilibrium $N_1 < N_0$. In (a), $U_1 > kT$. In (b), $U_1 \ll kT$.

the Boltzmann distribution (Fig. 15-9)

$$\frac{N_1}{N_0} = e^{-h\nu/kT} \tag{15-22}$$

This equation indicates that $N_1 < N_0$ for any temperature T. If $h\nu \ll kT$, the two populations are almost equal but N_1 is still slightly smaller than N_0.

Next we will relate the populations of these two levels to the absorption coefficient and then examine the relative importance of absorption and stimulated emission (assuming that only two levels are involved and that the upper one has a bandwidth of unity). These are competing processes; each is governed by the same coefficient B and each is proportional to the energy density. The rate at which energy is absorbed is proportional to $\hbar\omega N_0 B u_\omega(\omega)$ and the rate at which energy is emitted by the stimulated process is proportional to $\hbar\omega N_1 B u_\omega(\omega)$. The change in energy density is given by the difference, or

$$\frac{du_\omega(\omega)}{dt} = \hbar\omega(N_1 - N_0)Bu_\omega(\omega) \tag{15-29}$$

When a flux from a source is incident on this material, both absorption and stimulated emission will take place. To calculate the absorption coefficient α, we want to consider the variation of $u_\omega(\omega)$ with position rather than with time. Making use of the relation

$$\frac{d}{dt} = \frac{d}{dx}\frac{dx}{dt} = c\frac{d}{dx} \tag{15-30}$$

and dividing both sides by $u_\omega(\omega)$ we write Eq. (15-29) as

$$\frac{1}{u_\omega(\omega)}\frac{du_\omega(\omega)}{dx} = \frac{\hbar\omega}{c}(N_1 - N_0)B \tag{15-31}$$

or in terms of the flux $S_\omega(\omega)$

$$\frac{1}{S_\omega(\omega)}\frac{dS_\omega(\omega)}{dx} = \frac{\hbar\omega}{c}(N_1 - N_0)B \tag{15-32}$$

This can be integrated to yield

$$S_\omega(\omega) = S_0 e^{-\alpha x} \tag{15-33}$$

where α, the absorption coefficient, is

$$\alpha = \frac{\hbar\omega}{c}(N_0 - N_1)B \tag{15-34}$$

Normally, for transitions in the visible, we have $N_0 \gg N_1$ and α is positive; therefore $S_\omega(\omega)$ decreases with distance. But as absorption takes place, the populations N_0 and N_1 change. For weak sources and high concentrations N_0, this effect can be neglected, and α is called the absorption "constant." However, when a small concentration of atoms is irradiated by a large flux, there is a measurable change in the population of the excited state. This lowers the magnitude of α and reduces the rate of absorption, so that for a sufficiently large flux and a small sample, the absorption can become saturated. At this point $N_1 = N_0$; the number of upward transitions equals the number of downward transitions and the sample becomes transparent. It is not possible to illuminate the sample with enough flux at the transition frequency so that $N_1 > N_0$. The most that can be achieved with a large flux is to equalize the populations. (See Fig. 15-10.)

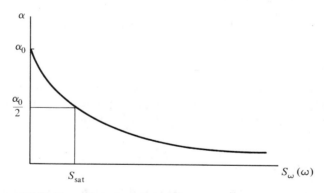

Figure 15-10 The variation of the absorption coefficient α with flux $S_\omega(\omega)$ irradiating a sample. The function is given by Eq. (15-39).

Let us calculate the relationship between energy density and the population of the upper level. Letting N_T equal the total number of atoms per unit volume in the sample, we have

$$N_0 = N_T - N_1 \tag{15-35}$$

With this substitution, Eq. (15-32) becomes

$$\frac{1}{S_\omega(\omega)}\frac{dS_\omega(\omega)}{dx} = -\frac{\hbar\omega}{c}B(N_T - 2N_1) \tag{15-36}$$

The time rate of change of N_1 is

$$\frac{dN_1}{dt} = \frac{S_\omega(\omega)}{c} B(N_T - 2N_1) - N_1 A \tag{15-37}$$

For steady-state conditions $dN_1/dt = 0$, or

$$N_1 = \left(\frac{N_T}{2}\right) \frac{1}{1 + (Ac/2BS_\omega(\omega))} \tag{15-38}$$

Using $A = \tau^{-1}$, where τ is the lifetime of the upper level, and substituting Eqs. (15-38) and (15-35) into Eq. (15-34) yields

$$\alpha = \frac{\hbar\omega BN_T}{c} \left[\frac{1}{1 + (2\tau BS_\omega(\omega)/c)} \right]$$

$$= \alpha_0 \left[\frac{1}{1 + (2\tau BS_\omega(\omega)/c)} \right] \tag{15-39}$$

where

$$\alpha_0 = \frac{\hbar\omega BN_T}{c}$$

The saturation value of $S_\omega(\omega)$ is taken as that value for which α decreases to $\alpha_0/2$ (after a period of time in which equilibrium has been achieved), and this condition is satisfied when

$$S_\omega(\omega) = \frac{c}{2\tau B} \tag{15-40}$$

Problem 15-7

Assume that at $t = 0$ a constant flux S illuminates a thin slab of absorbing material. Plot N_1 vs t for two different values of S near the saturation value. After reaching a steady-state condition, the flux is turned off. Plot N_1 vs t after the flux is off.   ▬

15-4c   Population Inversion

From Eqs. (15-33) and (15-34), we see that if $N_1 > N_0$, α would be negative, and the flux $S(\omega)$ would grow with distance. Hence, we would have more light coming out of the material than went in. This would constitute a light amplifier. N_1 can be greater than N_0 in a sample with three or more levels, with equilibrium populations of the energy levels as shown in Fig. 15-11. The populations shown in Fig. 5-11 are achieved as follows. The atoms are irradiated by an intense flux at a frequency ω_2 equal to $(U_2 - U_0)/\hbar$; this equalizes the population of these two states, giving $N_2 = N_0$. When this happens, there is an *inversion* in the population either between states 2 and 1 or between states 1 and 0; which one occurs depends on the relative lifetimes of states 1 and 2. If the lifetime of state 1 is shorter than that of state 2, the

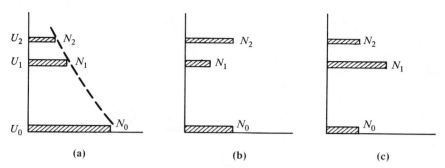

Figure 15-11 In (a), the population of a three-level atomic system is shown under equilibrium conditions. Irradiating the system with an intense flux of photons of energy, $U_2 - U_0$ equalizes the populations N_0 and N_2. A population inversion must exist between 2 and 1, shown in (b), or between 1 and 0, as shown in (c).

atoms in this level quickly decay to the ground state (state 0) and the inversion exists between states 2 and 1; but if it is longer than the lifetime of state 2, atoms accumulate in 1, making $N_1 > N_0$. This method of changing the populations of the energy levels is called *optical pumping*. We will investigate the application of these principles and other techniques for achieving a population inversion in a number of common laser systems in the following sections.

15-5 The Helium-Neon Laser

The majority of existing lasers use a mixture of helium and neon, typically with power outputs on the order of milliwatts. An important feature of the laser output is that its beam divergence $\Delta\theta$ can be diffraction-limited ($\Delta\theta = 1.22\lambda/d$). For a beam with diameter $d = 2$ mm and $\lambda = 0.7$ μm, most of the energy is confined to a cone of angular half-width equal to 4.2 milliradians. This points out one difference between lasers and spectral lamps. Spectral lamps with strong emission lines can also emit about a milliwatt in a line, but the power is radiated over 4π steradians. Thus a laser is many orders of magnitude brighter than any thermal source.

The laser transition occurs between energy levels in the neon, and the helium is used to pump the neon and obtain a population inversion. Neon will exhibit laser action on at least 130 different transitions, although usually only one operates at a time. The three strongest transitions are 632.8 nm, 1.15 μm, and 3.39 μm.

A He-Ne laser operating at 1.15 μm was the second laser system to be developed, appearing in 1960 shortly after the ruby laser. The most common type, however, operates at 632.8 nm in the visible region and is continuous

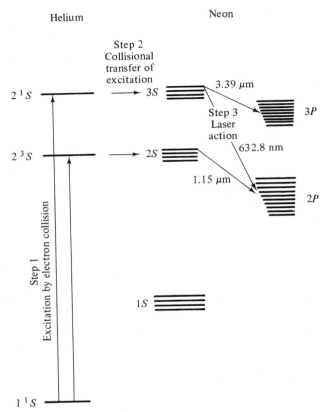

Figure 15-12 The energy-level diagrams for helium and neon. The 2^3S and 2^1S levels on helium are metastable. Laser action takes place between neon levels as shown.

(CW) as well as low in cost. Commercial units sell for around $150. We shall study the He-Ne laser first for pedagogic reasons; it provides a convenient illustration of cavity modes.

The successful development of the He-Ne laser was due to the accidental coincidence between energy levels in helium and neon. Amazingly enough, there is not one but two such coincidences (Fig. 15-12). The 2^3S and 2^1S levels in He are metastable, that is, they have a long lifetime because the transition from these states to the ground state is actually forbidden by the quantum mechanical selection rules.[16-6] When an electric current passes through helium, the atoms are raised to excited states by collision with free electrons, and then they cascade down the energy levels. Those that arrive in the 2^3S and 2^1S levels remain there for a long period. Atoms gradually accumulate in these levels, which then acquire a large population. When an

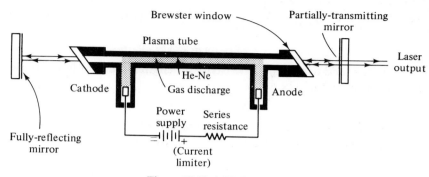

Figure 15-13 A He-Ne laser.

excited helium atom collides with a unexcited neon atom, the excitation is transferred to neon.

A typical laser shown in Fig. 15-13 consists of a pyrex tube which is 35 cm long and 2 mm in diameter with electrodes on the side and fused silica windows set at Brewster's angle on the ends of the tube. Setting the windows like this eliminates unwanted reflections from their surfaces for one polarization of light. Consequently, the output of a laser with Brewster angle windows is polarized. The laser tube is first evacuated and then filled with helium at about 2.5 torr and neon at 0.5 torr. The electrodes in the tube are connected to a high-voltage source of about 4000 volts dc. Mirrors are placed outside the tube and aligned perpendicular to the axis to within a few seconds of arc. The active medium is thus inside of a Fabry-Perot cavity whose mirrors have very high reflectivities. One mirror usually has the maximum reflectivity of 99.9%, and the other is the output mirror with $R \simeq 99.0\%$. They cannot have metallic surfaces, which absorb at least 4% of the incident radiation; this is too high for the low gain of the He-Ne system. Multiple-layer dielectric films must be used, and they are deposited on optically polished mirror substrates. Such films afford precise control of the value of the reflectivity and have almost no absorption. The surfaces of the mirrors must be held to a tolerance of $\lambda/20$ to $\lambda/200$.

15-5a The Gain of a Laser Medium

Equations (15-33) and (15-34) indicate that when a population inversion exists, so that $N_1 > N_0$, then α is negative, and an incident flux is amplified. Then α is the *gain* of the medium.

In the Fabry-Perot cavity formed by mirrors, there are finite losses. Some of the loss is due to absorption and unwanted reflections, but most of the loss is due to the finite transmission of the output mirror. Ignoring absorption, which we can include later by adjusting the numerical value of the loss parameter, the light flux in the cavity is reduced by the factor R_1R_2 for a single round trip, where R_1 and R_2 are the mirror reflectivities. The loss can be described by a parameter γ, defined as

$$\gamma = -\ln(R_1R_2) \tag{15-41}$$

For a single round trip in an active cavity of length l, the final flux S is given in terms of the initial flux S_0 by

$$S = S_0 e^{-\alpha l - \gamma} \tag{15-42}$$

Remembering that α is negative, we see that a steady value of S is maintained if the amplification just compensates for the loss, or

$$\alpha = \frac{\gamma}{l} \tag{15-43}$$

This is the *threshold condition* and it can be used to calculate the amount of inversion necessary to achieve laser action.

15-5b Radiation Modes

The simplest model of a laser cavity is a Fabry-Perot interferometer with a suitable light-emitting medium between the mirrors (Fig. 15-14). A wave

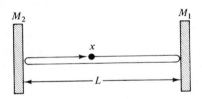

Figure 15-14 Simplified model of a laser cavity. A light wave is emitted at x, undergoes reflections at M_1 and M_2, and returns to x after being amplified by the medium.

starting at position x is partially reflected by mirrors M_1 and M_2 and returns to x after one round trip. The condition for constructive interference of the partial waves at x is given by

$$m\lambda = 2nl \tag{15-44a}$$

where m is an integer. The index of refraction n for gases may be taken as unity for our calculations, and a typical length l of a laser cavity is on the order of 35 cm.

A standing wave or mode is set up in the cavity at each of these wavelengths. Since at least one mirror is made partially transmitting, radiation is emitted at the wavelengths of these cavity modes. The frequency separation of the modes can be found by rearranging Eq. (15-44) to obtain

$$\frac{1}{\lambda} = \frac{m}{2l} \tag{15-44b}$$

or

$$\nu = \frac{c}{\lambda} = \frac{mc}{2l} \tag{15-44c}$$

The mode separation is dv/dm, or

$$\frac{dv}{dm} = \frac{c}{2l} \qquad (15\text{-}45)$$

For $l = 35$ cm, the mode separation is about 430 mHz. These are termed *principal modes* or *axial modes* and correspond to TEM_{00m} modes in electromagnetic theory.[5-2] Typical Doppler-broadened fluorescent line widths are on the order of 1000 MHz for transitions in the visible, giving several values of m which satisfy Eq. (15-44a); these correspond to wavelengths which lie within the line width.

Figure 15-15 shows the Doppler-broadened line width; the Fabry-Perot resonances are shown under the curves, and the laser can oscillate only at

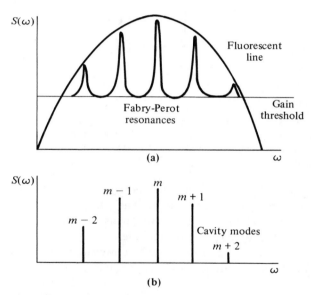

Figure 15-15 The fluorescence line is shown with the Fabry-Perot cavity resonances above the gain threshold. The spectrum of the laser output is shown below.

frequencies which are inside these resonances. Because of gain in the laser, the line width of the oscillation in each resonance is narrowed to a few hertz.

In a cavity with plane mirrors, i.e., a rectangular cavity, the propagation vector $\mathbf{k}$ has a magnitude given by

$$k^2 = k_x^2 + k_y^2 + k_z^2 \qquad (15\text{-}46)$$

The boundary conditions are $k_x A = \pi p$, $k_y B = \pi q$, and $k_z l = \pi m$, so that k^2 is

$$k^2 = \pi^2 \left(\frac{p^2}{A^2} + \frac{q^2}{B^2} + \frac{m^2}{l^2} \right) \qquad (15\text{-}47)$$

as shown in Eq. (15-4). Each index p, q, m denotes the number of nodes in the standing-wave pattern along the directions x, y, z, respectively. A TEM_{00m} mode has no nodes in a plane $z =$ constant. A TEM_{10m} has one node in the x direction and none in the y direction. Some of the mode patterns are illustrated in Fig. 15-16.

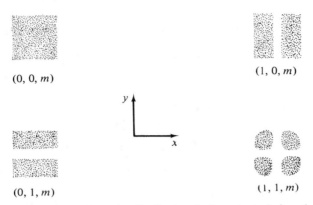

Figure 15-16 The field intensity distributions in the output of a laser beam. The plane shown is normal to z, the axis of the laser tube and the average direction of propagation of the beam. Higher intensities are shaded. Modes for a rectangular cavity are shown.

Early lasers used plane mirrors, but most gas lasers now have spherical mirrors because they are easier to align. The analysis of mode structure in cavities with spherical mirrors uses Huyghens' principle to find the self-reproducing field patterns of a wave making a round trip inside the cavity.

Boyd and Kogelnik and Goubau and Schwering obtained the solutions for the amplitude distributions for the case of a cylindrical cavity and spherical mirrors. An excellent summary of their results along with solutions for other resonator configurations is given by Smith and Sorokin[15-1]. The solutions take the form of narrow beams. The reflection surfaces intersect the beam along equiphase fronts, and the direction of propagation is taken along the z axis. The solution for the electric field is given by

$$E(r, \varphi, z) = E_0\left[\frac{r}{w}\sqrt{2}\right]^q L_q^p\left(2\frac{r^2}{w^2}\right) \exp\left[\frac{(-r^2 \cos q\varphi)}{w^2}\right] \sin kz \qquad (15\text{-}48)$$

where L_p^q are Laguerre polynomials and w is the spot size for which the field has dropped to $1/e$ of its maximum value, with w given by

$$w^4 = \left(\frac{\lambda}{\pi}\right)^2 \frac{R^2 l}{(2R - l)} \qquad (15\text{-}49)$$

where l is the length of the laser cavity and R is the radius of curvature of the mirrors. For constructive interference of partially reflected waves, the phase change after a round trip in the cavity must be an integer times 2π, and it is expressed by

$$m2\pi = 2kl - 2(2p + q + 1) \text{ arc cos } (1 - l/R) \qquad (15\text{-}50)$$

If the mirrors have different radii of curvature, the quantity $(1 - l/R)$ in Eq. (15-50) is replaced by the quantity $(1 - l/R_1)^{1/2}(1 - l/R_2)^{1/2}$.

Here again, the three letters p, q, and m characterize a mode, but we confine our interest to transverse electromagnetic modes, TEM_{pqm}. Axial modes occur when $q = p = 0$, and their frequency separation is just $c/2l$, exactly as obtained in the simple Fabry-Perot model. When q or p does not equal zero, the corresponding modes are off-axial. The letter p designates the number of radial nodes in the output spot. As an example, $p = 1, q = 0$ refers to a node at the center of the pattern and is called the *doughnut mode*. The letter q designates one-half the number of angular nodes. A mode with $p = 0, q = 2$ would appear to have a cross in the mode pattern. The letter m is the same quantity that appeared in Eq. (15-45). Some field distributions for modes in a cylindrical cavity are shown in Fig. 15-17. The frequency spacing of the off-axial modes may be calculated from the details of the laser cavity using Eq. (15-50).

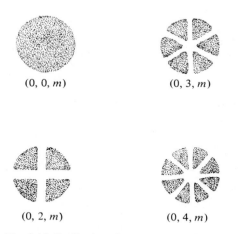

$(0, 0, m)$ $(0, 3, m)$

$(0, 2, m)$ $(0, 4, m)$

Figure 15-17 The field distributions for number of modes in a cylindrical cavity.

Problem 15-8

Assume that the Doppler-broadened line width of the He-Ne laser is 1000 MHz. What length should the cavity be for the laser to have only one axial or TEM_{00m} mode? ■

15-5c *Measurement of Mode Structure by the Beating Technique*

The mode structure of the laser can be deduced by measuring the intermode beat frequencies with a square-law detector and an electronic spectrum analyzer[15-2]. This not only provides valuable insight into the operation of the laser, but also permits measurement of the width of the line supporting laser action.

Optical detectors, as we have seen, measure the square of the electric field. If the light incident on a detector consists of two closely spaced frequencies each with a coherence time (coherence length divided by velocity of light) longer than the response time of the detector, the output will be a function of time varying at the beat frequency. To illustrate this, assume that two plane waves of frequencies ω_1 and ω_2 and amplitudes E_1 and E_2 are incident on the surface of the detector. The direction of propagation of the waves is along the normal to the surface. The total field E_T is given by

$$E_T = E_1 \cos \omega_1 t + E_2 \cos \omega_2 t \qquad (15\text{-}51)$$

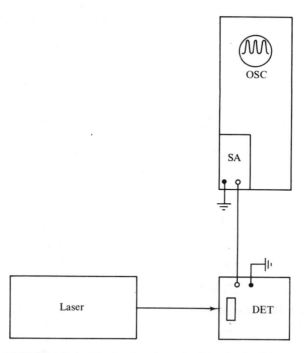

Figure 15-18 Experimental setup for measuring laser mode structure. The multimode output beam from a continuous laser falls on the surface of a silicon photodiode detector (DET). The detector output is fed to a spectrum analyzer (SA), which is an oscilloscope (OSC) plug-in unit.

and the square of the field will have a component $E_T^2 \propto E_1 E_2 \cos{(\omega_1 - \omega_2)t}$ which varies at the difference frequency $(\omega_1 - \omega_2)$. The detector output then varies at this frequency provided its response time is less than the reciprocal of $(\omega_1 - \omega_2)$.

The equipment required for measuring the difference frequency consists of a spectrum analyzer, a detector, and a laser (Fig. 15-18). The spectrum analyzer is commercially available as an oscilloscope plug-in unit, and the detector is a silicon photodiode with a response time of less than 1 ns. This detector can therefore measure beyond 1000 MHz. The photosensitive surface is mounted opposite a hole drilled in a minibox. The wiring diagram for the detector circuit is shown in Fig. 15-19. Since the biasing battery is mounted inside the box, the detector unit is self-contained.

The output beam of the laser to be investigated is directed perpendicularly onto the center of the surface of the detector. The analyzer is scanned over its range from a few MHz to a thousand or more MHz, and a strong signal appears at each of the intermode beat frequencies.

The effect of mirror alignment on modes can be observed in a laser with

Figure 15-19 Wiring diagram for a silicon photodiode.

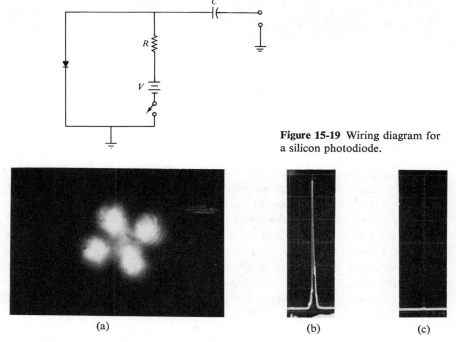

(a)   (b)   (c)

Figure 15-20 A laser with externally adjustable mirrors is set so that only TEM_{02m} modes are present. In (a), an expanded beam is shown. Beating between these modes produces a signal at 250 MHz (b), but not elsewhere; for example, not at 83 MHz (c).

adjustable mirrors. We shall show results of a measurement on a laser with two spherical mirrors of 120 cm radius separated by a distance of 60 cm. The alignment of the mirrors was varied while the projected spot pattern was observed, causing different modes to become active. First, the mirrors were aligned so that only TEM_{02m} modes were present (Fig. 15-20(a)). Then the spectrum analyzer was swept across the entire frequency range, and as predicted from Eqs. (15-48)–(15-50), beat frequencies were found at 250 MHz but not elsewhere [Figs. 15-20(b) and 15-20(c)]. Next, the mirrors were tilted slightly and both the TEM_{00m} and TEM_{02m} modes were observed (Fig. 15-21). Under these conditions a new signal at 83 MHz appeared.

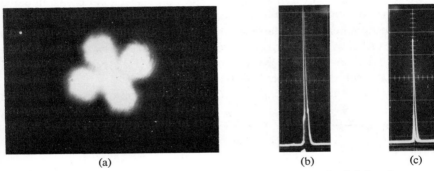

(a)   (b)   (c)

Figure 15-21 A laser with externally adjustable mirrors is slightly misaligned so that TEM_{00m} and TEM_{02m} modes are present. In (a), an expanded beam is shown. Beating between modes produces a signal at 250 MHz (b) and at 83 MHz (c).

15-6 The Ruby Laser

A widely used solid-state laser has ruby for the active medium. Such lasers generally have higher power outputs than do gas lasers. Population inversion is obtained in ruby by optical pumping.

The output of a ruby laser consists of pulses; heroic efforts must be made to obtain CW operation. Normally the pulse duration is 10^{-6} second and the energy per pulse is 10^{-2} joules, the average power in each pulse being on the order of 10 kilowatts. With special Q-switching techniques (described in Section 15-6a), the pulse duration can be reduced two orders of magnitude to 10^{-8} second and the energy per pulse increased to 10^{-1} joule, giving an average power in the pulse of 10 megawatts. Even narrower pulses as short as 10^{-11} second can be produced by synchronizing the phases of the modes in a multimode laser; these ultrashort pulses are useful in investigating fundamental processes in materials. At the high power densities available from the output of a ruby laser, the nonlinearities in the response of materials become significant, and effects such as the generation of optical harmonics, parametric amplification, and stimulated Raman emission become strikingly visible.

Ruby consists of a sapphire (Al_2O_3) host lattice into which a small amount of Cr_2O_3 has been introduced as an impurity. The red color of ruby is due to the Cr^{+3} ion. In the Al_2O_3 lattice, Cr^{+3} substitutes for Al^{+3} and gives the crystal two absorption bands—one in the green and the other in the blue. The depth of the red color depends on the concentration of Cr^{+3} ions; light-pink ruby has a Cr^{+3} concentration of 0.05% or less, but dark ruby has a Cr^{+3} concentration up to 1%. When white light is incident upon a ruby crystal, blue and green are absorbed while the red is reflected. In addition to these absorption bands, which account for the color, there are two narrow fluorescent lines at 694.3 nm and 692.8 nm. These are labeled R_1 and R_2 (R for the German word *rot* meaning "red"). The spectroscopic data for ruby appears in Fig. 15-22.

The laser transition (694.3 nm at room temperature) generally takes place between the R_1 level and the ground state. The ω^{-3} dependence of the B coefficient gives a higher gain for the R_1 than for the R_2 transition. Using special mirrors having high reflectivity at 692.8 nm and low reflectivity at 694.3 nm, the laser will operate on R_2 line. The R lines are about 0.4 nm wide at room temperature. Cooling the ruby shifts both lines to shorter wavelengths and narrows them.

The ruby laser is optically pumped using flash lamps filled with xenon. A high-current pulse is sent through the flash lamp; this heats the gas to several thousand degrees K, at which point much of the blackbody radiation takes place at wavelengths corresponding to the absorption bands. When this light is absorbed by the Cr^{+3} ions, they are excited to the band levels. The ions quickly decay to the R levels by creating a phonon (lattice vibration). The R levels are metastable and R_1 has a lifetime of 4.3×10^{-3} second. As the pumping continues, the ions accumulate in the R levels, and a large population inversion develops; thus laser action follows.

A diagram of a ruby laser is shown in Fig. 15-23. In the configuration illustrated, the Xe flash lamp is linear and is located along one focus of a cylindrical elliptical cavity. The ruby rod, typically 5 cm long by 1 cm in diameter, is placed along the other focus. The cavity focuses almost all of the light from the flash lamp into the ruby rod. This provides an energy density in the rod equal to that in the source. Most lamps cannot operate continuously at the high temperature needed to pump the ruby and are therefore pulsed. One mirror is replaced by a rooftop which is cut into one end of the rod and polished. The half-angle of the rooftop is greater than the critical angle. Light propagating along the axis of the rod and incident on the rooftop is reflected 180° and back down the axis. The other end of the rod is cut at Brewster's angle, so that light of one polarization experiences no reflection at this surface. The output mirror is mounted externally.

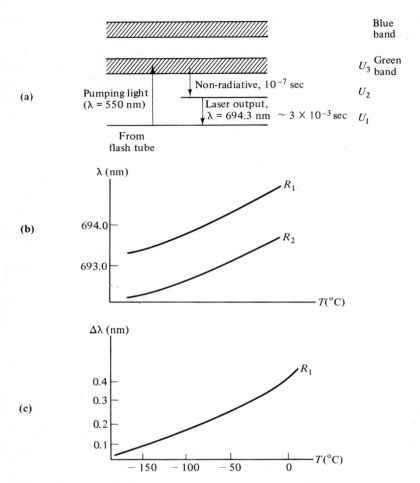

Figure 15-22 The energy-level diagram from ruby (a) shows the pumping scheme. The wavelengths (b) of the R_1 and R_2 lines change with temperature. The fluorescent width of the R_1 line is 0.4 nm at room temperature. This decreases to 0.005 nm at the temperature of liquid nitrogen (c).

We will now calculate the temperature that must be reached in the flash lamp in order to pump the ruby at a fast enough rate to overcome losses. We know that, with any optical system, light from a source cannot be focused to a higher energy density than that of the source. If this could happen, it would mean that energy was flowing from a lower temperature region to a higher temperature region, and this would violate the second law of thermodynamics. Therefore we will assume that the energy density in the rod is the same as that in the lamp.

To obtain an exact answer to the question "How hot should the flash lamp

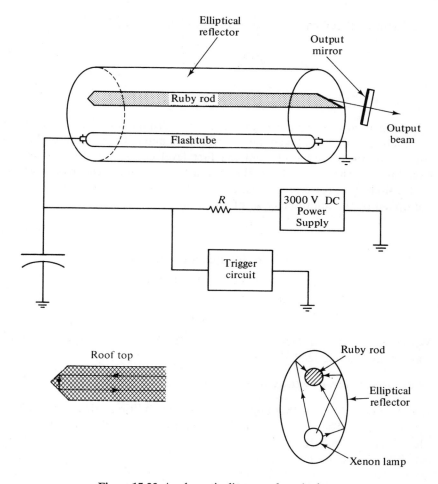

Figure 15-23 A schematic diagram of a ruby laser.

be?" we must take into account all of the processes which produce transitions between any of the levels in ruby. This leads to a set of rate equations, one equation for each level, and this set of coupled equations is difficult to solve. We shall be satisfied with a less-than-exact answer and will consider a simpler requirement for pumping.

In the three-level pumping scheme shown in Fig. 15-20, a necessary condition for laser action is that the rate of absorption for the $0 \rightarrow 2$ transition must be greater than the rate of spontaneous emission from the R_1 level. If this does not hold, the population in the R levels will never build up. This condition is

$$B_{02} u(\omega_2) > A_{10} \tag{15-52}$$

The quantity $Bu(\omega)$ can be obtained from Planck's law, Eq. (15-26),

$$\frac{A_{20}}{e^{\hbar v/kT} - 1} > A_{10} \qquad (15\text{-}53\text{a})$$

or

$$\frac{A_{20}}{A_{10}} > e^{\hbar v/kT} - 1 \qquad (15\text{-}53\text{b})$$

In our treatment we have neglected the multiplicities or degeneracies of the levels, and we have taken the line widths as unity. The second level really consists of twelve sublevels and the first level consists of four sublevels. Consequently the total rate of absorption for the combined upper levels is twelve times that for one level, and each of the A's in Eq. (15-53b) must be multiplied by the multiplicity, g, of the level, or

$$\frac{g_2 A_{20}}{g_1 A_{10}} = e^{hv/kT} - 1 \qquad (15\text{-}53\text{c})$$

Maiman measured the following quantities for ruby

$$hv = 3.6 \times 10^{-19} \text{ joules}$$
$$A_{20} = 3 \times 10^5 \text{ s}^{-1}$$
$$A_{10} = 232 \text{ s}^{-1}$$

With these values, Eq. (15-53c) gives $T = 3880°K$, and the temperature used was actually about 3200°K.

15-6a   Giant Pulses by Q-Switching

Equation (15-43) states that under steady-state conditions, the gain in a laser equals the loss. The population in the upper laser level builds up until threshold is reached and thereafter remains at that value. Larger power outputs can be attained if the output mirror is closed off while the pumping is taking place; then the population inversion will build up past the threshold value that corresponds to the condition for an unobstructed mirror. Now when the mirror is opened to the cavity, the excess inversion gives a very high initial gain, as indicated by Eq. (15-34), and a large pulse is produced as the inversion is reduced to the steady-state condition. This technique is called *Q-switching* because the Q factor (energy stored/energy lost per second) of the cavity is switched from a low value to a high value. The process can be described by solving a set of coupled equations such as Eqs. (15-36) and (15-37).

One way to switch the Q of the cavity is to mount the mirror on a rotating shaft. Each time the mirror is perpendicular to the axis of the rod, a giant pulse appears in the output. Pulse widths on the order of 100 nanoseconds and peak powers on the order of 100 megawatts can be produced in this manner.

A second way to obtain giant pulsing is to place a saturable absorber in the cavity. Initially, the absorption coefficient is α_0. Light from the ruby rod then bleaches the absorber making it transparent. The rod then "sees" the high-reflectivity mirror, and its inversion is well above the threshold. The excess inversion is quickly dumped into a pulse, giving it a high peak power.

Problem 15-9

A Q-switched ruby laser emits a pulse of 0.1 joule in 100 nanoseconds. The output beam is 1 cm in diameter and its beam divergence is diffraction-limited. If it is focused by an $f = 5$ cm lens, calculate the electric field strength at the focus and compare it to the interatomic fields in a crystal. ■

15-7 The CO_2 Laser

Carbon dioxide (CO_2) is a molecular gas, and molecules have energy levels which correspond to the vibration and rotation of the nuclei. Since the mass of the nucleus is much greater than that of an electron, the vibrational and rotational energy levels are much lower than the electronic energy levels. The CO_2 laser transition is in the middle infrared—normally at 10.6 μm—but the wavelength of the output can be adjusted from 9.2 to 10.8 μm by the insertion of a dispersing element such as a prism or grating into the cavity. This enables the laser to be tuned from one vibration-rotation level to another. These levels lie about 1 cm^{-1} apart and are about 50 MHz wide at a pressure of 10 torr. A CO_2 laser 1 m in length operating at several frequencies produces about 80 watts of CW power, but pulsed outputs in excess of megawatts have been achieved with special configurations.

The CO_2 molecule is linear, with a center of symmetry, and has three fundamental modes of vibration (Fig. 15-24). These energies are quantized; the rotational energies are small and also quantized. Each of the vibrational levels has many rotational levels associated with it (Fig. 15-25). Each vibrational mode can have one or more quanta, the state of a molecule being determined by the number of quanta in each mode. The state is designated

Symmetric stretch	Bending	Antisymmetric stretch

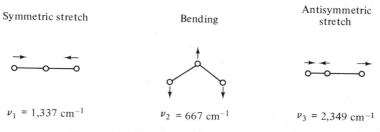

$\nu_1 = 1{,}337$ cm^{-1}	$\nu_2 = 667$ cm^{-1}	$\nu_3 = 2{,}349$ cm^{-1}

Figure 15-24 The three vibrations modes of CO_2.

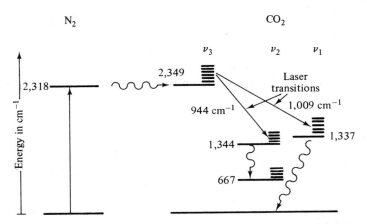

Figure 15-25 The energy levels of N_2 and CO_2. Each vibrational level has rotational sublevels.

by the set of three numbers (v_1, v_2, v_3). The CO_2 molecules are pumped by nitrogen. The excited state of the N_2 molecule is metastable and is at 2318 cm^{-1}, very nearly the energy of the (0, 0, 1) state of CO_2. As a current passes through N_2, the excited state is populated by electron collisions. When excited N_2 molecules collide with CO_2 molecules, a resonant transfer of the excitation to CO_2 takes place. This produces a population inversion between the (0, 0, 1) state and the (1, 0, 0) and (0, 2, 0) states in CO_2. Helium is added to the discharge to depopulate the (0, 1, 0) level, which has a long lifetime and acts as a bottleneck for CO_2 molecules. A typical gas mixture is 10 Torr He, 1 Torr N_2, and 1 Torr CO_2.

Flowing systems are the easiest to build. The CO_2 tends to decompose into CO and O which removes the active molecules from the discharge, and CO further decomposes into C and O; the carbon deposits on the electrodes and walls of the discharge tube. In sealed-off systems, water vapor is added to drive the $CO_2 \rightarrow CO + O$ reaction backward. Platinum electrodes act as a catalyst and also help drive the reaction toward CO_2. Sealed systems have operated for several thousand hours but are more difficult to construct than flowing systems.

A schematic diagram of a flowing CO_2 laser is shown in Fig. 15-26. The mirrors and windows must be made of a material which is transparent at 10.6 μm. NaCl, GaAs, and Ge are often used as window materials. A diffraction grating ruled on an aluminum substrate is used for one mirror. This enables the output to be tuned over a wide range of rotation-vibration transactions (Fig. 15-27).

For the pressures stated earlier, Doppler broadening is greater than pressure broadening. The Doppler-broadened linewidth is on the order of 60 MHz. For a cavity length of 1 meter, Eq. (15-25) indicates that the TEM$_{00m}$

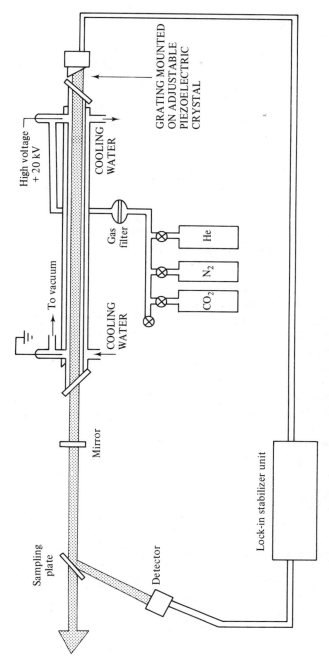

Figure 15-26 A length-controlled CO_2 laser. The grating selects the rotational transition on which the laser operates.

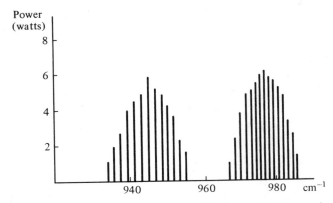

Figure 15-27 The output spectrum of the CO_2 laser. With the grating in the cavity, the output is confined to one line.

modes are separated by 150 MHz. In general, one of these modes will not fall within the line width of a specified transition. To operate on a particular transition, the length of the cavity must be adjusted. To sustain operation on the same transition, compensation must be made for temperature-induced changes in the length of the supports for the cavity. A piezoelectric crystal on which the mirror or grating is mounted will enable fine adjustments to be made of the cavity length. Electronic circuits stabilize the length against temperature changes.

A more efficient pumping of the N_2-CO_2-He system is achieved by means of a transverse discharge (called a T-laser). (See Fig. 15-28.) The discharge is pulsed at a rate of about 20 s^{-1} and produces power outputs of hundreds of kilowatts. A discharge cannot be sustained all along two long parallel bars. At the point where the discharge first starts, the plasma is created which has

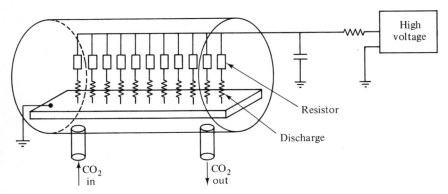

Figure 15-28 The transverse-discharge configuration provides more efficient pumping of CO_2.

lower resistance than the rest of the gap. That path is preferred, and all the current flows along it. To attain multiple discharge paths, a line of 1-watt carbon resistors is connected to the positive high voltage. Each resistor limits the current through the path, and a hundred or more paths are produced. The T-laser operates at atmospheric pressure.

References

15-1 W. V. Smith and P. P. Sorokin, *The Laser*, McGraw Hill, New York (1966).

15-2 R. A. Phillips and R. D. Gehrz, "Laser Mode Structure Experiments for Undergraduate Laboratories" *Am. J. Phys.* **38**, 429 (1970).

General References

A. Bloom, "Optical Pumping," *Scientific American* (October, 1960).

J. E. Faller and E. J. Wampler, "The Lunar Laser Reflector," *Scientific American* (March, 1970).

D. R. Herriot, "Applications of Laser Light," *Scientific American* (May, 1967).

B. A. Lengyel, *Introduction to Laser Physics*, Wiley & Sons, New York (1966).

F. M. Morehead, "Applications of Laser Light," *Scientific American* (September, 1968).

R. H. Pantell and H. E. Puthoff, *Fundamentals of Quantum Electronics*, John Wiley & Sons, New York (1969).

C.K.N. Patel, "High Power Carbon Dioxide Lasers," *Scientific American* (August, 1968).

M. V. Ratner, *Spatial, Temporal, and Spectral Properties of Lasers*, R. A. Phillips, translation ed., Plenum, New York (1972).

A. Schawlow, "Optical Masers," *Scientific American* (June, 1961).

A. Schawlow "Advances in Optical Masers," *Scientific American* (July, 1963).

16

DETECTORS

16-1   Introduction

Interference, diffraction, and polarization, and the wave properties of light were the subjects of earlier chapters. These phenomena can only be observed by the process of detecting light. In this chapter the main types of optical detectors will be considered, and the fundamental physical process taking place in each type will be described. The relative advantages and optimum spectral range of each will be stated and a few applications will be illustrated.

Most detectors exhibit *quantum effects*. That is, their operation indicates that light travels in discrete packets of energy and is localized. Quantum effects are most clearly appreciated when low-level light signals are detected, but in this regime the noise generated by the detector is quite important.

Although we normally think of a detector as some sort of electronic or solid-state device, the most common form is the *human eye*, which is truly a remarkable instrument. For example, its working range is 100 million to 1 in intensity, and no fabricated detector comes even close to this performance. For our purposes here, we are interested only in the structure of the *retina*, which is the detecting surface at the rear of the organ; details on how the eye functions will be found in the book of LeGrand.[16-1]

The retina consists of about 12×10^8 rod-shaped cells and 6×10^6

conical cells, the cell size being about 2.5 μm. The rods, which are located mostly on the periphery of the retina, function at low light levels, and the cones operate at high levels. Hence, daytime vision, known as *scotoptic* vision, is different from low-level or *photoptic* vision, and it is the former type that is considered in Chapter 4 in connection with the standard luminous efficiency curve (Fig. 4-13). The cones are the seat of color vision. There are three sets (which cannot be distinguished under a microscope), and each set has a different spectral response: the blue receptors peak at 450 nm, the green at 540 nm, and the red at 570 nm. It is believed that the eye can perceive a flash when two rods receive only one photon each.

Turning next to man-made detectors, we shall consider a number of types; these are photoemissive, photoconductive, photovoltaic, thermal, and photographic.

16-2 Photoemission Detectors

Electrons can be ejected from a solid body upon which radiant energy impinges. In particular, if the radiation is in the visible region, then this *photoelectric effect* occurs for alkali metals such as cesium, sodium, and potassium. To see what pertinent property of the solid is involved, consider the photoelectric detector—or *photocell*—in Fig. 16-1. The cathode is coated with

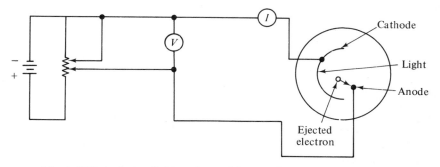

Figure 16-1 A photocell. The voltage giving a null current is proportional to the frequency of the incident light. The energy of a photoelectron therefore is $U = h\nu - \varphi$.

one of the alkali metals. A photon with energy U as given by Planck's relation

$$U = h\nu \qquad (16\text{-}1)$$

can strike the cathode and cause an electron to be ejected. If the electrostatic potential difference between cathode and anode is positive, electrons are attracted to the anode and a current is detected in the ammeter. Reducing the value of the applied potential to zero and then reversing the polarity, we reach a value $-V_0$ at which the current just vanishes. At this point, conservation of energy tells us that the most energetic electrons—the ones which just have enough energy to escape from the solid cathode and reach the anode—must obey the relation

$$hv = \varphi + |-eV_0| \qquad (16\text{-}2)$$

where φ is the energy barrier which must be surmounted to leave the solid. This quantity is the *photoelectric work function*, and Eq. (16-2) is known as the *Einstein photoelectric equation*. It is evident that the work function must be properly chosen for the frequency range of the radiation being detected. Some typical spectral response curves are shown in Fig. 16-2, where the ver-

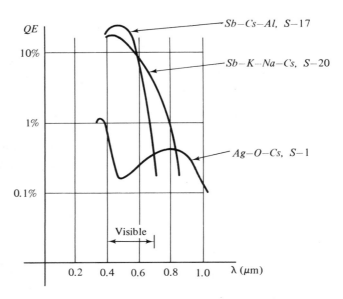

Figure 16-2 The spectral response of several photoemissive surfaces.

tical scale is expressed in terms of *quantum efficiency QE*, defined as the number of electrons released per incident quantum. The designations S-1, S-17, etc. are arbitrary designations used in the industry; an S-1 surface, for example, responds beyond the visible region. Further information on photocell theory will be found in Bauer[16-2] and in van der Ziel[16-3]; specific details on optical and electrical characteristics can generally be obtained from manufacturers' literature.

The output of a photocell can be significantly enhanced by taking advantage of *secondary emission*, which is the ejection of several electrons from a solid bombarded with a single electron at high velocity. The initial photoelectron can be accelerated by a potential so that it achieves a high velocity. Then, when it strikes the next stage, it will cause several secondary electrons to be emitted. *Photomultipliers* are devices which multiply a single photo-electron by five or six orders of magnitude. In addition to a photoemissive cathode, they have several *dynodes* which are maintained at successively higher potentials. Generally, the potential difference is increased in about 100-volt steps from one dynode to the next. The potential difference accelerates the electrons so that when they strike the surface of the dynode, each one causes three or four secondary electrons to be ejected. These secondary electrons are then accelerated toward the next dynode and the process repeated. In this manner, a single electron at the cathode can be multiplied to one million electrons at the anode after ten stages. The last dynode, maintained at a high positive voltage with respect to the cathode, is grounded so that the output will be at low voltage (Fig. 16-3).

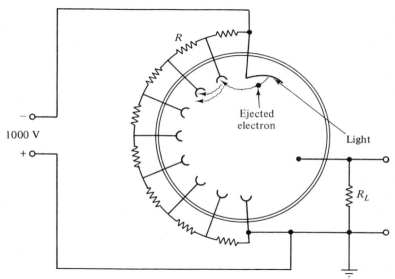

Figure 16-3 Circuit diagram for a photomultiplier tube; R is typically 100,000 Ω.

The load resistor R_L together with the capacitance C in the output stage and cables determines the response time τ of the circuit

$$\tau = R_L C \tag{16-3}$$

The photomultiplier is a current multiplier.

For *cw* or steady state operation when the signal is longer than τ, the voltage V developed across the load resistor is given by

$$V = IR_L \tag{16-4}$$

On the other hand, when the duration of a light pulse is less than τ, the output voltage is

$$V = \frac{Q}{C} \tag{16-5}$$

where Q is the amplified charge of the pulse. This is equal to the charge on a single photoelectron times the number of photoelectrons times the gain of the tube. The following two examples illustrate how the output voltage is calculated taking into account the relation between the pulse duration and response time of the circuit.

To appreciate the sensitivity of a photomultiplier tube, consider the typical calculation which follows.

Example 16-1

A photocathode intercepts two thousand photons in one microsecond from a blue source. The load resistor R_L is 1 kΩ, the gain of the tube is 10^6, the quantum efficiency is 10% in this region, and the output capacitance C is 100 pF. Find the maximum voltage developed across R_L.

First we note that the pulse duration is longer than the response time τ, which we obtain from the usual relation for an RC circuit as

$$\tau = R_L C \tag{16-3}$$

so that

$$\tau = 10^3 \times 10^{-10} = 10^{-7} \text{ s}$$

The output voltage V for this situation should then be calculated from Eq. (16-4). ΔQ is the charge corresponding to the amplified output. This charge, in turn, is the product of the number of photons, N, the quantum efficiency QE, the gain G, and the charge on the electron e, or

$$\Delta Q = (2 \times 10^3)(0.10)(10^6)(1.6 \times 10^{-19})$$
$$= 3.2 \times 10^{-11} \text{ coulombs}$$

Since $I = \Delta Q / \Delta \tau$,

$$I = \frac{\Delta Q}{\Delta \tau} = \frac{3.2 \times 10^{-11}}{10^{-6}} = 3.2 \times 10^{-5} \text{ amperes}$$
$$V = IR = 3.2 \times 10^{-5} \times 10^3$$
$$V = 3.2 \times 10^{-4} \text{ volts}$$
$$= 320 \ \mu V$$

Thus, 2000 photons, producing approximately 32×10^{-12} coulombs of charge, yield a current which is readily measurable. ▄

Example 16-2

One hundred photons of blue light are incident on the above surface in 10^{-8} sec. The load resistor has been changed to 10 kΩ. Find the peak voltage across R_L.

First we note that the pulse duration 10^{-8} s^{-1} now is much less than the time constant.

$$\tau = RC = 10^4 \times 10^{-10} = 10^{-6} \text{ sec}$$

The voltage should then be calculated from Eq. (16-5).

$$Q = N \times Q.E. \times g \times C = 1.6 \times 10^{-12} \text{ coulombs.}$$

$$V = \frac{Q}{C} = \frac{1.6 \times 10^{-12}}{10^{-10}} = 16 \text{ millivolts}$$

Further details on photomultipliers will be found in References 16-2 and 16-3; a discussion which outlines the problems associated with their use at extremely low photon levels is given by Keyes and Kingston.[16-4]

Problem 16-1

If the maximum steady-state anode current for the above tube is 1 milliampere, what is the maximum flux of blue light in watts that can illuminate the photocathode? (There is also a maximum safe cathode current, so one must take care in focusing a beam onto a photocathode not to exceed the tube ratings.)

Under the heading of photoemissive radiation detectors, we should also consider briefly some very advanced devices used as *image tubes*. For example, the *image orthicon* used in television cameras is shown in Fig. 16-4. Light is focused onto the photocathode. The light causes the emission of electrons from the left side of the photocathode. They are accelerated by an electric charge and strike the glass membrane, where they are held briefly by a positive charge augmented by the fine-mesh screen that is 50 microns from the membrane. A beam of electrons from the gun at the left scans the membrane and carries off the image. The return beam is amplified at the left to yield the output of the tube.

Another device in this category is the *image dissector* tube, used to convert optical images to digital form. The tube, shown in Fig. 16-5, has an end-on photocathode. An image is projected onto the photocathode. A pinhole is located a distance behind the photocathode, and magnetic deflection coils surround the region between the cathode and pinhole. The coils are operated so that electrons emitted from a selected spot on the cathode pass through the pinhole and into the dynode chain where they are amplified. The current in the coils is varied, so the effective area scans across the cathode in a programmed manner. The photomultiplier output corresponds to a point-by-point reading of the image intensity. A typical tube has a 1-inch-square cathode which is broken down into 1000×1000 resolution elements.

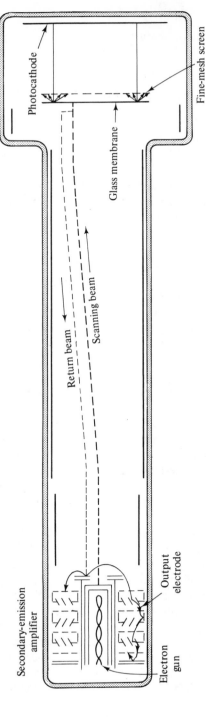

Photocathode

Fine-mesh screen

Glass membrane

Return beam

Scanning beam

Secondary-emission
amplifier

Output
electrode

Electron
gun

Figure 16-4

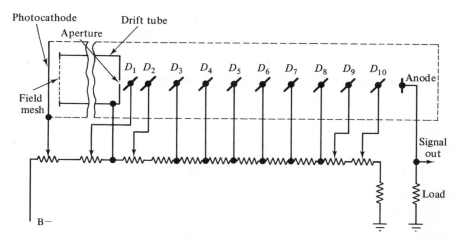

Figure 16-5

Problem 16-2

An argon laser with an output of 100 milliwatts in the blue region at 4880 Å is used to illuminate a one-inch-square positive transparency. The image of the transparency is projected onto the surface of an image dissector. There are to be 1000 × 1000 resolution elements. What is the maximum permissible flux in photons/second into one element? The output of the dissector tube is to be digitized by assigning it to the nearest of eight levels. The levels are to be in geometric progression (i.e., proportional to 0: 1: 2: 4: 8: 16: 32: 64) since the eye responds to the logarithm of the intensity. The signal corresponding to each element is to be integrated for a sufficient period of time so that it will be unlikely that random fluctuations in the signal will cause the wrong level to be read. What number of photons must correspond to each of the levels? If the electronics can be made to operate as fast as we wish, how long will it take for the system to scan one transparency, reading the level of illumination at each element? Is the argon laser too powerful for the dissector tube? ■

16-3 Noise

When we are dealing with detectors which are sensitive to a very low photon flux, such as the photomultipliers just considered, then any signals in the output which are generated in the cell or its associated electronic circuits will result in detection errors. Such signals are known as *noise*. Noise theory is a rather involved subject, and our discussion here is designed merely to give an appreciation of the way in which it enters into considerations in the choice of a radiation detector. An elementary introduction to noise theory applied to semiconductor and electronic devices is given by van der Ziel.[16-3]

In general, three types of noise sources are recognized. Every current-carrying circuit has associated with it what is known as *thermal* or *Johnson* noise, due to the finite energy of the electrons at any temperature. We may visualize intuitively how this noise originates by considering the total capacity in a given circuit, either deliberately introduced or due to circuit elements in close proximity. When this capacity is charged to a potential V, it possesses a stored energy equal to $\frac{1}{2}CV^2$. However, this energy is not constant—even with the applied potential fixed—because the associated electronic charge given by $Q = CV$ is fluctuating due to thermal agitation. This fluctuating charge, passing through any resistances in the circuit, generates a noise voltage. It is obvious that every electronic device must have inherent thermal noise; hence this effect represents the lower limit on the sensitivity of a detector.

The next type of noise to consider is *shot noise* which is due simply to the discrete nature of electronic processes. For example, the emission of electrons from the cathode of a vacuum tube does not take place at a uniform rate, and there is a noise signal due to this random effect.

Finally, each device has noise due to the particular mechanism responsible for its operation, and all of these phenomena are usually lumped together and appear as $1/f$ *noise*, because this noise generally falls off with frequency.

In addition to the noise signals produced by the phototube, there is noise associated with the incident radiation. *Photon noise* arises from the statistical distribution of photons in a light beam. If there are N photons or photoelectrons in a given time interval, then statistical theory[16-5] shows that the fluctuation in that number is $\sqrt{N}$. Approximately two-thirds of the measurements will be in the interval $N \pm \sqrt{N}$. This arises from the fact that the emission process is a random event. For example, if we try to measure an average of 4 photoelectrons in a series of equal time intervals, then the fluctuation would be 2. This is 50% of the signal and is therefore large. Nothing can be done about this source of noise for it is a fundamental property of light. Only by increasing N, such as by integrating over a longer time interval, can we reduce this percentage.

The *dark current*—the current with no applied light energy—is noise which arises from the random emission of electrons from the cathode. The energy of electrons in a metal is distributed over a range which depends on temperature. There are always a few that have energies greater than the work function φ. If these electrons are physically near the surface, they can escape and will be amplified just as the photoelectrons are amplified. Inexpensive photomultipliers have a dark current at the anode of about 1 milliampere. Refinements in the construction of photomultipliers have produced tubes with the dark current at the cathode equal to a few photoelectrons per second. This can be reduced even further by cooling the cathode and by reducing its effective area. The dark current is approximately proportional to the cathode area. A tube with a few photoelectrons per second and a gain of 10^6 has a dark current at the anode of approximately 10^{-12} amperes. An example of such a low-noise tube is the ITT Corp. FW-130. In this tube the effective area of the photocathode is $\frac{1}{4}''$ in diameter. The few noise pulses generated in this tube are believed to be caused by cosmic rays. This tube can count light signals as weak as one photon per second.

16-4 Photoconductive and Photovoltaic Detectors

Another class of detectors depends on the fact that there are many solids whose electrical resistance decreases substantially under illumination. Such materials are known as *semiconductors*; for a complete treatment of their theory in both an elementary and a rigorous fashion, we refer the reader to a previous book.[16-6] An alternate approach is given by van der Ziel.[16-3] We shall confine ourselves here to a very intuitive discussion of this aspect of solid-state physics.

We start by considering the elements in column IVA of the periodic table: carbon (in diamond form), silicon, germanium, and tin (in the gray modification), all of which have four electrons in their outer or valence shell. These elements crystalize in such a way that every atom shares a valence electron with each of its four nearest neighbors. This results in complete shells effectively having 8 electrons around every interior atom of the crystal, since each atom is sharing a pair of electrons with its immediate neighbors. The resulting material is mechanically rigid and electrically inert; the valence electrons are tightly bound. Hence, very pure column-IVA crystals are electrical insulators. The *binding energy* of an electron in these crystals—the amount of energy required to remove an electron from one of the shared bonds—is on the order of 1 eV, where an *electron volt* (eV) is the energy required to move an electron (of charge 1.6×10^{-19} C) through a potential difference of 1 volt.

Using Eq. (16-1) and the relation

$$\nu\lambda = c$$

which connects the frequency ν and wavelength λ of a wave, we obtain the useful formula

$$U(\text{eV}) = \frac{1.24}{\lambda}(\mu\text{m}) \tag{16-6}$$

for predicting the effect of light of a given wavelength on a semiconductor. For example, the diamond crystal requires about 5.6 eV to free an electron from a bond. This corresponds to a wavelength given by

$$\lambda = \frac{1.24}{5.6} = 0.22 \ \mu\text{m} \tag{16-7}$$

Thus, light of any wavelength above 220 nm will not have enough energy to release electrons. This means that the entire visible spectrum (400–700 nm) passes through diamond without being absorbed, accounting for the fact that it is transparent and colorless. Hexagonal sulfur (in column VIA) requires only about 2.0 eV to produce free electrons (this quantity is called the *energy gap* in semiconductor theory), giving $\lambda = 1.24/2.0 = 0.6$ μm. A wavelength of 600 nm lies in the orange part of the visible region; thus the red, orange, and yellow is passed or reflected and the green, blue, and violet portion is strongly absorbed, explaining the characteristic straw-yellow color of sulfur. Finally, germanium has an energy gap of 0.75 eV, giving $\lambda = 1.6$ mμ. Thus, germanium absorbs all visible light, the energy being used to produce free electrons, and only infrared frequencies are passed. The relatively large increase in the number of charge carriers which are free to conduct is responsible for the transformation from the insulating to the conducting condition.

It is possible to achieve photoconductive cells with special characteristics by mixing two materials. As an example, consider the problem of detecting infrared radiation with a wavelength of about 10 μm. This has important applications, since it corresponds roughly to the surface temperature of the earth and is useful for mapping and surveillance. The energy gap of the materials necessary for use in a detector, by Eq. (16-6), is 0.12 eV. There is no suitable semiconducting element or compound with such a gap. However, CdTe has a gap of 1.50 eV and HgTe is metallic in nature—it contains enough free electrons to be a conductor. Since these two materials are similar in structure, it would be expected that the proper proportions of the two would result in a single crystal whose gap could be adjusted to obtain the right value. As it turned out, the metallurgical problems were formidable, but a 10-μm photocell was finally developed having a composition of the form $Hg_{1-x} Cd_x Te$, where x corresponds to about 10%–15% CdTe.[16-7],[16-12]

Photoconductive detectors involving the optical activation process described above have been made from an enormous number of elemental and compound semiconductors. A description of many others will be found in Kruse et al.[16-8] and in Willardson and Beer.[16-12]

As with all electronic devices, noise is a limiting factor in the sensitivity of a photoconductive detector. To specify noise in a quantitative way, a number of definitions have been proposed for a suitable figure of merit. One of these is the *noise equivalent power* (*NEP*), which is the radiation power capable of generating a signal equal to the noise in the detector. That is, the signal-to-noise ratio is equal to unity. The radiation is assumed to come from a blackbody at some specified temperature, and it must have a specified modulation frequency and bandwidth. Hence, it is customary to designate noise equivalent power as, for example, *NEP* (500K, 900, 1) where the second figure denotes a 900-hz chopping frequency and the last one is a 1-hz bandwidth. The reciprocal of the *NEP* is called the *detectivity D*, so that

$$D = \frac{1}{NEP} \tag{16-8}$$

Another common figure of merit is the *noise equivalent input (NEI)*, which is the radiant power per unit area of detector (the quantity called the irradiance in Chapter 4) required for a signal-to-noise ratio of unity. Multiplying its reciprocal by the square root of the area, we obtain the relation

$$D^* = \frac{1}{(NEI)A^{1/2}} = DA^{1/2} \tag{16-9}$$

where D^* (which has no name) is expressed in units of m-hz$^{1/2}$-W^{-1}. Some typical values of this sensitivity specification are shown in Fig. 16-6.

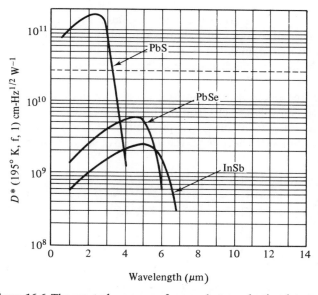

Figure 16-6 The spectral response of some photoconductive detectors.

More elaborate uses of photoconductive detection are represented by the *vidicon* TV camera tube. One of the newest is the *plumbicon*, as developed by Phillips (Fig. 16-7). A glass window has on its inside a transparent layer of stannic oxide and a photoconducting layer of lead monoxide. Light striking the stannic oxide is conducted as a current by the lead monoxide. The photoconducting layer is scanned by an electron beam that is accelerated by the anode.

Another type of detector which involves a different aspect of semiconductor physics is the *photovoltaic cell*, and again we refer the reader to a previous book[16-6] for a fuller explanation. The photoconductivity of semiconductors

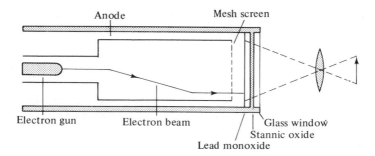

Anode   Mesh screen

Electron gun   Electron beam   Glass window
Stannic oxide
Lead monoxide

Figure 16-7

as we have described it involves inherent or *intrinsic* properties of the material. It is also possible to convert an insulator into a semiconductor by
adding very small amount of column-V impurities. For example, a phos-
phorous atom in a germanium crystal can be bonded with four out of five of
its valence electrons; the fifth electron is then free to conduct, and we have
impurity conduction. Such a material is said to be *N-type*, since the excess
charges are free electrons. If, however, we use a column-III element such as
boron, there is a missing electron in the bonding structure. We can see by
analogy that this vacant place is equivalent to a free *positive* charge, for a
stone dropped in water falls under the influence of gravity, whereas a bubble—

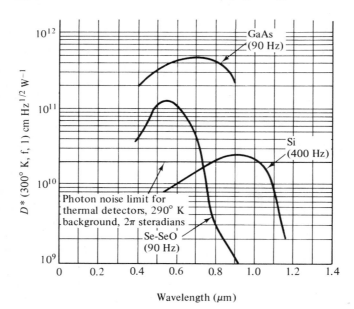

Wavelength (μm)

Figure 16-8 The spectral response of some photovoltaic detectors.

which is also a vacancy in the structure—rises. In the same way, the absent electron permits the bonding vacancy, called a *positive hole* or a *hole* for short, to move in a direction opposite to that in which the electrons move. Material in which this hole conduction process is dominant is said to be *P-type*.

Now if we create a *PN junction* by having an *N* region and a *P* region exist in the same crystal, then we can generate a potential by optical excitation. The presence of free electrons on one side of the junction and free holes on the other side sets up an electric field. Incoming photons produce free electrons and free holes in pairs. The original field causes the electrons to diffuse in one direction and the holes in another, creating a potential difference which can be detected by a sensitive meter. As an example, polycrystalline selenium has been used for some years in the photoelectric cells which are part of photographic light meters. Characteristics of some typical photovoltaic detectors are shown in Fig. 16-8.

16-5 Thermal Detectors

Thermal detectors operate on the principal that, when light is absorbed by a substance, its temperature increases. Measuring a property that depends on temperature, such as resistance, is a way to measure the intensity of light.

A *thermocouple* consists of a junction of dissimilar metals. Such an arrangement thus involves metals of different work functions, and the difference between these work functions is a measure of the amount of energy required to transfer an electron across the junction, thereby establishing a potential difference.[16-6] Thermocouples used as radiation detectors are coated with lampblack and absorb almost all of the incident radiation. Their spectral response is essentially flat over the entire visible and near infrared range, but the response time is quite slow because of the delay in heating a relatively large mass.

Other types of thermal radiation cell[16-8] are the *bolometer*, which utilizes the fact that the resistance of a metal or a semiconductor depends on its ambient temperature, the *thermopile*, a large number of thermocouples in series, and the *Golay cell*, which involves the expansion of a heated gas.

16-6 Photographic Film

Everyone is familiar with the use of film in photography, X-ray diagnosis, and radiation detection. Less well known is the availability of infrared-sensitive film and other special types. At the present times the precise mechanism of image formation is not fully understood. We shall discuss here the Gurney-Mott theory[16-9]; for a very thorough and clear description of what is known, refer to the book by Brown.[16-10]

The types of film mentioned above generally consist of microcrystals of a silver halide, such as AgBr or AgBr-AgI, suspended in a gelatine emulsion. The emulsion is coated on a transparent base and exposed to radiation. It is believed that an incident photon then releases an electron-hole pair from the silver halide, which is a semiconducting material. The electron diffuses to a spot on a microcrystal which is called a *sensitivity center* and is trapped there. Once the electron has become stationary, it is possible for a silver ion in the emulsion to migrate to the same point and be neutralized and thus converted into a free silver atom. This two-stage process is repeated for subsequent photons, and eventually a speck of metallic silver is formed. The development process removes the emulsion and its associated alkali halides, leaving behind the free silver to form the image on the negative. The characteristics of some commercial films are shown in Fig. 16-9.

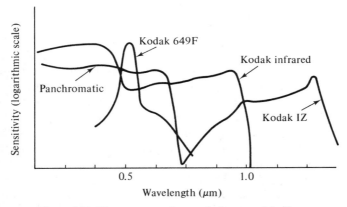

Figure 16-9 The response of several photographic films.

A color photograph contains information about the subject at many wavelengths. After a photograph has been taken, it can be filtered so that each spectral region is examined separately. A photograph of a small island surrounded by water of various depths can be filtered so that each layer of water is examined separately. The deeper the layer, the bluer is its color. Looking at bluer spectral sections amounts to looking at deeper layers. In this way one can see beneath the surface and enhance underwater objects.

The idea of multispectral analysis can be used with black and white infrared film. Several identical cameras are mounted pointing in the same direction. Each camera has a distinct narrow pass-band filter in front of its lens. All cameras are triggered simultaneously, and each samples a different portion of the infrared and visible spectrum. The visual analysis of the photographs can be facilitated by false color reconstruction. Each photograph representing a different band is projected onto a screen through a colored

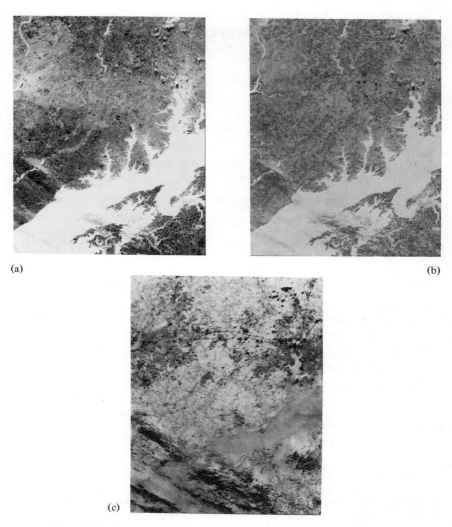

(a)   (b)

(c)

Figure 16-10 A scene of Chesapeake Bay reconstructed from an electronic multispectral 13 band scanner in Skylab. Three of the bands are shown. The outlines of the bay and tributaries are most pronounced in the .46 to .51 band (a), whereas the clouds, roads, bridge, and water variations are most pronounced in 1.55 to 1.75 (c). (*Courtesy Honeywell Radiation Center*)

filter. Differences between the photograph taken at different wavelengths are heightened. See Fig. 16-10 which shows a section of the Chesapeake Bay.

Newer photographic processes as surveyed by Tubbs,[16-11] eliminate both wet processing (which is inconvenient) and the expense of silver. An example is the well-known Xerox system, using electrostatic image formation from a selenium photon detector. Although these newer systems are the basis of commercial photocopy uses, they are not regarded as detectors in the conventional sense.

References

16-1 Y. LeGrand, *Light, Colour, and Vision*, Chapman and Hall, London (1968).

16-2 G. Bauer, *Measurement of Optical Radiations*, Focal Press (1965).

16-3 A. van der Ziel, *Solid State Physical Electronics*, 2nd ed., Prentice-Hall, N.J. (1968).

16-4 R. J. Keyes and R. H. Kingston, *Physics Today*, **24**, 48 (1972).

16-5 A. W. Drake, *Fundamentals of Applied Probability Theory*, McGraw-Hill (1967).

16-6 A. Nussbaum, *Electromagnetic and Quantum Properties of Materials*, Prentice-Hall, Englewood Cliffs, N.J. (1966).

16-7 P. W. Kruse, M. D. Blue, J. H. Garfunkel, W. D. Saur, *Infr. Phys.* **2**, 53 (1962).

16-8 P. W. Kruse, L. D. McGlauchlin, R. B. McQuistan, *Elements of Infrared Technology*, John Wiley & Sons, New York (1963).

16-9 R. W. Gurney and N. F. Mott, *Electronic Processes in Ionic Crystals*, Oxford University Press, New York (1938), 16-10.

16-10 F. C. Brown, *The Physics of Solids*, W. A. Bergman, Inc., New York (1967).

16-11 M. R. Tubbs, *Phys. Educ.* **6**, 227 (1971).

16-12 R. K. Willardson and A. C. Beer, eds., *Semiconductors and Semimetals*, *Vol. 5, Infrared Detectors*, Academic Press, New York (1970).

General References

R. N. Colwell, "Remote Sensing of Natural Resources," *Scientific American* (January, 1968).

R. C. Jones, "How Images Are Detected," *Scientific American* (September, 1968).

APPENDIX

A

FOURIER INTEGRALS

A-1 Fourier Series

The square pulse shown in Fig. A-1 can be defined by the expression

$$y = 1 \quad \text{when} \quad -\frac{\pi}{2} \le x \le \frac{\pi}{2}$$

$$y = 0 \quad \text{when} \quad \frac{\pi}{2} < |x| \le \pi \tag{A-1}$$

It is sometimes convenient, however, to have an analytic equivalent of Eq. (A-1), and this may be obtained if y is expressed in terms of the sine and

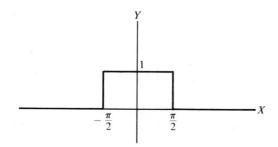

Figure A-1

473

cosine functions as

$$y = a_0 + a_1 \cos x + a_2 \cos 2x + \cdots + b_1 \sin x + b_2 \sin 2x + \cdots$$

$$\text{(A-2)}$$

or

$$y = \sum_{n=0}^{\infty} (a_n \cos nx + b_n \sin nx) \tag{A-3}$$

Such an expression is known as a *Fourier series*.

To determine the coefficients a_n and b_n, we can multiply Eq. (A-3) by $\cos mx$ or $\sin mx$, where m is a positive integer, and integrate over the interval $(-\pi, \pi)$. For example, to determine a_m,

$$\int_{-\pi}^{\pi} y \cos mx \, dx = \int_{-\pi}^{\pi} \sum (a_n \cos nx + b_n \sin nx) \cos mx \, dx$$

To evaluate this expression, we use the definite integrals

$$\int_{-\pi}^{\pi} \sin mx \sin nx \, dx = \begin{cases} \pi & \text{if} \quad m = n \\ \\ 0 & \text{if} \quad m \neq n \end{cases}$$

$$\int_{-\pi}^{\pi} \cos mx \cos nx \, dx =$$

$$\int_{-\pi}^{\pi} \sin mx \cos nx \, dx = \quad 0$$

The latter may easily be demonstrated by means of the identity

$$\sin mx \cos nx = \tfrac{1}{2}[\sin (m + n)x + \sin (m - n)x]$$

As a result, we see that multiplying y by $\cos m_x$ and integrating from $-\pi$ to π has the effect of picking out the coefficient a_m, since only this term is non-zero when the definite integral is evaluated:

$$\int_{-\pi}^{\pi} y \cos mx \, dx = \int_{-\pi}^{\pi} \cos^2 mx \, dx + 0 + 0 + \cdots = \pi a_m$$

This may be rewritten as

$$a_m = \frac{1}{\pi} \int_{-\pi}^{\pi} y \cos mx \, dx \qquad (m \neq 0) \tag{A-4}$$

Similarly

$$b_m = \frac{1}{\pi} \int_{-\pi}^{\pi} y \sin mx \, dx \tag{A-5}$$

When $m = 0$, then $b_m = 0$, but $\int_{-\pi}^{\pi} y \, dx = \int_{-\pi}^{\pi} a_0 \, dx = 2\pi a_0$ or $a_0 = (1/2\pi) \int_{-\pi}^{\pi} y \, dx$. If we replace a_0 in Eq. (A-2) by a new constant $a_0/2$, then Eq. (A-4) becomes

$$a_n = \frac{1}{\pi} \int_{-\pi}^{\pi} y \cos nx \, dx \qquad (n = 0, 1, 2, \ldots) \tag{A-6}$$

and in the same way

$$b_n = \frac{1}{\pi} \int_{-\pi}^{\pi} y \sin nx \, dx. \qquad (n = 0, 1, 2, \ldots) \qquad \text{(A-7)}$$

For the example of Fig. A-1, we find that

$$a_0 = \frac{1}{\pi} \int_{-\pi}^{\pi} y \, dx = \frac{1}{\pi} \int_{-\pi/2}^{\pi/2} (1) \, dx = 1$$

$$a_1 = \frac{1}{\pi} \int_{-\pi}^{\pi} y \cos x \, dx = \frac{1}{\pi} (\sin x) \Big|_{-\pi/2}^{\pi/2} = \frac{2}{\pi}$$

$$b_1 = \frac{1}{\pi} \int_{-\pi}^{\pi} y \sin x \, dx = -\frac{1}{\pi} \cos x \Big|_{-\pi/2}^{\pi/2} = 0$$

$$a_2 = \frac{1}{\pi} \int_{-\pi}^{\pi} y \cos 2x \, dx = \frac{1}{2\pi} (\sin 2x) \Big|_{-\pi/2}^{\pi/2} = 0$$

$$b_2 = 0$$

$$a_3 = \frac{1}{\pi} \int_{-\pi/2}^{\pi/2} \cos 3x \, dx = \frac{1}{3\pi} \sin 3x \Big|_{-\pi/2}^{\pi/2} = -\frac{2}{3\pi}$$

etc.

so that

$$y = \frac{a_0}{2} + a_1 \cos x + a_2 \cos 2x + \cdots + b_1 \sin x + \cdots$$

$$= \frac{1}{2} + \frac{2}{\pi} \cos x - \frac{2}{3\pi} \cos 3x + \cdots$$

$$= \frac{1}{2} + \frac{2}{\pi} \left[\cos x - \frac{1}{3} \cos 3x + \frac{1}{5} \cos 5x - \cdots \right] \qquad \text{(A-8)}$$

The term in square brackets can be pictured as shown in Fig. A-2. Figure A-2(a) at the top is $\cos x$, and this curve is also repeated at the bottom. To this, we add $(-1/3) \cos 3x$ of Fig. A-2(b) to obtain Fig. A-2(d). Then we combine the curve in Fig. A-2(d) with the curve for $(1/5) \cos 5x$ of curve (c) in the figure to obtain the curve in Fig. A-2(e), and so on. The resulting square pulse has a height of $2(\pi/4)$, for at $x = 0$

$$\cos x - \frac{1}{3} \cos 3x + \frac{1}{5} \cos 5x - \cdots = 1 - \frac{1}{3} + \frac{1}{5} - \cdots = \frac{\pi}{4}$$

as may be demonstrated if we numerically sum the series for the first few terms. Taking $2/\pi$ times this result and shifting the entire square pulse up a distance $y = 0.5$ then gives Fig. A-1, as it should.

The function expressed by Eq. (A-1) and shown in Fig. A-1 is a single square pulse which is nonzero only from $-\pi/2$ to $\pi/2$. The Fourier series expression in Eq. (A-8), however, does not vanish outside this interval, because $\cos x$, $\cos 3x$, ... are periodic functions which repeat themselves

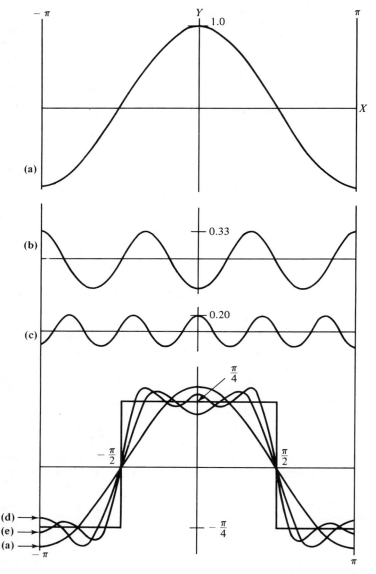

Figure A-2

to the right and to the left of the vertical boundaries of Fig. A-2. Hence, Eq. (A-8) describes the function of Eq. (A-1) only in the interval $(-\pi, \pi)$. However, it *is* an analytic approximation to the periodic function

$$f(x) = \begin{cases} 1 & \text{when} \quad |x| \leq \dfrac{\pi}{2} \\ 0 & \text{when} \quad \pi > |x| > \dfrac{\pi}{2} \end{cases}$$

$$f(x \pm 2\pi) = f(x)$$

and this train of square waves is shown in Fig. A-3.

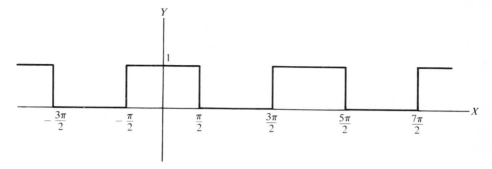

Figure A-3

Equation (A-3) as applied to an arbitrary periodic function $f(x)$ is then

$$f(x) = \frac{a_0}{2} + \sum_{n=1}^{\infty} (a_n \cos nx + b_n \sin nx) \tag{A-9}$$

It is convenient to change x to a new variable z, defined as

$$z = kx = \frac{2\pi x}{\lambda}$$

where k is the propagation constant of Eq. (5-2). The Fourier series of Eq. (A-9) is then

$$f(z) = \frac{a_0}{2} + \sum_{n=1}^{\infty} (a_n \cos nz + b_n \sin nz) \tag{A-10}$$

or

$$f(x) = \frac{a_0}{2} + \sum_{n=1}^{\infty} (a_n \cos nkx + b_n \sin nkx) \tag{A-11}$$

where the coefficients, by Eqs. (A-6) and (A-7), become

$$\left. \begin{aligned} a_n &= \frac{1}{\pi} \int_{-\pi}^{\pi} f(z) \cos nz \, dz = \frac{2}{\lambda} \int_{-\lambda/2}^{\lambda/2} f(x) \cos nkx \, dx \\ b_n &= \frac{1}{\pi} \int_{-\pi}^{\pi} f(z) \sin nz \, dz = \frac{2}{\lambda} \int_{-\lambda/2}^{\lambda/2} f(x) \sin nkx \, dx \end{aligned} \right\} \tag{A-12}$$

The series for $f(x)$ just given applies to a function whose wavelength is 2π units. For a function of arbitrary wavelength, which we shall designate as $2L$, the substitution $\lambda = 2L$ converts (Eq. (A-11) to)

$$f(x) = \frac{a_0}{2} + \sum_{n=1}^{\infty} \left(a_n \cos \frac{n\pi x}{L} + b_n \sin \frac{n\pi x}{L} \right) \tag{A-13}$$

where

$$\left.\begin{aligned} a_n &= \frac{1}{L} \int_{-L/2}^{L/2} f(x) \cos \frac{n\pi x}{L} \, dx \\ b_n &= \frac{1}{L} \int_{-L/2}^{L/2} f(x) \sin \frac{n\pi x}{L} \, dx \end{aligned}\right\} \tag{A-14}$$

Using

$$\cos \theta = \frac{e^{i\theta} + e^{-i\theta}}{2}$$

$$\sin \theta = \frac{e^{i\theta} - e^{-i\theta}}{2i}$$

we convert Eq. (A-11) into the complex form

$$f(x) = \sum_{n=-\infty}^{\infty} c_n e^{inkx} \tag{A-15}$$

where

$$c_n = \frac{a_n - ib_n}{2} \qquad (n = 1, 2, \ldots)$$

$$c_0 = \frac{a_0}{2} \qquad (n = 0)$$

$$c_n = \frac{a_n + ib_n}{2} \qquad (n = -1, -2, \ldots)$$

Since

$$\int_{-\pi}^{\pi} e^{i(m-n)x} \, dx = \begin{cases} 2\pi & \text{if} \quad m = n \\ 0 & \text{if} \quad m \neq n \end{cases} \tag{A-16}$$

then

$$c_n = \frac{1}{\lambda} \int_{-\lambda/2}^{\lambda/2} f(x) e^{-inkx} \, dx \tag{A-16a}$$

A-2 The Fourier Integral and the Fourier Transform

We shall replace the variable x in Eqs. (A-11) and (A-12) by a new variable y (since the symbol assigned to the variable of integration in a definite integral is arbitrary), and the coefficients are then

$$\left.\begin{aligned} a_n &= \frac{2}{\lambda} \int_{-\lambda/2}^{\lambda/2} f(y) \cos nky \, dy \\ b_n &= \frac{2}{\lambda} \int_{-\lambda/2}^{\lambda/2} f(y) \sin nky \, dy \end{aligned}\right\} \tag{A-17}$$

Substituting Eq. (A-17) into Eq. (A-16) gives

$$f(x) = \frac{1}{\lambda} \int_{-\lambda/2}^{\lambda/2} f(y)\, dy + \sum_{n=1}^{\infty} \frac{2}{\lambda} \int_{-\lambda/2}^{\lambda/2} f(y) \cos \{nk(x - y)\}\, dy \quad \text{(A-18)}$$

where we have used the well-known identity for cos $(a - b)$.

Now let

$$k_n = \frac{2\pi n}{\lambda} = nk$$

so that

$$k_n - k_{n-1} = \Delta k_n = \frac{2\pi n}{\lambda} - \frac{2\pi(n - 1)}{\lambda} = \frac{2\pi}{\lambda}$$

If we let λ approach infinity, then the second term on the right of Eq. (A-18) becomes

$$\frac{1}{\pi} \sum \Delta k_n \int_{-\lambda/2}^{\lambda/2} f(y) \cos \{k_n(x - y)\}\, dy = \frac{1}{\pi} \int_0^{\infty} dk \int_{-\infty}^{\infty} f(y) \cos \{k(x - y)\}\, dy$$

where the summation from $n = 1$ to $n = \infty$ is assumed to be approximately equal to the integral over the continuous variable w from 0 to ∞. We further assume that the integral in the first term on the right of Eq. (A-18) has a finite value so that the entire term vanishes as $\lambda \longrightarrow \infty$, and we have

$$f(x) = \frac{1}{\pi} \int_0^{\infty} dk \int_{-\infty}^{\infty} f(y) \cos \{k(x - y)\}\, dy \quad \text{(A-19)}$$

This is the *Fourier integral theorem.*

If we write this as

$$f(x) = \frac{1}{\pi} \int_0^{\infty} \cos (kx)\, dk \int_{-\infty}^{\infty} f(y) \cos (ky)\, dy$$

$$+ \frac{1}{\pi} \int_0^{\infty} \sin (kx) \int_{-\infty}^{\infty} f(y) \sin (ky)\, dy \quad \text{(A-20)}$$

we may look on the new form as an extension of Eq. (A-3) to an interval $(-\infty, \infty)$, where the term $\int_{-\infty}^{\infty} f(y) \sin (ky)\, dy$ may be regarded as the coefficient $a(k)$ and $\int_{-\infty}^{\infty} f(y) \cos (ky)\, dy$ is $b(k)$.

Using

$$\cos \{k(x - y)\} = \frac{e^{ik(x-y)} + e^{-ik(x-y)}}{2}$$

we convert Eq. (A-19) into

$$f(x) = \frac{1}{\pi} \int_{-\infty}^{\infty} f(y) \left\{ \int_0^{\infty} \left(\frac{e^{ik(x-y)} + e^{-ik(x-y)}}{2} \right) dk \right\} dy$$

The inner integral in this expression may be written

$$\int_0^{\infty} (e^{ik(x-y)} + e^{-ik(x-y)})\, dk = \int_0^{\infty} \frac{e^{ik(x-y)}}{2}\, dk + \int_{-\infty}^{0} \frac{e^{ik(x-y)}}{2}\, dk$$

from which we obtain

$$f(x) = \frac{1}{2\pi} \int_{-\infty}^{\infty} f(y) \left\{ \int_{-\infty}^{\infty} e^{ik(x-y)} \, dk \right\} dy$$

or

$$f(x) = \frac{1}{2\pi} \int_{-\infty}^{\infty} \left\{ \int_{-\infty}^{\infty} f(y) e^{ik(x-y)} \, dk \right\} dy \qquad \text{(A-21)}$$

This is the *complex form* of the Fourier integral theorem.

We next define the *Fourier transform* $\mathfrak{F}[f(x)] = F(k)$ of the function $f(x)$ as

$$\mathfrak{F}[f(x)] = \int_{-\infty}^{\infty} f(x) e^{-ikx} \, dx = F(k) \qquad \text{(A-22)}$$

Comparing Eqs. (A-21) and (A-22) shows that the *inverse Fourier transform* will be

$$f(x) = \frac{1}{2\pi} \int_{-\infty}^{\infty} \left\{ \int_{-\infty}^{\infty} f(y) e^{-iky} \, dy \right\} e^{ikx} \, dk$$

$$= \frac{1}{2\pi} \int_{-\infty}^{\infty} F(k) e^{ikx} \, dk = \mathfrak{F}^{-1}[F(k)] = \mathfrak{F}^{-1}\mathfrak{F}[f(x)] = f(x) \qquad \text{(A-23)}$$

We define a *pair of Fourier transforms* $f(x)$ and $F(k)$ as two functions which satisfy both Eqs. (A-22) and (A-23).

Let us identify x in Eq. (A-22) or (A-23) as one coordinate of an arbitrary point whose position vector is $\mathbf{r}$ and k as the x component of the propagation vector $\mathbf{k}$ for an arbitrary wave described by the Fourier integral theorem. Then the three-dimensional analogue of Eq. (A-22) is

$$F(\mathbf{k}) = \mathfrak{F}[f(\mathbf{r})] = \int_{-\infty}^{\infty}\int_{-\infty}^{\infty}\int_{-\infty}^{\infty} f(\mathbf{r}) e^{-i\mathbf{k}\cdot\mathbf{r}} \, dx \, dy \, dz \qquad \text{(A-24)}$$

and the inverse is

$$f(r) = \mathfrak{F}^{-1}[F(k)] = \frac{1}{(2\pi)^3} \int_{-\infty}^{\infty}\int_{-\infty}^{\infty}\int_{-\infty}^{\infty} F(\mathbf{k}) e^{i\mathbf{k}\cdot\mathbf{r}} \, dk_x \, dk_y \, dk_z \qquad \text{(A-25)}$$

The linear differential equation of the form

$$A\frac{d^2 f(t)}{dt^2} + B\frac{df(t)}{dt} + Cf(t) = g(t) \qquad \text{(A-26)}$$

or

$$A\ddot{f}(t) + B\dot{f}(t) + Cf(t) = g(t) \qquad \text{(A-27)}$$

governs the behavior of damped oscillating systems. For example, if we apply a potential $v(t) = g(t)$ to a resistor, an inductor, and a capacitor in series, then the resulting current $i(t) = f(t)$ is obtained by solving Eq. (A-26). In fact, we call $g(t)$ the *excitation* or *forcing function* and $i(t)$ is the *response function*.

To solve Eq. (A-26) we shall take the Fourier transform of both sides. First, we replace x in Eq. (A-22) by the time t, so that k is replaced by the angular frequency ω of the applied potential. Equation (A-22) becomes

$$F(\omega) = \int_{-\infty}^{\infty} f(t)e^{-i\omega t}\, dt \tag{A-28}$$

and Eq. (A-23) is now

$$f(t) = \frac{1}{2\pi} \int_{-\infty}^{\infty} F(\omega)e^{i\omega t}\, d\omega \tag{A-29}$$

Next, we integrate by parts to obtain

$$\int_{-\infty}^{\infty} \frac{df}{dt}\, e^{-i\omega t}\, dt = fe^{-i\omega t}\Big|_{-\infty}^{\infty} + i\omega \int_{-\infty}^{\infty} fe^{-i\omega t}\, dt \tag{A-30}$$

where

$$v = e^{-i\omega t}, \qquad dv = \frac{df}{dt}\, dt$$

Since the magnitude of $e^{-i\omega t}$ is unity, the integrated term will vanish if we require $f(t)$ to vanish when $|t|$ is large. We arrange for the current $i(t) = f(t)$ to start at $t = 0$, so that it has a value of zero for $t < 0$, and we know that it decays in a damped circuit; thus, these conditions are satisfied. Hence

$$\mathcal{F}\left[\frac{d}{dt}f(t)\right] = i\omega \mathcal{F}[f(t)] \tag{A-31}$$

and similarly, we may show that

$$\mathcal{F}\left[\frac{d^2}{dt^2}f(t)\right] = (i\omega)^2 \mathcal{F}[f(t)] \tag{A-32}$$

Then Eq. (A-27) becomes

$$\{(i\omega)^2 A + (i\omega)B + C\}F(\omega) = G(\omega) \tag{A-33}$$

where

$$G(\omega) = \mathcal{F}[g(t)] \tag{A-34}$$

or

$$F(\omega) = \frac{G(\omega)}{(i\omega)^2 A + (i\omega)B + C} \tag{A-35a}$$

The polynomial in $(i\omega)$ of the form

$$H(\omega) = \frac{1}{(i\omega)^2 A + (i\omega)B + C} \tag{A-36}$$

is called the *transfer function*. Since Eq. (A-35a) becomes

$$F(\omega) = G(\omega)H(\omega) \tag{A-35b}$$

where $F(\omega)$ and $G(\omega)$ are the transforms of the current and voltage, respectively, then $H(\omega)$ must be the admittance $Y(\omega)$ (i.e., the reciprocal of the impedance) of the series circuit expressed as a function of ω.

To convert back to time as a variable, we use the definitions

$$h(t) = \mathcal{F}^{-1}[H(\omega)]$$
$$f(t) = \mathcal{F}^{-1}[F(\omega)] \tag{A-37}$$

so that

$$F(\omega) = G(\omega)H(\omega) = \int_{-\infty}^{\infty} g(\alpha)e^{-i\omega\alpha}\, d\alpha \int_{-\infty}^{\infty} h(\beta)e^{-i\omega\beta}\, d\beta$$

$$= \int_{-\infty}^{\infty}\int_{-\infty}^{\infty} g(\alpha)h(\beta)e^{-i\omega(\alpha+\beta)}\, d\alpha\, d\beta$$

Let

$$\alpha + \beta = t$$

and integrate first over β, holding α constant. This gives

$$F(\omega) = \int_{-\infty}^{\infty} \left\{ \int_{-\infty}^{\infty} g(\beta)h(t-\alpha)e^{-i\omega t}\, dt \right\} d\alpha$$

Then we reverse the order of integration to obtain

$$F(\omega) = \int_{-\infty}^{\infty} \left\{ \int_{-\infty}^{\infty} g(\alpha)h(t-\alpha)\, d\alpha \right\} e^{-i\omega t}\, dt$$

By Eq. (A-29) and Eq. (A-37), this result is equivalent to

$$f(t) = \mathcal{F}^{-1}[F(\omega)] = \int_{-\infty}^{\infty} g(\alpha)h(t-\alpha)\, d\alpha \qquad (A\text{-}38a)$$

which is known as the *convolution theorem,* and we say that *the response of the circuit is the convolution of the excitation with the transfer function.* This is written as

$$f(t) = g(t) * h(t) \qquad (A\text{-}38b)$$

and we realize that the Fourier transform of the convolution of two functions is the product of the individual transforms, using Eq. (A-35b).

As a concrete example, consider a resistance of 4 ohms in series with an inductance of 2 henries. An exponential voltage specified by the relation

$$v(t) = \begin{cases} 0 & \text{for } t \le 0 \\ 10e^{-t} & \text{for } t \ge 0 \end{cases} \qquad (A\text{-}39)$$

is applied to this circuit. The differential equation to be solved is then

$$v = L\frac{di}{dt} + iR \qquad (A\text{-}40)$$

Since there is a response only for $t \ge 0$, we write the equation as

$$2\frac{di}{dt} + 4i = 10e^{-t}$$

and integrate only from $t = 0$ to $t = \infty$, obtaining the Fourier transform of $v(t)$ as

$$V(\omega) = \int_{0}^{\infty} (10e^{-t})e^{-i\omega t}\, dt = 10\frac{e^{-(1+i\omega)t}}{-(1+i\omega)}\bigg|_{0}^{\infty}$$

$$= 10\left(\frac{1}{1+i\omega}\right)$$

Combining this with Eq. (A-31) and Eq. (A-40) gives

$$[(i\omega)2 + 4]I(\omega) = 10\left(\frac{1}{1 + i\omega}\right)$$

or

$$I(\omega) = 5\frac{1}{(2 + i\omega)(1 + \omega)}$$

To find the inverse transform, we can write the function on the right as

$$\frac{1}{(2 + i\omega)(1 + i\omega)} = \frac{1}{1 + i\omega} - \frac{1}{2 + i\omega}$$

But we have just seen that the Fourier transform of the function given by Eq. (A-39) is e^{-t}, and we similarly can show that the transform of the function

$$f(t) = \begin{cases} 0 & \text{for } t \le 0 \\ Ae^{-at} & \text{for } t \ge 0 \end{cases}$$

is

$$F(\omega) = \frac{A}{a + i\omega}$$

Using

$$\mathfrak{F}^{-1}[\mathfrak{F}[f(t)]] = f(t)$$

then gives

$$\mathfrak{F}^{-1}\left[\frac{1}{a + i\omega}\right] = e^{-at}$$

Hence

$$i(t) = \mathfrak{F}^{-1}[I(\omega)]$$

$$= 5\mathfrak{F}^{-1}\left[\frac{1}{1 + i\omega} - \frac{1}{2 + i\omega}\right]$$

$$= \begin{cases} 0 & t \le 0 \\ 5(e^{-t} - e^{-2t}) & t \ge 0 \end{cases}$$

To summarize, we have obtained the following results:

	Time Domain		Frequency Domain
Excitation	$v(t) = \begin{cases} 0 & t \le 0 \\ 10e^{-t} & t \ge 0 \end{cases}$		$V(\omega) = \dfrac{10}{1 + i\omega}$
Response	$i(t) = \begin{cases} 0 & t \le 0 \\ 5(e^{-t} - e^{-2t}) & t \ge 0 \end{cases}$		$I(\omega) = \dfrac{5}{(2 + i\omega)(1 + i\omega)}$
Transfer Function	$y(t) = \dfrac{1}{2}e^{-2t}$		$Y(\omega) = \dfrac{1}{2(2 + i\omega)}$

$$i(t) = v(t) * y(t)$$
$$I(\omega) = V(\omega)Y(\omega)$$

A-3   The Meaning of Convolution Processes

The convolution operation is of such practical importance in optics and engineering that it is worthwhile to examine its properties and its physical significance. A very clear discussion of these points has been given by Healy[A-1] and the material which follows is extracted from his article.

Transformers of a certain type are delivered by two companies, which will be called A and B, in lots of three and four, respectively. In a lot from company A, there will be zero, one, two, or three defective transformers. The probability that each of these numbers of defects, represented by $a(a = 0, 1, 2, 3)$, will occur is assumed to be

a	$P(a)$
0	0.4
1	0.3
2	0.2
3	0.1

Similarly, the probability of $b(b = 0, 1, 2, 3, 4)$ occurrences is taken as

b	$P(b)$
0	0.3
1	0.2
2	0.2
3	0.2
4	0.1

in a shipment from company B. The problem is to find the probabilities of the total number of defects in two shipments, one from each company. This sum or total number of defects will be indicated as $c(c = 0, 1, 2, 3, 4, 5, 6, 7)$.

The probability that $c = 0$ is the probability that $a = 0$ and $b = 0$; that is, there are no defects in either shipment. This is written as

$$P(c = 0) = P(a = 0, b = 0)$$

If the events $a = 0$ and $b = 0$ are independent of each other, which we assume here and which is necessary if the solution is to be a convolution, then this probability reduces to the product

$$P(c = 0) = P(a = 0) \times P(b = 0)$$
$$= 0.4 \times 0.3$$
$$= 0.12$$

The probability that we have a total of one defect is the probability that the shipment from A has one defect and the shipment from B has none, or vice versa. That is,

$$P(c = 1) = P[(a = 1, b = 0) \text{ or } (a = 0, b = 1)]$$
$$= P(a = 1, b = 0) + P(a = 0, b = 1)$$
$$= P(a = 1)P(b = 0) + P(a = 0)P(b = 1)$$
$$= (0.3 \times 0.3) + (0.4 \times 0.2)$$
$$= 0.17$$

Similarly,

$$P(c = 2) = P[(a = 2, b = 0), (a = 1, b = 1), \text{ or } (a = 0, b = 2)]$$
$$= P(a = 2)P(b = 0) + P(a = 1)P(b = 1) + P(a = 0)P(b = 2)$$
$$= (0.2 \times 0.3) + (0.3 \times 0.2) + (0.4 \times 0.2)$$
$$= 0.20$$

Continuing in this way, we can obtain the probabilities for all eight possible values of c. The result is given as

c	$P(c)$
0	0.12
1	0.17
2	0.20
3	0.21
4	0.16
5	0.09
6	0.04
7	0.01

Although it is possible to solve this problem in the manner just described, it is highly desirable to find a shortcut to determine the probabilities of the values of c. Notice how the entries for a and b are used to find those for c. The first entry for c is the product of the first entries for a and b. The second entry for c is entry 1 of b times entry 2 of a plus entry 2 of a times entry 1 of b. The third entry of c is the sum of cross terms 1 and 3, 2 and 2, 3 and 1 of a and b.

We can systematize the process of finding the entries for c in the following way. Write the probabilities for a and b as sequences A and B

$$A = [0.4, 0.3, 0.2, 0.1]$$
$$B = [0.3, 0.2, 0.2, 0.2, 0.1]$$

Reverse the order of one of the sequences, say B. Call the reversed sequence B_{inv}, where

$$B_{inv} = [0.1, 0.2, 0.2, 0.2, 0.3]$$

Position sequence A and the inverted sequence B_{inv} so that the first right-hand term of B_{inv} is under the first left-hand term of A

$$[0.4, 0.3, 0.2, 0.1]$$

$$[0.1, 0.2, 0.2, 0.2, 0.3]$$

The probability of zero defects is the product of the overlapping numbers 0.4 and 0.3. Now shift the inverted sequence one position to the right:

$$[0.4, 0.3, 0.2, 0.1]$$

$$[0.1, 0.2, 0.2, 0.2, 0.3]$$

The probability of *one* defect is the sum of the overlapping products 0.2×0.4 and 0.3×0.3. The remaining terms in c are obtained by shifting the inverted sequence one step at a time to the right, and for each step summing the overlap products.

The process of inverting a sequence, sliding it one step at a time to the right, and summing the overlap products is called *discrete convolution* or *serial multiplication*.

The asterisk (*) is generally used to indicate convolution. Thus, we write

$$C = A * B \qquad\qquad (A\text{-}41)$$

or

$$C = [0.4, 0.3, 0.2, 0.1] * [0.3, 0.2, 0.2, 0.2, 0.1]$$
$$= [0.12, 0.17, 0.20, 0.21, 0.16, 0.09, 0.04, 0.01]$$

where the last term in C is the first product, the next-to-last term is the sum of the second and third products, and so on.

Using this example as a guide, we see that the convolution of two sequences A and B, where

$$A = [a_0, a_1, \ldots, a_i, \ldots]$$
$$B = [b_0, b_1, \ldots, b_i, \ldots]$$

to obtain a sequence

$$C = [c_1, c_2, \ldots, c_i, \ldots]$$

has the following properties:

(a) The convolution process is commutative. That is

$$A * B = B * A \qquad\qquad (A\text{-}42)$$

(b) The ith term of C is given by the relation

$$c_i = \sum_{j=0}^{i} a_i b_{i-j} \qquad\qquad (A\text{-}43)$$

(c) The number of terms n_C in C is

$$n_C = n_A + n_B - 1 \qquad\qquad (A\text{-}44)$$

where n_A and n_B are the number of terms in A and B, respectively.

(d) The product of the sum of the elements of A and the sum of the elements in B is the sum of the elements in C, or

$$(\sum_{i=0}^{n_A-1} a_i)(\sum_{j=0}^{n_B-1} b_j) = \sum_{k=0}^{n_A+n_B-1} c_k \tag{A-45}$$

Returning to Eq. (A-38a),

$$f(t) = \int_{-\infty}^{\infty} g(\alpha)h(t-\alpha)\,d\alpha \tag{A-38a}$$

where $f(t)$ is a *continuous convolution*. To see why, let $g(t)$ and $h(t)$ be two arbitrary functions whose appearance is sketched in Fig. A-4(a). Let α be some value of t at which we shall evaluate $h(t)$, so that $h(-\alpha)$ represents what happens when we take the function $f(t)$ and fold or convolve it about the vertical axis to obtain its mirror image. Then we slide it back to the right somewhat (Fig. A-4(b)) by changing the argument to $(-\alpha+t)$, so that we end up with $h(t-\alpha)$ as shown. We multiply this function by $g(t)$ evaluated at the same value of α to obtain the shaded area in Fig. A-4(c), and then we "add" all these products—i.e., integrate over all possible values of α—to obtain the convolution in Fig. A-4(d). This process is essentially what we did in the discrete case: one of the two sets of variables is reversed, the overlaping values are computed, and then the overlaps are totaled. In fact, the dummy variable of integration α is analogous to the dummy summation index j. We have a further parallel which comes from letting

$$t - \alpha = \beta$$

Then

$$f(t) = g(t) * h(t) = -\int_{+\infty}^{-\infty} g(t-\beta)h(\beta)\,d\beta = h(t) * g(t) \tag{A-46}$$

which is the analogue of Eq. (A-42). We can find the relation which corresponds to Eq. (A-45) by integrating Eq. (A-38a), obtaining

$$\int_{-\infty}^{\infty} f(t)\,dt = \int_{-\infty}^{\infty}\int_{-\infty}^{\infty} g(\alpha)h(t-\alpha)\,d\alpha\,dt$$

$$= \int_{-\infty}^{\infty} g(\alpha)\left\{\int_{-\infty}^{\infty} h(t-\alpha)\,dt\right\}d\alpha$$

$$= \int_{-\infty}^{\infty} g(t)\left\{\int_{-\infty}^{\infty} h(t)\,dt\right\}dt$$

$$= \int_{-\infty}^{\infty} g(t)\,dt \int_{-\infty}^{\infty} h(t)\,dt \tag{A-47}$$

where we have used the fact that $h(t-\alpha)$ and $h(t)$ take on the same set of values if we are integrating from $-\infty$ to ∞. Equation (A-47) states that the product of the areas under the curves for two functions is equal to the area produced by this convolution.

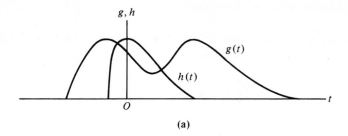

(a)

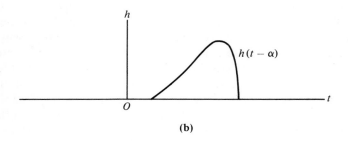

(b)

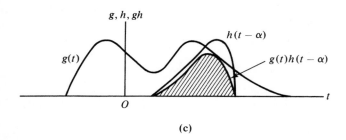

(c)

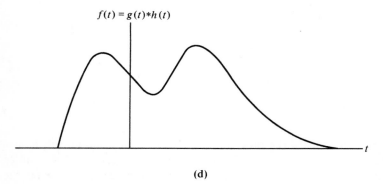

(d)

Figure A-4

488

A-4 The Delta Function

The function $f(t)$ defined by

$$f(t) = \frac{1}{\sqrt{2\pi a^2}} e^{-t^2/2a^2} \tag{A-48}$$

is called the *Gaussian function*. The constant $1/\sqrt{2\pi a^2}$ makes the area under the curve in the range $-\infty < x < \infty$ have a value of unity, as may be verified by a table of definite integrals. The quantity a in the exponent is a measure of the width of the symmetric Gaussian curve. Figure A-5 shows these curves

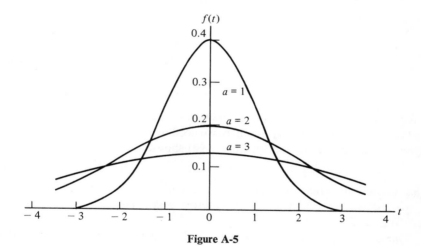

Figure A-5

for $a = 1$, 2, and 3. If we consider the width $2w$ as being specified by the values of $f(t)$ which are $e^{-1} = 0.37$ times the value at $t = 0$, then Eq. (A-48) indicates that

$$w = \pm\sqrt{2a} \tag{A-49}$$

The Fourier transform of the Gaussian function is

$$F(\omega) = \int_{-\infty}^{\infty} \frac{1}{\sqrt{2\pi a^2}} e^{-t^2/2a^2} e^{-i\omega t}\, dt = e^{-2\pi^2 a^2 \omega^2} \tag{A-50}$$

This function is also Gaussian, as shown in Fig. A-6.

Let us consider the effect on $f(t)$ and $F(\omega)$ if a is permitted to become indefinitely small. As Fig. A-6 indicates, the curve for $f(t)$ will become higher and narrower. If we let $a \to 0$, we have, in fact, a width of zero and a height of infinity, but the area under the curve remains equal to unity. Under these conditions, we call $f(t)$ the *Dirac delta function* $\delta(t)$, and specify it by the

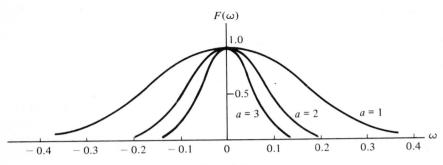

Figure A-6

relations

$$\delta(t) = 0 \qquad t \neq 0 \tag{A-51}$$

$$\delta(t) = \infty \qquad t = 0 \tag{A-52}$$

$$\int_{-\infty}^{\infty} \delta(t)\, dt = 1 \tag{A-53}$$

It is not really correct to call $\delta(t)$ a function—it is properly known as a *distribution*[A-2]—but for our purposes here we need not worry about mathematical rigor. Taking the same limit in Eq. (A-50) shows that

$$F(\omega) = 1 \tag{A-54}$$

That is, we have obtained a Gaussian curve of infinite width. Figure A-7(a) shows the way in which the delta function might be represented and Fig. A-7(b) is its Fourier transform.

One of the useful properties of the delta function is obtained if we consider

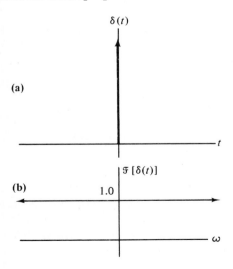

Figure A-7

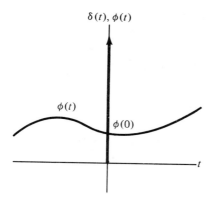

$\delta(t), \phi(t)$

$\phi(t)$

$\phi(0)$

t

Figure A-8

its effect on an arbitrary function $\varphi(t)$. Let $\varphi(t)$ have the general behavior shown in Fig. A-8 and consider the integral

$$I = \int_{-\infty}^{\infty} \varphi(t)\delta(t)\, dt \qquad (A-55)$$

Since the product $\varphi(t)\delta(t)$ vanishes for $t \neq 0$, the only value of $\varphi(t)$ which has any effect on the value of the integral I is $\varphi(0)$. But $\varphi(0)$ is a constant and can be placed outside the integral. Hence

$$I = \varphi(0) \int_{-\infty}^{\infty} \delta(t)\, dt = \varphi(0) \qquad (A-56)$$

We can generalize this result by shifting the origin of coordinates to $t = a$. Then Eq. (A-53) becomes

$$\int_{-\infty}^{\infty} \delta(t - a)\, d(t - a) = \int_{-\infty}^{\infty} \delta(t - a)\, dt = 1 \qquad (A-57)$$

and the arguments just given lead to the relation

$$\int_{-\infty}^{\infty} \varphi(t)\delta(t - a)\, dt = \varphi(a) \qquad (A-58)$$

This equation expresses the *sampling property* of the delta function. We can visualize it as shown in Fig. A-9; placing $\delta(t)$ at $t = a$ picks out the corresponding value $\varphi(a)$ of $\varphi(t)$ as we integrate from $t = -\infty$ to $t = \infty$.

The sampling property may also be obtained from Fourier transforms involving the delta function. We have seen from Eq. (A-50) and Eq. (A-54) that

$$F(\omega) = \mathfrak{F}[\delta(t)] = 1 \qquad (A-59)$$

and the inverse of this is

$$\mathfrak{F}^{-1}[1] = \delta(t) \qquad (A-60)$$

Using Eq. (A-29), we see that Eq. (A-60) is equivalent to the relation

$$\frac{1}{2\pi} \int_{-\infty}^{\infty} e^{i\omega t}\, d\omega = \delta(t) \qquad (A-61)$$

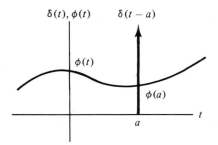

$\delta(t), \phi(t)$ $\delta(t-a)$

$\phi(t)$

$\phi(a)$

a

Figure A-9

or replacing t by $t - t'$

$$\frac{1}{2\pi} \int_{-\infty}^{\infty} e^{i\omega(t-t')}\, d\omega = \delta(t - t') \qquad (A\text{-}62)$$

Returning to Eq. (A-21), let $x = t$, $y = t'$, and $k = \omega$ to obtain

$$f(t) = \frac{1}{2\pi} \int_{-\infty}^{\infty}\int_{-\infty}^{\infty} f(t')e^{i\omega(t-t')}\, d\omega\, dt' \qquad (A\text{-}63)$$

By Eq. (A-62)

$$f(t) = \int_{-\infty}^{\infty} f(t')\delta(t - t')\, dt' \qquad (A\text{-}64)$$

which is equivalent to Eq. (A-58), since

$$\delta(t - t') = \delta(t' - t) \qquad (A\text{-}65)$$

This result follows from the fact that the area defined by the delta function $\delta(x)$ is unity for any value of x.

If we rewrite Eq. (A-16) as

$$\frac{1}{2\pi} \int_{-\pi}^{\pi} e^{i(m-n)x}\, dx = \delta_{mn} \qquad (A\text{-}66)$$

where

$$\delta_{mn} = \begin{cases} 1 & \text{if} \quad m = n \\ 0 & \text{if} \quad m \neq n \end{cases}$$

is the Kronecker delta, then Eq. (A-66)—which involves the integers m and n—is analogous to the continuous relation of Eq. (A-62) and explains the source of the name given to $\delta(t)$.

The use of the delta function in the convolution process leads to interesting results. Since the sampling property as illustrated in Fig. A-9 simply picks out the value of $\varphi(t)$ at $t = a$, it follows that the convolution, $\varphi(t) * \delta(t - a)$ will first reverse the delta function (which leaves it unaltered) and then reproduce each of the values of φ in a new position. That is, $\varphi(t)$ is converted into $\varphi(t - a)$ as shown in Fig. A-10. We express this analytically, from Eq. (A-46), as

$$\varphi(t) * \delta(t - a) = \delta(t - a) * \varphi(t) = \int_{-\infty}^{\infty} \delta(\alpha - a)\varphi(t - \alpha)\, d\alpha \qquad (A\text{-}67)$$

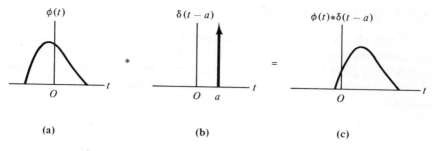

Figure A-10

The sampling property will reduce the integral to the value of φ for which $\alpha = a$. Hence Eq. (A-67) becomes

$$\varphi(t) * \delta(t - a) = \varphi(t - a) \qquad \text{(A-68)}$$

as we have already deduced.

Consider finally a series of evenly spaced delta functions, known as a *Dirac comb* (Fig. A-11a). This function is expressible as

$$\text{comb }(t) = \sum_{n=-\infty}^{\infty} \delta(t - na) \qquad \text{(A-69)}$$

The convolution of $\varphi(t)$ and the comb function then repeats $\varphi(t)$ at equally spaced intervals of $\pm a$, $\pm 2a$, $\ldots$, as shown in Fig. A-11b.

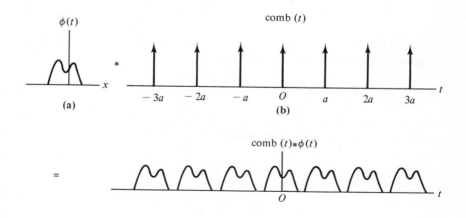

Figure A-11

A-5 The Fourier Spectrum

A periodic function can be expanded in a Fourier series of trigonometric functions as specified by Eq. (A-11) or in a series of complex exponential functions as given by Eq. (A-12). To express the series, we need merely know the set of coefficients a_n, b_n or the set c_n and the value of k. As an example, consider a generalization (Fig. A-12) of the pulse train in Fig. A-3. Let

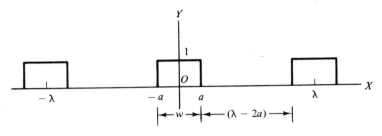

Figure A-12

the pulses have a width $2a$ and a spacing $(\lambda - 2a)$. To determine the c_n, we use Eq. (A-14) and obtain

$$c_n = \frac{1}{\lambda} \int_{-\lambda/2}^{\lambda/2} f(x)e^{-inkx}\, dx$$

$$= \frac{1}{\lambda} \int_{-a}^{a} (1)e^{-inkx}\, dx = \left. \frac{e^{-inkx}}{\lambda(-ink)} \right|_{-a}^{a}$$

$$= \frac{2 \sin nka}{\lambda nk} = \frac{\sin nka}{\pi n} = \frac{2a}{\lambda} \frac{\sin nka}{nka} = \frac{\omega}{\lambda} \operatorname{sinc}(nka) \qquad \text{(A-70)}$$

This result does not hold for $n = 0$, since c_0 has the form $0/0$. However, using a Taylor series for $\sin nka$, we find that

$$c_0 = \lim_{n \to 0} \frac{nka + \frac{1}{3}!(nka)^3 + \cdots}{\pi n} = \frac{ka}{\pi} = \frac{\omega}{\lambda}$$

We now wish to find the dependence of c_n on nk. To do this, we must specify a and λ numerically. For example, if $\lambda = 6$ and $a = 1$ so that

$$a = \frac{\lambda}{6}$$

then

$$c_0 = \frac{ka}{\pi} = \frac{2a}{\lambda} = \frac{1}{3}$$

$$c_1 = \frac{\sin(\pi/3)}{\pi} = \frac{\sqrt{3}}{2\pi}$$

$$c_2 = \frac{\sin (2\pi/3)}{2\pi} = \frac{\sqrt{3}}{4\pi}$$

$$c_3 = \frac{\sin (3\pi/3)}{2\pi} = 0$$

and so on, producing the graph in Fig. A-13(a). This dependence of c_n on nk is called the *Fourier spectrum* and, as stated above, provides a compact summary of information such as that in Figs. A-2(a), (b), and (c). That is, it gives the amplitudes and wavelengths of the constituent parts of the periodic function. Note that the ordinates are labeled with the values of n.

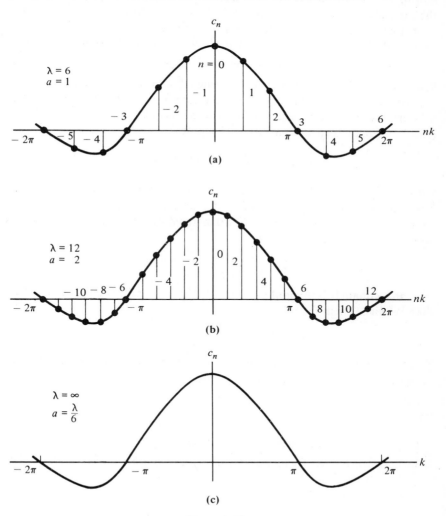

Figure A-13

If the pulse train has its wavelength doubled by letting

$$\lambda = 12, \qquad a = 2$$

then the ratio a/λ is unchanged; therefore c_0 is unchanged. A condition that

$$\sin nka = 0$$

is that

$$nka = \pi$$

In the case $\lambda = 6$, this is equivalent to $n = 3$, and when $\lambda = 12$, this gives $n = 6$. Hence, we obtain the spectrum shown in Fig. A-13(b), which has the same shape as the curve above it, but twice as many points.

It then is clear that as λ increases, we approach the curve of Fig. A-13(c) as a limit. That is, we have the Fourier spectrum for the function

$$f(x) = 1 \quad \text{for} \quad -\infty \leqq x \leqq \infty \qquad (A\text{-}71)$$

This constant function contains all values of k in its spectrum, rather than a comparatively small number of discrete values. Also, we cannot talk about a Fourier series for this function; instead, we have a Fourier integral. By Eq. (A-23), we may express an arbitrary function as

$$f(x) = \frac{1}{2\pi} \int_{-\infty}^{\infty} F(k)e^{ikx}\, dk$$

where $F(k)$ may be interpreted as the infinite set of Fourier coefficients. It is, however, the Fourier transform of $f(x)$. Hence, the limit of the Fourier spectrum as λ becomes infinite is the Fourier transform. In fact, for the function of Eq. (A-71), we have

$$F(k) = \frac{1}{2\pi} \int_{-\infty}^{\infty} (1)e^{-ikx}\, dx$$

We can evaluate this integral by simply letting the limits be very large, rather than infinity. Then

$$F(k) = \frac{\sqrt{2}\,\sin kx}{\sqrt{\pi}\,k}$$

This is the function shown in Fig. A-13(c), thus verifying the statement that the Fourier transform is the limiting spectrum.

References

A-1   T. J. Healy, *IEEE Spectrum*, **6**, 87 (1969).

A-2   J. Arsac, *Fourier Transforms and the Theory of Distributions*, Prentice-Hall, Englewood Cliffs, N.J. (1966).

INDEX

INDEX